Functional Gradient Materials and Surface Layers Prepared by Fine Particles Technology

NATO Science Series

A Series presenting the results of scientific meetings supported under the NATO Science Programme.

The Series is published by IOS Press, Amsterdam, and Kluwer Academic Publishers in conjunction with the NATO Scientific Affairs Division

Sub-Series

I. **Life and Behavioural Sciences**	IOS Press
II. **Mathematics, Physics and Chemistry**	Kluwer Academic Publishers
III. **Computer and Systems Science**	IOS Press
IV. **Earth and Environmental Sciences**	Kluwer Academic Publishers

The NATO Science Series continues the series of books published formerly as the NATO ASI Series.

The NATO Science Programme offers support for collaboration in civil science between scientists of countries of the Euro-Atlantic Partnership Council. The types of scientific meeting generally supported are "Advanced Study Institutes" and "Advanced Research Workshops", and the NATO Science Series collects together the results of these meetings. The meetings are co-organized bij scientists from NATO countries and scientists from NATO's Partner countries – countries of the CIS and Central and Eastern Europe.

Advanced Study Institutes are high-level tutorial courses offering in-depth study of latest advances in a field.
Advanced Research Workshops are expert meetings aimed at critical assessment of a field, and identification of directions for future action.

As a consequence of the restructuring of the NATO Science Programme in 1999, the NATO Science Series was re-organized to the four sub-series noted above. Please consult the following web sites for information on previous volumes published in the Series.

http://www.nato.int/science
http://www.wkap.nl
http://www.iospress.nl
http://www.wtv-books.de/nato-pco.htm

Series II: Mathematics, Physics and Chemistry – Vol. 16

Functional Gradient Materials and Surface Layers Prepared by Fine Particles Technology

edited by

Marie-Isabelle Baraton

University of Limoges,
Faculty of Sciences,
Limoges, France

and

Irina Uvarova

Institute for Problems of Materials Science (IPMS),
Kiev, Ukraine

Kluwer Academic Publishers

Dordrecht / Boston / London

Published in cooperation with NATO Scientific Affairs Division

Proceedings of the NATO Advanced Study Institute on
Functional Gradient Materials and Surface Layers Prepared by Fine Particles
Technology
Kiev, Ukraine
June 18–28, 2000

A C.I.P. Catalogue record for this book is available from the Library of Congress.

ISBN 0-7923-6924 6 (HB)
ISBN 0-7923-6925-4 (PB)

Published by Kluwer Academic Publishers,
P.O. Box 17, 3300 AA Dordrecht, The Netherlands.

Sold and distributed in North, Central and South America
by Kluwer Academic Publishers,
101 Philip Drive, Norwell, MA 02061, U.S.A.

In all other countries, sold and distributed
by Kluwer Academic Publishers,
P.O. Box 322, 3300 AH Dordrecht, The Netherlands.

Printed on acid-free paper

Printed in the Netherlands.

Table of Contents

vi

Preface

The NATO Advanced Study Institute on "Functional Gradient Materials and Surface Layers Prepared by Fine Particles Technology" was held in Kiev (Ukraine) on June 18-28, 2000 where more than 90 participants, ranging from Ph.D. students to experienced senior scientists, met and exchanged ideas.

This meeting was aimed at stimulating the research work across traditional disciplinary lines by bringing together scientists from diverse research areas related to functional gradient materials and surface layers. It also intended to give opportunities for initiating collaborative works between scientists from NATO and Partner countries and to trigger fruitful and exciting discussions between experienced and young researchers. In this respect, this NATO-ASI has been quite successful.

The term of functional gradient materials which originates from Japan in the 1980's describes a class of engineering materials with spatially inhomogeneous microstructures and properties (*MRS Bulletin*, 1995, **20**, N°1). These materials can be successfully utilized in various applications like electronic devices, optical films, anti-wear and anti-corrosion coatings, thermal barrier coatings, biomaterials, to name only a few. Although these functional gradient materials are not fundamentally new, the use of nanoparticles in their fabrication and in surface layers as well has greatly improved their performances to meet challenging requirements for industrial applications. However, the technology of nanoparticles is not straightforward and requires a particular knowledge and know-how. Processing methods and characterization techniques should be adapted to the nanometer scale for optimized results. Therefore, the research works on functional gradient materials and surface layers are very diverse, covering a broad range of scientific disciplines.

This diversity was reflected in the talks presented during this NATO Advanced Study Institute. From synthesis methods to various applications, many aspects of functional gradient materials were addressed. New approaches for solving problems were thoroughly discussed during the sessions, pursuing during the breaks. The processing of nanomaterials, which is far from being straightforward, was presented under different angles, allowing better understanding of complex phenomena. As an important part, characterization and properties evaluation were also addressed.

This NATO-ASI publication contains most of the invited lectures and selected contributions as well, presented at the meeting. All the published papers were peer-reviewed.

The directors of this NATO-ASI acknowledge the contribution of the Organizing Committee: Dr J.A. Akkara (The National Science Foundation, USA), Prof. R. Andrievski (Institute for Problems of Chemical Science, Russia), Prof. V. Skorokhod

(Institute for Problems of Materials Science, Ukraine), Dr L. Kolomiets (Institute for Problems of Materials Science, Ukraine).

The directors gratefully acknowledge the major financial support offered by the NATO Science Committee and the co-sponsorships of the Office of Naval Research International Field Office, the European Office of Aerospace Research and Development, the US Army Research Laboratory - European Research Office, the Ministry of Science and Education of Ukraine, the Science and Technology Center in Ukraine, and the National Science Foundation under the ASI Travel Awards Programme. The directors also would like to thank many others, specially in Ukraine, who provided efficient assistance and help in the preparation and organization of this successful meeting.

Marie-Isabelle Baraton   and   Irina Uvarova
ASI director for NATO countries   ASI co-director for Partner Countries
University of Limoges (France)   IPMS, Kiev (Ukraine)

November 21, 2000

FROM NATURE THROUGH ART TO FINE PARTICLE TECHNOLOGY

I. UVAROVA
I.N. Frantsevich Institute for Problems of Materials Science,
Ukrainian Academy of Science,
3, Krzhizhanovsky Str.
03142, Kiev, Ukraine

For a long time the scientists of all the world tried to achieve a perfection for materials as bases of high qualitative production studying the nature of plants and animals. The several examples, bonded with the achievements of materials science and the nature have been discussed in this lecture. Thus inverse image of the first self-grinding material was the self-sharpening tooth of beaver and the other animals and the structure of bamboo and other plants was the foundation for composite materials.
The second stage for the development of materials science had been the art of of different high qualitative goods preparation.
And only successive study of all the details for preparation of different materials had been resulted to development of materials science. The aspiration to improve the physical, chemical and technological properties of different materials have resulted in the creation of the composite and gradient functional materials. Last few years special attention has been devoted to nano-sized materials and composite nanophased coatings. These questions will be discussed in this lecture.

1. The studying of the nature as the foundation for the composite and gradient materials

Remembering the history of the Institute for problems of Materials Science which began their life as a small laboratory located on the territory of ancient building in Kiev, namely Kiev Pecherska Lavra with the young scientist I.N.Frantsevich as the head. Thinking about future, I.N.Frantsevich gathered in his institute the specialists in different branch of chemistry, physics, physical chemistry. In his lecture, which he read at 1974 year on the 4-th International Powder Metallurgy Conference in Czech Republic [1], I.N.Frantsevich made an attempt to compare the properties of different materials with the various functions of objects in the nature. It is known that the scientists were trying to achieve the level of nature objects always.

The studies of composite materials and their unique properties had been begun many years ago. The first objects of composite materials knowledge were the natural materials as the foundation for living organism namely the animals and plants. The comparative structure of bamboo and glass fiber materials is shown on the Fig. 1 [1]. In the both cases, the structures represented by the matrix media reinforced by fiber with the higher strength characteristics.

1

M.-I. Baraton and I. Uvarova (eds.),
Functional Gradient Materials and Surface Layers Prepared by Fine Particles Technology, 1–15.
© 2001 *Kluwer Academic Publishers. Printed in the Netherlands.*

2

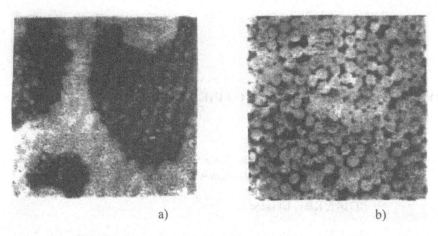

a)   b)

Figure 1. The structure of bamboo (a) and glass fiber (b)

The most extraordinary example of nature composite materials can be radiolyariya (Fig. 2).Their plasma having the mechanical characteristics nearest to solid and fiber reinforced skillet based on silicic acid, silicium and amorphous quarts correspond the optimum demands concerning with load-carrying ability. The sea ice which strain hardened by salt is more stronger than river ice.The comparative description of cell wall of a plant such as Alstonia spathulata (a) and the soft glass filament materials (b) are represented on the Fig. 3 [2].

Figure 2. Skeletal forms of radiolyariya

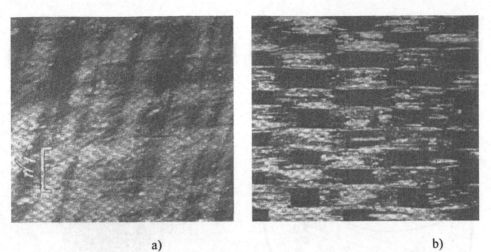

a)   b)

Figure 3. The cell wall of such plant as Alstonia spathulata (a) and soft glass filament materials (b)

In our works [3-4], we used the different methods of filament netting as the metal foundation for supported catalysts using at the purification from harmful additions (Fig.4).

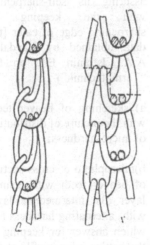

Figure 4. The different methods of the filament netting.

Using various methods of metal filament netting, with the nets of different diameters (0,03-0,3 mm) it is possible to construct the active catalytic structures, consisted of metal filament nets with deposited on them barrier electrophoretic partially crystallized glass of boro-barium composition, alumina and zirconium oxide coatings and upper catalytic layer. Such compositions are very successful as the catalysts in the reactions of CO and NO conversion.

The most exact examples of gradient functional nature materials are the teeth of animals (Fig.5). The self-sharpening teeth of beaver and other animals consist of hard external and soft internal layers. Under the external hard, strong and thin enamel there is anothe stronger layer but more less hard, passed into ductile mass bonded this organ of animal with another organs. As can be seen from Fig.5, the highest microhardness is on the upper layer, then it decreases.

4

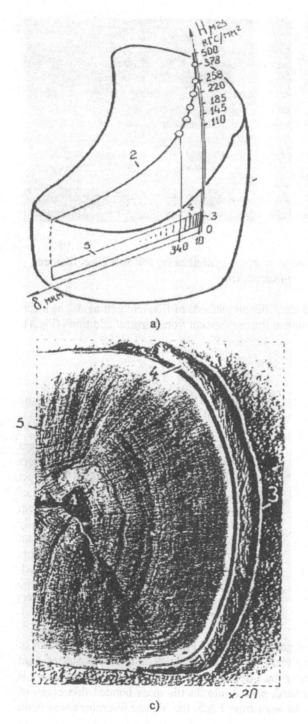

a)

b)

Figure 5. The main elements of beaver teeth structure, assuring its self-sharpening with the keeping of sharpening edge of cater [the data obtained by Kaydalov A.A., Istomin E.I. on the "Termotechnik"]

a) sketching of beaver teeth with the picture of distribution of micro-hardness;

b)microphoto of cross section of beaver teeth with enamel layer (3), intermediate layer with alternating hardness (4), which answer for keeping of the cutting edge profile at the process of self-sharpening and dentin foundation (5)

c) microphoto of cross-cut of front part of beaver teeth

2. The Art as the First Step in the Creation of New Technology

The man always aspired to construct materials to work in different conditions. Early the construction of similar materials was bonded with the art of ordinary man. For many years, the idea to construct hard, strong and ductile materials have been bonded with the production of good cutting tool such as sword, sabre and safety equipment such as armor. When Alexander Makedonski went to India he discovered that the of Tsar Por's armour did not damaged by arrows and spears. All Indian weapon were unique. Aristotel named the Indian steel as a "white iron". The ancient masters had hidden the secret production of such materials. The master worked with one blade for more than a month or even a year. If he have retreated a small from prescriptions the blade would have been destroyed and his work would have been useless.

In Russia, similar material had been named as "bulat" and the goods from it – "bulats". There are several examples from the popular science book of Russian scientist Yu. Gurevich [5] (Fig.6-8)

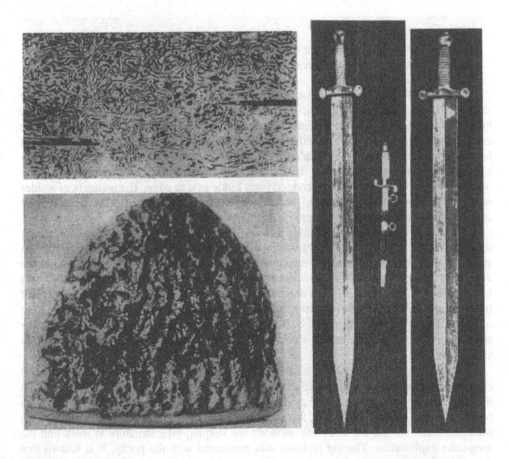

Figure 6.The figure of welding "bulat" and "bulat" alloy

Figure 7.The swords prepared from "bulat" steel

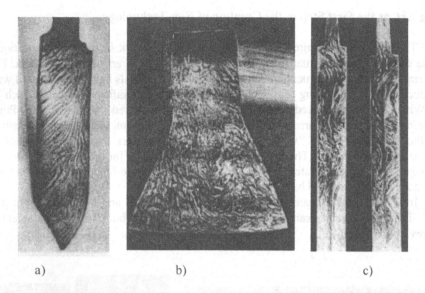

a)   b)   c)

Figure 8. The figures of toots prepared from "bulat" with carbon layers (a), on
hatchet from "bulat" steel (b), on the edge of hanger from "bulat" steel (c)

"Bulat" must have the definite figure, and produced clear and long sound at the stroke and may deform by strong bending. The "bulat" blade had been sharpened to improbable edge. It's known the many methods for preparation of figure blades. For example, the strips and wire bits with the different content of carbon and having different hardness have been wrung as a cable. This cable have been welded under pressing by hammer. By changing the manner of wringing, the masters prepared the different figure blades. The carbon content have increased from blade axis to edge. The manners of oxides removal, the welding temperature, the rate of welding the manner of strips and wire connection were secrets of the ancient masters, which they had hidden.

The scientists studied the structure and the properties of "bulat" steel for a long time and now known that the "bulat" is a high carbon steel having non equilibrium structure with macro and micro- inhomogeneity. "Bulat" is the layer steel in which the very hard layers alternate with the layers having the smaller carbon content with less hardness and biggest ductility. The alternation of soft and very hard places on the edge of blade supply its extraordinary sharpness and self-sharpening. "Bulat" is the combination of two different components, one is ductile and the other hard and strong. By studying the structure of "bulat", the scientists established also the preliminary importance of grain size.

Today, the layer materials are well known and used in different branches of chemistry, electrotechnics, in medicine and so on. Special significance are materials with the nanometer-sized structure. The implementation of the nanometer-sized materials began with the Art too. We know, that the first time the scientists learned to prepare the these materials but did not know the methods for keeping their structure at work and the long-time exploitation. The big problem was connected with the purity. It is known that the most of nanometer-sized materials are the materials with the very far from equilibrium structure.

3.   **Nanometer-sized Materials**

The nanometer-sized materials are the materials with a microstructure with characteristic size on the order a few (1-100) nanometers. As established by investigations of the physics and chemistry of solids, the most properties of solids depend on the microstructure. The microstructure have been determined by chemical composition, the atomic structure and the size of individual grains and their arrangement. Today it's established and known many physical parameters which can change when the grain sizes are reduced to 1-100 nm.

All the nanosized materials may be divided into three categories, as it was proposed in the recent review of H.Gleiter [6].

3.1. THREE CATEGORIES OF NANOMETER-SIZED MATERIALS [6].

The first category materials with nanometer-sized particles, thin wires or thin films can be prepared by CVD, PVD, inert gas condensation, various aerosol techniques, precipitation from vapor, from supersaturated liquids. These methods were known for a long time and used until appearance of conception of nanometer-sized particles. They have been used in catalysts with the big specific surface area, and for semiconductor devices utilizing single or multilayer quantum structures.

The second category is the materials in which the nanometer-sized microstructure is limited to a thin surface region of a bulk materials. The main methods for their preparation are the same CVD and PVD in addition to ion implantation and laser beam treatments. The such methods allow to create the chemical composition and atomic structure on the solid surfaces. These surfaces have the more high corrosion resistance, hardness, wear resistant and protective action. This category of materials are expected to play a key role in the production of the next generation of electronic devices. Today, these materials play important roles such as highly integrated circuits, terrabit memories, single electron transistors, quantum computers, etc. The development of such high technologies is the foundation of future more effective devices.

The third category is the bulk solids with the nanometer-scale microstructure. In this case, the chemical composition and the size of blocks (crystallites and atomic/molecular groups) vary on the length scale of a few nanometers throughout the bulk. These materials are prepared by quenching a high-temperature (equilibrium) structure to low temperatures at which the structure is far from equilibrium and have a nanometer-sized blocks and crystallites. In this case, these blocks may differ in their atomic structure, crystallographic orientation and chemical composition. These heterogeneous structure distinguishes them from glasses and gels that are microstructurally homogeneous. Materials with a nanometer-sized microstructure are called "Nanostructured Materials". Such materials have a significant future as the materials with controlled microstructural characteristics needed for the creation of new devise concept and manufacturing methods.

As I said, the early the scientists who worked in the such branches of chemistry as catalysts and adsorbents aspired to prepare the powders with the biggest specific surface area. The big specific surface area explained not only the small particles of powder but ramification of their surface, roughness and other defects. And only the appearance of conception about nanometer-sized microstructures resulted the scientists to idea that the physical, chemical and technological properties can be changed enough at the reducing of the characteristic size of microstructure blocks to the point where critical length scales (e.g. the free paths of electrons or phonons) became the comparable with the characteristic size of microstructure blocks. Thus, if the thickness of the surface layers is comparable to the wavelength of the electrons at the Fermi edge , discrete levels for electrons are formed in the quantum wells. Such effects resulted in change of the mechanical and optical properties.

Thus by combining the different layers with the particles of different sizes, we can produced the gradient materials with the expected properties. There may be the combination of different methods of coating precipitation such as galvanic and plasma deposition and heat treatment using laser, induction, and heat-chemical techniques and the following precipitation on the surfaces of these coatings and thin films by PVD or CVD methods. Durability, hardness and wear resistance of ultrafine and nanocrystalline bulk materials are several times higher than that of conventional materials. For ultrafine and nanocrystalline monolayer and especially multilayer films and coatings, these properties are higher than that of bulk. In multicomponent systems, the corrosion resistance and resistance to high temperatures can be increased.

Now, I would like to deserve one method for coating precipitation, namely electrolytic composite coatings and the methods for preparation of disperse powders, using in these coatings as a fillers. These coatings have been prepared by co-deposition of metals from iron group and dispersed fillers such as molybdenum, molybdenum and tungsten carbides and so on. These coatings have a high hardness which isin 3-5 times higher the wear and abrasive resistance in aggressive medium and a long-time work [7, 8].

3.2. THE LOW TEMPERATURE REDUCTION AS THE ONE OF THE METHODS OF FINE PARTICLE PREPARATION

One of the most known and old method for production of metals is the reduction of metals from oxides. Not stopping on the principles of producing of the different metals by reduction I would like to pay your attention on one problem how we can increase the dispersivity of metals and the different compounds based on their metals preparing by the reaction of heterogeneous reduction. Studying the processes of reduction we had concluded that we can prepare the nanometer-sized particles especially the powders of refractory metals by hydrogen reduction at decreased initial temperature of reduction and use a combination of systems.

3.2.1. *The Influence of Reaction Temperature on the Possible Change of the Reduction Mechanisms.*

As shown in our previous works [9-13], we can decrease the possibility of particle coalescence by changing the reaction mechanism at low initial temperatures of reduction. We can create the such conditions at which it is possible for momentary nucleation and followed by homothetical nucleus growth resulting in pseudomorphism transformation. In the other words, we change the ordinary mechanism of nucleus growth with the constant rate by pseudomorph transformation which decrease the possibility of particle coalescence at the covering of growing nucleus. At the low temperature reduction, we observed the small number of nucleus per one grain of initial oxide phase (Table 1). By increasing the temperature, the number of nucleus per one grain of initial oxide phase increase, and the rate of nucleus growth increase with temperature too. The relationship between the nucleation rate and the nucleus growth rate is determined by mechanism of reaction.

Let us compare the several examples of molybdenum reduction at the different temperatures. For this study, we had chosen the molybdenum oxide with the big initial specific surface area. This oxide have been reduced at the temperatures of 320 and 350 °C in according to the mechanism of momentary nucleation and the homothetical nucleus growth. At temperatures 400 °C and higher, the reaction mechanism changed and the nucleus growth with the constant rate. As can be shown from data presented on the Fig.9, the specific surface area decreases 8-10 times by changing the mechanism of reaction.

TABLE 1. The rate of nucleation and nucleus growth [9].

Initial phase	$t, °C$	Nucleation rate $V_i \bullet$ 10^7, cm/min	Nucleus growth rate $Vg \bullet 10^{-7}$, cm^{-2}/min^{-1}	Number of nucleus per grain of initial oxide phase
WO_3	500	0,47	74,8	23
	600	2,6	830	28
$S=4,5$ m^2/g	650	3,6	5700	180
MoO_3	500	6,4	0,019	2
	575	8,6	-	20
$S=0,2$ m^2/g	600	9,8	4,0	100
$MoO_3+0,1\%Pd$	200	-	-	0,4
	300	-	-	1,2
$S=0,8$ m^2/g	350	5,78	-	-
	400	25,6	-	1,75
	450	224	-	-

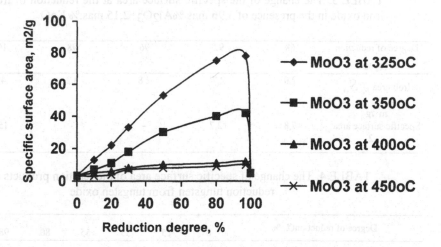

Figure 9. The change of specific surface area by the reduction of molybdenum from molybdenum trioxide at the different of initial temperatures.

3.2.2. The Influence of Additions on the Change of Specific Surface Area of the Reduction Products

The second method to increase the specific surface area by the reduction of metals from oxides is the use of additions of different non reduced oxides. Compare the specific surface area in two systems which are ordinarily used as a catalyst for ammonia synthesis. The first example (Table 2) is the reduction iron from iron oxide. The second case is the reduction of iron from iron oxide in the presence of additions of non reduced oxides, such as Al_2O_3, K_2O (Table 3).

As can be seen, the maximum specific surface area was achieved at the 97 % reduction degree. Then the specific surface area decreased to 1 m^2/g. By the presence of non reduces oxides, the specific surface area increased not only for the intermediate but also for the final products.

This effect increase at the long-time exploitation at the higher temperatures. Similar trends are observed for the reduction of molybdenum and tungsten from oxides (Table 4, 5).

TABLE 2. The change of the specific surface area at the reduction of iron from iron oxide

Degree of reduction α, %	0	5	35	50	80	97	100
Iron area $_{Fe}S$, m^2/g	0,03	0,3	0,3	0,3	0,4	0,6	1,0
Specific surface area S, m^2/g	0,1	0,3	1,0	1,2	1,6	1,9	1,0

TABLE 3. The change of the specific surface area at the reduction of iron from iron oxide in the presence of 3,96 mas.%Al_2O_3+2,15 mas.% K_2O

Degree of reduction α, %	68	92	96	98	100
Iron area $_{Fe}S$, m^2/g	2,0	2,7	2,8	3,2	4,0
Specific surface area S, m^2/g	7,8	12,7	-	-	15,8

TABLE 4. The change of specific surface area of the reaction products at the reduction tungsten from tungsten oxide

Degree of reduction α, %	0	15	35	55	80	98	100
Specific surface area S, m^2/g at 600 ^{o}C	3,5	8,0	11	16	20	25	5,0
Specific surface area S, m^2/g at 650 ^{o}C	3,5	6,8	9,5	13	15	19	4,0

TABLE 5. The change of specific surface area of the reaction products at the reduction tungsten from tungsten oxide in the presence of Al_2O_3

Degree of reduction α, %	0	15	30	50	65	98	100
Specific surface area S, m^2/g at 600 ^{o}C	8	25	26	27	28	29	20
Specific surface area S, m^2/g at 650 ^{o}C	8	18	20	20	20	20	15

3.2.3. *The Reduction in Complex Systems*

The third method to increase the specific surface area by the reduction of metals from oxides is the use of complex oxide systems with two and more reduced oxides. In this case, the effect depends on the quantity of reduced phases. Let us consider the reduction of molybdenum in the presence of the small additions of other reduced phase (Fig. 10).

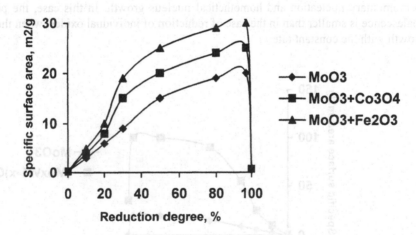

Figure 10. The change in specific surface area by the reduction of molybdenum from molybdenum trioxide. The initial temperature of reduction is 350 ^{o}C

As can be seen from data presented on the Fig. 10, the specific surface area by the reduction molybdenum from molybdenum trioxide achieve a maximum of 20 m^2/g at 350°C. At the same temperature of reduction in the presence of small additions (0,5%) of iron, nickel or cobalt oxides it can be achieved the maximum specific surface area of 30 m^2/g.

By the joint reduction of refractory metals from oxides, for example, molybdenum and rhenium at the more lower temperature (300 ^{o}C) it can be achieved the maximum specific surface area of 150 m^2/g at the presence of 0,5 % rhenium and 200 m^2/g at the presence of 30 % rhenium (Fig.11).

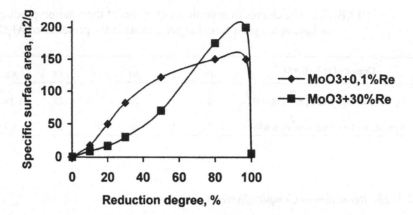

Figure 11. The change of specific surface area at the reduction of molybdenum from molybdenum trioxide. The initial temperature of reduction is 300 °C

The same trend have been observed by the joint reduction molybdenum and tungsten from trioxides (Fig. 12, 13).

The mechanism of joint reduction of refractory metals from oxides corresponds to momentary nucleation and homothetical nucleus growth. In this case, the process of coalescence is smaller than in the case of reduction of individual oxides, when the nucleus growth with the constant rate.

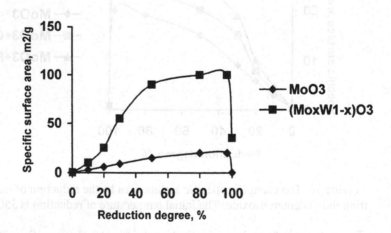

Figure 12. The change of specific surface area by the reduction of molybdenum from molybdenum trioxide and molybdenum and tungsten (50:50)from mixture. The initial temperature of reduction is 350 °C

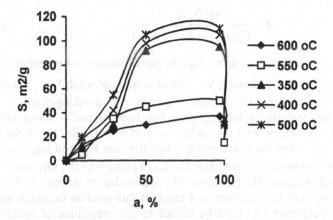

Figure 13. The change of specific surface area by the joint reduction of molybdenum and tungsten from mixture trioxides (50:50) at the different initial temperatures.

Taking the advantage of this low temperature reduction, we prepared the disperse carbides of molybdenum, tungsten and their mixture. These highly dispersed powders had been used as the fillers in composite electrolytic coatings, for metalization of ceramics and in polish paste.

3.2.4. *The Non-Isothermal Reduction*

It is very interesting the process for the non-isothermal reduction with the constant rate of temperature rise [14]. In this case, the calculation carrying out with the help of computer modeling had established the pseudo-branched chain mechanism. Comparing the computer and experimental data for the change of the specific surface area, we observed an interesting dependence between kinetic parameter "A" and the magnitude of the specific surface area. Changing the reaction conditions we can also change directionally the specific surface area.

$$\alpha = 1 - \exp^{-A(chBt-1)}, \tag{1}$$

α - the degree of transformation; $A=0{,}693/(chBt-1)$; $t_{0,5}$ – time of half transformation; B – the function of the formation of primary nucleus and growth of the second which bonds the kinetic parameter of with the rates of nucleation, growth and breaking-down of chains. Not stopping on the calculation in detail it is interesting to demonstrate the final results which tell about possibility to control the dispersity of preparing powder in dependence of kinetic parameter of non isothermal reduction or other heterogeneous reactions..

Based on the computer modeling studies for the nickel reduction, the following formula for calculation of parameter "A" was proposed

14

$$A = \frac{\varphi_f^1 N_o K_{g_1}}{V_s (K_{g_{1_s}} - K_{g_1} - K_{g_0})} \quad , \tag{2}$$

φ_f^1 - a factor of form: N_O- the initial number of potential nucleus; Kg_1 – the specific constant of nucleation; Vs – initial volume of solid; Kg_S – the constant of the rate of growth chains; Kg_o – the constant of the rate of its breaking-down.

An analysis of this equation has been shown a linear relationship for the parameter "A" bonds with the surface area of new phase (Ni) in the volume of initial phase (NiO). When the constant Kg_S equal the sum Kg_1 and Kg_o, the nucleus growth is minimum. The process of reduction has taken place with the formation of the big number of the small nucleus. The parameter "A" approaches to infinity. In this case, the specific surface area must be maximum as a result of this reaction the small nucleus are formed. Thus the parameter "A" must be related to the magnitude of specific surface area of straight relation as the bigger parameter "A" related to the increased the specific surface area is also increased(Fig. 12).

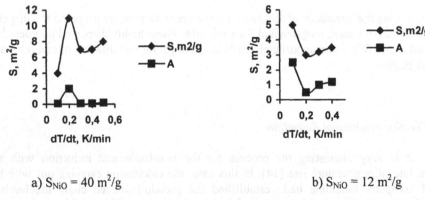

a) $S_{NiO} = 40$ m^2/g   b) $S_{NiO} = 12$ m^2/g

Figure 14. Computer modeling of reduction kinetic

Thus, by changing the reduction conditions, namely the rate of temperature rise, we can change the kinetic parameter "A" and as a result the specific surface area of the reaction products.

4. Conclusions.

1. To make novel materials, we must study the possibility of living organism and aspire to learn from them.
2. Prepare the multilayer gradient coatings and films for a hard, durable, wear and corrosion resistant upper layer.
3. To use the possibility of nano-sized materials in composite electrolytic and galvanophoretic coatings.

4. At the creation of nano-sized powder materials aspire to decrease the influence of the processes of coalescence.
5. To study the mechanism of heterogeneous and solid reaction for direct preparing the nano-sized powders.
6. To use the composite systems to decrease the level of particle coalescence during the preparation of metals, alloys and complex compounds based on refractory metals.

5. References

1. Frantsevich I.N. (1974) Position of powder metallurgy in present material science and engineering, in *The 4-th International Powder Metallurgy Conference in CSSR*, Dom Techniky Publishers, Kosice, pp. 18-46.
2. Felix R (1974) *Plants – genial engineers of nature*, Duseldorf – Vienna Puplisher.
3. Rutcovskyi, A., Chekhovskyi, A., Vlasenko, V. at al. (1999) Development and study of properties of composite catalysts based on metallic netted frame, in *The International conference "Advanced material"* Kiev, pp. 91.
4. Chekhovskyi, A., Podsossonnyi, V. Rutkovskyi, A. (1999) Manufacturing of thin, hollow metallic fibers by thermal decomposition of organometallic compounds used to produce high performance filtering and catalytic structures, in *The International conference "Advanced material"* Kiev, pp. 274.
5. Gurevich Yu.G. (1985) *Secret of "Bulat"*, "Knowledge" Publisher, Moscow.
6. Gleiter H. (2000) Nanostructured materials: Basic concepts and microstructure, *Acta Materialia* **48**, pp 1-29
7. Uvarova I.V. (2000) Ultrafine and nanophased powders as the fillers in composite coatings, *J, Advanced Materials* **32 (2)**, pp.26-31.
8. Kostenko, V.K., Saviyak, M.P., Uvarova, I.V. and Guslienko, Yu.A. (1992) Manufacture of composite electrophoretic coatings using very finely divided Mo and Mo_2C powders, *Powder metallurgy and Metal Ceramics* **12**, 1010-1014.
9. Skorokhod, V.V., Solonin, Ju.M. and Uvarova, I.V. (1990) *Chemical, diffusion and reological processes in powder metallurgy technology*, Naukova Dumka, Kiev.
10. Uvarova, I.V. (1990) Phenomenological aspects of dispersion of the products in topochemical reactions of reduction of metal from oxides (review), *Powder Metallurgy and Metal Ceramics* **2**, pp. 141-146.
11. Uvarova, I. (1994) Processes of Surface Energy Relaxation in Topochemical Reactions, *Science of Sintering*, **26 (3)**, pp. 245-250.
12. Uvarova, I.V. (1998) Preparation of Fine-Particle Materials through Topochemical Reactions *Inorganic materials*, **34 (8)**, pp. 911-923.
13. Uvarova, I.V. (1998) Heterogeneous and solid state reactions of reduction and carbidization, *J of Materials Processing & Manufacturing Science*, pp.251-265.
14. Stepanjuk, S.A., Uvarova, I.V., Savjak, M.P. and Ragulya, A.V. (1996) Simulation of the processes of solid-phase transformation during reduction of Ni from NiO under nonisothermical conditions, *Powder metallurgy and Metal Ceramics*, **1/2**, pp.1-4.

NEW SUPERHARD MATERIALS BASED ON NANOSTRUCTURED HIGH-MELTING COMPOUNDS: ACHIEVEMENTS AND PERSPECTIVES

R.A. ANDRIEVSKI

Institute of Problems of Chemical Physics, Russian Academy of Sciences
Chernogolovka, Moscow Region 142432, RUSSIA

1. Introduction

In recent 10-15 years, there has been increased interest in nanostructured materials (NMs). In the USA, the National Nanotechnology Initiative will be the top priority in science and technology in FY2001 with a significant increase in funding [1]. Note at least three factors accompanying the burst of research work in the field of nanotechnology as a whole and in the field of NMs, in particular. First, the interest is connected with the hope to realize a high level of the physical, mechanical, and chemical properties (and, therefore, potential service characteristics) in the nanocrystalline state. Some NMs such as nanostructured hard alloys, nanostructured Ni foil, the magnetic alloy Finemet, etc., are successfully produced on an industrial scale [2, 3]. Second, the topic of nanotechnology and NMs is really an interdisciplinary problem for physicists, chemists, material scientists, biologists, engineers, and specialists in the Earth Sciences. Third, this topic has revealed some gaps in both our understanding of the nature of the nanocrystalline state and its practical realization. This led to a wide scope of studies, numerous conferences, and a great flow of publications.

The present review is focused on the possibilities of obtaining new types of superhard materials (SMs), which can be designed via atomic-level structural control that tailors the high hardness properties. Some preliminary ideas on this topic have been discussed in the previous papers of the present author [4-7] (see also reviews [8,9]). It seems necessary to remind some general information on the SMs and hard materials.

2. Some features of materials with high hardness value

It is well known that SMs are of primary importance in modern science and technology. Table 1 shows hardness and elastic properties of some compounds and diamond. Notice that the information on hardness and elastic moduli values vary from different sources. However, leaving aside both a discussion on the methods of the hardness determination and the data on such exotic materials as $B_{22}O$ ($H_V\sim60$ GPa), fullerite C_{60} ($H_V=50-300$ GPa) and so on (see, for example [9, 10, 12, 13]), Table 1 seems to be quite realistic. The boundary between hard materials and SMs is very conditional and it is assumed that SMs have Vickers hardness, H_V,

17

M.-I. Baraton and I. Uvarova (eds.),
Functional Gradient Materials and Surface Layers Prepared by Fine Particles Technology, 17–32.
© 2001 *Kluwer Academic Publishers. Printed in the Netherlands.*

TABLE 1. Hardness (H_V), bulk modulus (K), and shear modulus (G) of intrinsic SMs and some hard materials

Material	H_V (GPa)	K (GPa)	G (GPa)
Diamond C	96±5	443	535
Cubic BN	63±5	400	409
Wurtzitic BN	50-78	410	330
BC$_2$N		408	445
Cubic C$_3$N$_4$		496	332
β-C$_3$N$_4$		437	320
B$_4$C	40±2	247	171
AlB$_{12}$	37±2	139	163
B$_6$O	35±5	228	204
WB	35±3		
TiB$_2$	33±2	244	263
SiO$_2$ (stishovite)	33±2	305	220
BP	33±2	169	174
Ta$_3$B$_4$	33±2		
TiC	30±2	288	198
β-Si$_3$N$_4$	30±2	236	122
CeB$_6$	30±2		
ZrB$_2$	30±2	218	221
VC$_{0.88}$	29±2	398	160
ZrN$_{0.75-0.78}$	27±2	267	160
SiC	26±2	240	188
TiN	20±2	292	160
Al$_2$O$_3$	20±2	250	160
WC	17±2	410	300

exceeding 40 GPa [9, 10] or, more precisely, 40-50 GPa.

It is common knowledge that hardness is defined as the material resistance against elastic and plastic deformation as well as fracture in the state of nonuniform volume compression. High values of intrinsic hardness and elastic moduli are connected with high cohesive energy, short bond length and covalent bonding. All these features result in a high level of Peierls - Nabarro stress and resistance to dislocation propagation and generation.

The example of successful calculations of C$_3$N$_4$ elastic properties from ab-initio and semi-empirical considerations is well known[14]. By now this compound has been synthesized only in film amorphous or partly crystalline forms [15, 16]. The highest value of Knoop hardness (load 0.15 N) of the CN$_X$ films (thickness of 1 μm) was observed by Fujimoto and Ogata [17] and equaled about 64 GPa. The experimental search for other SMs based on the boron-rich systems such as B-N-C, B-O-N, B-C-O, B-P-O, etc., is persistently going on [9, 10, 12].

3. Nanostructured SMs

3.1. SOME NEW RESULTS ON FRACTURE AND DEFORMATION OF NM BASED ON HIGH-MELTING COMPOUNDS

As mentioned in the Introduction, the design of nanoscale structures is now very popular. In this connection it cannot be doubted that an attempt to increase hardness at the cost of nanocrystalline structure seems to be realistic at an using of some hard compounds (Table 1) as a base. Analysis of numerous results in this area [2-7, 18, 19] shows that in the case of metals, alloys and intermetallics, the transition to a nanocrystalline state is accompanied by an increase of hardness by 4-6 times. In many cases, this increase is described by the known Hall-Petch relationship $Hv = kL^{-1/2} + h_O$, which is based on the dislocation - grain boundary interaction mechanism assuming a pileup at an interface. At the same time, in some cases, an inverse Hall-Petch-like behavior was also observed; therefore some other mechanisms of deformation may come into play. These features attract the attention of much research (see, for example, [19, 20]). However, the information on the mechanical properties of NM based on high-melting compounds is very limited and the mechanisms of their deformation and fracture need to be explored.

Some new recent results in this area [21-27] can be outlined:

1. The fracture surface and crack propagation in NM are intergranular. Figure 1 shows SEM micrographs of the fracture surfaces of nanostructured TiN bulk materials prepared by high-pressure sintering of the ultrafine powder. The intergranular fracture is evident and one can see also that the increase in the crystallite size results in the transition from intergranular fracture to transgranular one. This observation of the intergranular/ transgranular fracture seems to be important for understanding and explaining of the fracture toughness data for the one-phase and two-phase NMs.

2. In the case of the films with grain size of 3-15 nm, the majority of grain boundaries are to a great extent crystalline in nature and have a typical clear fringe contrast. Essentially amorphous phases and layers are not revealed in the grain boundaries.

3. Edge dislocations have been observed inside some crystallites (Fig. 2). The presence of dislocations inside of the NM crystallites is also expected by estimations of the characteristic length of dislocation stability in nanocrystals $\Lambda=0.04Gb/\sigma_{PN}$, where b is the Burgers vector and σ_{PN} is the Peierls-Nabarro stress [28]. Using the value of $\sigma_{PN} \sim 3.7$ GPa [29], it is possible to estimate the Λ value as of about 0.1 nm. So our dislocation observations seem to be very realistic especially regarding the availability of residual plastic deformation of TiN nanocolumns under indentation (Fig. 3) [30].

4. There are two types of deformation of nanostructured films: homogeneous (Fig. 3) and inhomogeneous deformation (Fig. 4). The first is connected with the presence of a columnar structure of the films. In the case of the clearly defined columnar structure, homogeneous deformation by slip on the boundary columns is dominant as for the TiN films. Localized inhomogeneous deformation is accompanied by the formation of shear bands and observed in the case of the TiB₂ and AlN films with partly columnar or stonelike (equiaxed) structure.

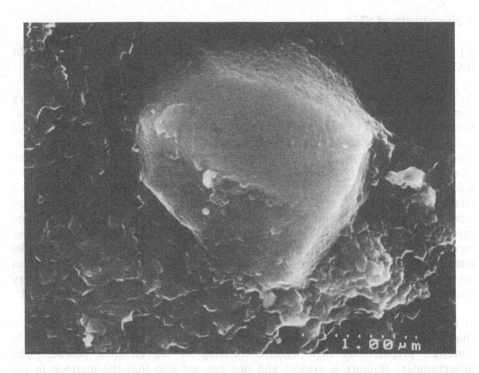

Figure 1. FE-SEM micrograph of the fracture surface of the TiN ultrafine powder (initial particle size about 80 nm with some large particles) consolidated at 1200°C.

3.2. HARD AND SUPERHARD NANOSTRUCTURED FILMS AND BULK MATERIALS. SYNTHESIS AND PROPERTIES

3.2.1. *Films*

Investigation of hard PVD nitride films started in the early seventies. Wide application of these films in tool and electronic materials was an impetus to the further development of the research. The TiN films are a particular leader both in applications and publications [5]. Undoubtedly, hardness measurements are most used for the film characterization because of the simplicity of the technique. However, in some cases the interpretation of the results is not so easy with respect to the known high structural sensitivity of hardness and its dependence on load as well as the test method used. So the absolute value of the film hardness needs to be discussed in every specific case.

Figure 5 shows the effect of crystalline size on TiN film hardness [31]. As compared with the H_V value for the conventional polycrystalline TiN bulk (Table 1), a hardness increase of 1.5-2 times is evident. Note also that the result may reflect not only the effect of crystalline size but also the influence of stoichiometry deviations, film thickness, substrate hardness, residual stresses, and the methods of film deposition and the hardness testing. It is probable that the above-mentioned increase of 1.5-2.0 is a lower level.

Table 2 shows hardness properties of monolayer and multilayer films;

Figure 2. HREM micrograph of the TiN film.

22

Figure 3. High-resolution FE SEM image of a fractured cross-section of the TiN film through an indentation imprint demonstrating the occurrence of microbuckling behavior at high indentation load (courtesy of Ma and Bloyce [30]).

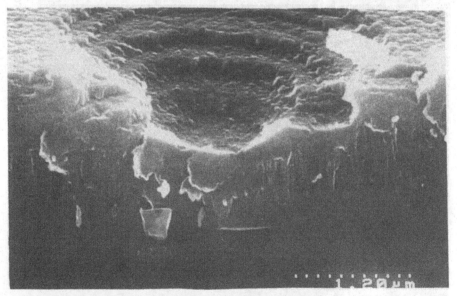

Figure 4. Fracture high-resolution FE SEM image through indentation on the Ti-B-N film deposited onto the Si substrate.

however, it includes only information on extremely high values of film hardness (see also reviews [5, 8, 9]). In connection with the properties of multilayer films, it seems necessary to point out the importance of the results obtained by Palatnik, Il'inskiy, Movchan, Bunshah, Holleck and their coworkers [47-51] (see also review [52]).

The following comments are useful for correct interpretation of the data presented in Table 2.

1. It is evident that in spite of different conditions of the hardness measurements, the values of hardness are on the level of that for SMs (see also Table 1). In some cases [25, 36], accounting for the effect of the film thickness and substrate material, the hardness of the film in itself (intrinsic hardness, H_{VO}) was estimated by the method proposed by Jonsson and Hogmark [53].

2. There is no doubt that the high values of hardness of the majority of crystalline monolayer and multilayer films are the result of the nanocrystalline structure, although the measurement of the grain size has not realized in all cases. The high hardness values of some amorphous B_4C and BC_4N films are also noteworthy.

3. Superhard films have been mainly prepared by magnetron sputtering. In addition, cathodic arc plasma deposition was used for preparation of nitride multilayer films [46]. The plasma CVD methods were also employed by some investigators [9, 39-43]. In addition, extremely hard TiN-Si_3N_4-$TiSi_2$ films were prepared from $TiCl_4$+SiH_4+H_2+N_2 (1:0.1:100:10) at 550°C by plasma CVD in abnormal d.c. glow discharge with the substrate connected as a cathode [43].

4. The high hardness values of crystalline films can be explained by the combination of various effects such as the image dislocation force at interfaces, the structural differences of the constituents, the availability of compression stresses and so on [5, 8, 52, 54]. Figure 6 shows the influence of the number of layers on the hardness of TiN/ZrN, TiN/NbN, and TiN/CrN films of similar total

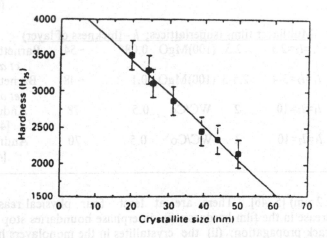

Figure 5. The variation in the microhardness of the TiN films with the crystallite size [31]. The load was 25 mN at nanoindentation testing.

TABLE 2. Vickers microhardness of some boride and nitride films

Content	Grain size (nm)	Thickness (µm)	Substrate	Load (N)	H_V (GPa)	Authors (year)
			Monolayer films			
Ti (B,C)$_x$	n.d.	6-12	HSS	0.5	~70	Knotek et al. [32,33;1990]
TiB$_{1.6}$N$_{0.6}$	n.d.	5	WC/Co	0.15-0.3	~60	Mitterer et al. [34;1990]
TiBN$_{0.5}$	n.d.	1-3	HSS	nanoindent. (δ=100 nm)	~55	Gissler et al. [35; 1994]
Ti(B,N,C,O)$_{2.5}$	n.d.	1.8	(100)TiB$_2$	0.3	45±4 (H_{VO}=50-52)	Andrievski et al. [36; 1995]
TiB$_{~2}$	n.d.	~4	Mo	0.1	~68	Mitterer et al. [37; 1998]
B$_{12}$C$_{2.9}$Si$_{0.35}$;					~63	Badzian et al.
Si$_3$N$_{2.2}$C$_{2.16}$					~65	[39; 1998]
n-TiN/BN					60-80	Veprek et al. [40; 1998]
Ti(B,N,Si,C,O)$_x$	2-4	3.5-4	HSS	0.25	70	Shtansky et al. [38; 1999]
a-SiB$_{0.9}$C$_{2.8}$N$_{1.1}$		1.2	(100)Si	0.01	22.3±4.4	Riedel et al. [41; 1999]
a-B$_4$C					50-70	Veprek et al. [42; 1992]
Ti(B,N,C,O)$_{~1.6}$	2-5	1	Si	0.2	34 (H_{VO}=70-80)	Andrievski et al. [25; 2000]
n-TiN/a-Si$_3$N$_4$/ (a+n)TiSi$_2$	3	3.5	HSS	0.1	~100	Veprek et al. [9,43; 1999]
			Multilayer films (superlattices; l_i - thickness of layer)			
TiN/VN	l_1=l_2=2.5	2.5	(100)MgO	0.08	~54	Barnett, Sundgren et al. [44;1987]
TiN/NbN	l_1=l_2=3-4	2.5-3	(100)MgO	0.1	~48	Barnett, Hultman et al. [45;1992]
TiN/NbN	l_1=l_2=10	2	WC/Co	0.5	78	Andrievski et al. [46;1992]
TiN/ZrN	l_1=l_2=10	2	WC/Co	0.5	70	Andrievski et al. [46;1992]

thickness (~ 2 µm) [5, 46]. There are at least four physical reasons for the observed increase in the film hardness: (i) interphase boundaries stop dislocation motion or crack propagation; (ii) the crystallites in the monolayers have a low size value (equal to, close to or smaller than the layer thickness); (iii)the dislocation density increases as a result of the layer lattices mismatch; and (iv) possibly a more

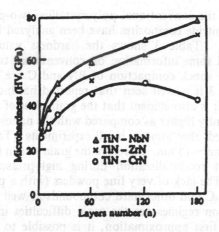

Figure 6. Influence of the number of layers on the film microhardness.

favorable situation exists with the residual strains in the multilayer films. The variable influence in the case of TiN/CrN films is determined by the (Ti,Cr)N solution formation for a number of layers greater than 100-120, which has been shown by an XRD analysis. This fact also confirms the main effect of the interfaces and layer lattice mismatch on the increase of hardness.

5. A particular emphasis is drawn upon high hardness values obtained by Veprek *et al.* [9, 40, 42, 43]. Veprek's main idea for the design of novel superhard composites is that it is necessary to form nanocomposites with strong interfaces such as nanocrystalline/amorphous (*n-/a-*) or nanocrystalline two-phase (*n-/n-*) composites with coherent boundaries as in the case of multilayer films. Such an approach can help to avoid grain boundary sliding and so achieves a high hardness. From general considerations, this concept seems to be quite realistic because it is known that two-phase composites have many benefits compared with one-phase those. Some features of *n-/a-* interfaces have been discussed by Glezer and the present author [18]. In addition, the selection of amorphous phases (*a*-Si_3N_4 and *a*-BN with high hardness in itself) was very successful and good results were reported [9, 40, 42, 43]. It is also important that these composites are characterized by high values of fracture toughness and elastic recovery.

3.2.2. *Bulk Materials*

The success in nanostructured bulk material development is more modest as compared with nanostructured films. The main difficulty in obtaining NM by powder technology methods is a retention of the nanocrystalline structure with full densification [55]. In this connection mainly high-energy consolidation methods such as hot forging, high static and dynamic pressures, etc., seem to be most useful for NM preparation. Only rate-controlled sintering results in a preparation of NM with the grain size of about 50-100 nm [56]. The present author analyzed this question three years ago [55] since the situation with nanostructured bulk materials based on one-phase hard materials varied only slightly. There are many data on the properties of the conventional polycrystalline two-phase composites. The properties of boride/nitride composites have been analyzed by J. Desmaison and M. Desmaison [57]. Table 3 shows the hardness values of one-phase nanocrystalline bulks and some information on conventional two-phase ceramic composites. The data on shock compaction of BN and C are also included for comparison. From Table 3 it is well seen that reported hardness values are far lower than those of SM. It is also evident that the grain sizes of the bulk material from Table 3 are significantly higher as compared with the nanostructured films in Table 2. It should be noted that almost in all experiments (Table 3), the initial particle size was higher than ~ 15 nm, therefore the grain size in bulk material was not so small because of recrystallization during high-pressure sintering and particularly during HIP. The lack of very fine powders (with a particle size below 5-10 nm) of TiB_2, TiN, B_4C and other hard compounds as well as the absence of data on their consolidation regimes are the main difficulties in obtaining novel SMs in bulk form.To a first approximation, it is possible to use the Veprek's concept [9] on the design of superhard NMs.

Note also that development of nanostructured SMs with high values of fracture toughness seems to be based on the features of the intergranular/ transgranular fracture of NMs as pointed out earlier. So their nanostructure must consist of at least two kinds of nanoparticles of the same or different nature with a different size and intergranular fracture.

3.2.3. *Stability and Degradation*

Essentially all NMs are far from the equilibrium state. Thermal activation results in enhancement of diffusion and recrystallization, with partial or total annihilation of the nanocrystalline structure, that decreases the level of properties such as hardness, strength, and so on. The study of a stability and degradation of NMs is considered to be very important in both scientific and practical aspects. From the general point, diffusion and recrystallization processes in NMs should occur intensively due to a substantial free-energy excess provided by extended interfaces and numerous lattices defects [18].

Figure 7 illustrates the effect of the vacuum annealing temperature on the hardness of various nitride monolayer and multilayer films [64]. It is evident that up to 800-1000°C all types of the TiN-NbN and TiN-ZrN films studied are stable. An anomalous hardness increase of alloyed (Ti,Zr)N is connected with spinodal decomposition in this system. The temperature of the hardness decrease for the multilayer TiN/NbN films qualitatively coincides with estimations based on interdiffusion studies of TiN and NbN in superlattice structures [65]. It is also interesting that the high hardness values of TiN/ZrN and TiN/NbN multilayer films change after a long period of about 1-2 years at room temperature [7]. Unfortunately, the exact time of the hardness decrease has not been settled and a more complete database will be required for understanding this phenomenon and developing the pertinent theory. Degradation of one-phase NMs could be caused not only by recrystallization but also by decomposition of the non-equilibrium phases. One reason for the two-phase NMs degradation may be also a superplasticity effect [4, 6, 19]. There are some other works on stability of NMs [5, 9, 38, 51, 66]. However, as a whole this problem is far from being settled and is needed for a fundamental understanding of the degradation mechanism.

TABLE 3. Vickers microhardness of some hard NMs

NM	Preparation method	Relative density	Grain size (nm)	H_V (GPa) at load (N)
TiN [21,58]	High-pressure sintering	~1.0	50-80	26 (0.3)
	(P=4 GPa, T=1400°C)	~1.0	30-40	29 (0.3)
TiN [56]	Rate-controlled sintering	~1.0	~50	26
TiN [24]	High-pressure sintering (P=4 GPa, T= 1200°C)	~1.0	~60	31.5 (0.5)
TiN/TiB$_2$ [24]	High-pressure sintering (P=4 GPa, T=1200°C)	~1.0	n.d.	34(TiN;0.5)
Si$_3$N$_4$ [59]	High-pressure sintering (P=8.5 GPa, T=1200°C)	0.99	500	32.5 (1)
SiC [60]	HIP (P=350 MPa, T=1500°C)	>0.97	~70	27
SiC$_{1+x}$ [61]	High-pressure sintering (P=4 GPa, T=1000°C)			40 (2-9)
TiB$_2$/AlN [57]	HIP (P=195 MPa, T=1900°C)	>0.99	n.d.	~38(5)
BN [62]	Sintering in shock waves	~0.96	25	43-80(5)
C [63]	Shock compaction	0.91		63-68
		0.94		56-65

4. Conclusions

In this review, an attempt has been made to analyze the current progress in the development of nanostructured SMs. As it is evident from the foregoing, there are several examples of synthesis of nanostructured monolayer and multilayer boride and nitride films which are characterized by high hardness values on the level of those of SM. At the same time, the results for bulk materials based on high-melting compounds are more modest as compared with the nanostructured films. It is believed that an use of very fine powders with the particle size lower than 10 nm may increase the hardness of bulk materials. The data on stability and features of degradation of superhard NMs are still incomplete, requiring intense studies in the immediate future in order to determine the best strategies for future technologies.

5. Acknowledgments

The author would like to express his deep appreciation to Drs. G.V. Kalinnikov and V.S.Urbanovich for their active help in this work. This research is sponsored by NATO's Scientific Affairs Division in the framework of the Science for Peace Programme (Project No 973529) and the Russian Programme "Integration".

28

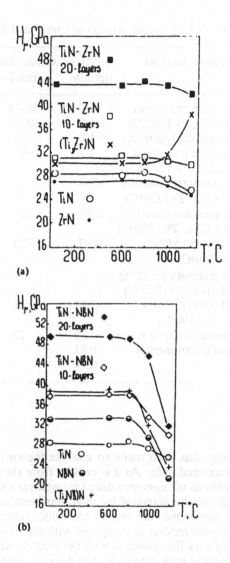

Fig.7. Microhardness of nitride films (thickness of 2 μm) deposited by cathodic arc deposition on WC/Co substrate as a function of annealing temperature: a) Ti-Zr-N films; b) Ti-Nb-N films.

References

1. Anonymous author (2000) National Nanotechnology Initiative (NNI) receives top priority in President's S&T budget proposal, *MRS Bulletin* **25**, 8-9.
2. Suryanarayana, C. (1995) Nanocrystalline materials, *Int. Mater. Rev.* **40**, 41-64
3. Robertson, A., Erb, U., and Palumbo, G. (1999) Practical applications for electrodeposited nanocrystalline materials, *Nanostruct. Mater.* **12**, 1035-1040.
4. Andrievski, R.A. (1994) Review - Nanocrystalline high melting compound-based materials, *J. Mater. Sci.* **29**, 614-631.
5. Andrievski, R.A. (1997) Review-Films of interstitial phases: synthesis and properties, *J. Mater. Sci.* **32**, 4463-4484.
6. Andrievski, R.A.(1999) The-state-of-the-art of high-melting point compounds, in Y.G. Gogotsi and R.A. Andrievski (eds.), *Materials Science of Carbides, Nitrides and Borides*, Kluwer Academic Publishers, Dordrecht, pp. 1-18.
7. Andrievski, R.A. (1999) New superhard materials are becoming a reality, *MRS Bulletin* **23**(5), 3.
8. Madan, A. and Barnett, S.A. (1999) Fundamental of nitride-based supperlattice thin films, in Y.G. Gogotsi and R.A. Andrievski (eds.), *Materials Science of Carbides, Nitrides and Borides*, Kluwer Academic Publishers, Dordrecht, pp. 187-204.
9. Veprek, S. (1999) The search for novel, superhard materials, *J. Vac. Sci. Technol.* **A17**, 2401-2420.
10. Teter, D.M. (1998) Computational alchemy: the search for new superhard materials, *MRS Bulletin* **23** (1), 22-27.
11. Andrievski, R.A. and Spivak, I.I. (1989) *Strength of High-Melting Compounds and Materials on Their Base,* Metallurgia, Cheliabinsk (in Russian).
12. Riedel, R. (1994) Novel ultrahard materials, *Adv. Mater.* **6**, 549-560.
13. Blank, V.D., Levin, V.M., Prokhorov, V.M., Buga, S.G., Duibitskiy, G.A. and Serebrianaja, N.R. (1998) Elastic properties of superhard fullerites, *J. Exper. Theoret. Physics* **114**, 1365-1374 (Engl. transl.).
14. Liu, A.Y. and Cohen, M.L.(1989)Prediction of new low compressibility solids, *Science* **245**, 841-842.
15. Sharma,A.K. and Narayan, J.(1997) Synthesis and processing of superhard ca-carbon nitride solids, *Int. Mater. Rev.* **42** (4), 137-153.
16. Korsounskii, B.L. and Pepekin, V.I. (1997) On the way to carbon nitride, *Russ. Chem. Rev.* **66**, 1003-1014 (Engl. transl.).
17. Fujimoto, F. and Ogata,K. (1993) Synthesis and properties of CN_x films, *Jpn. J. Appl. Phys.* **32**, L420-L423.
18. Andrievski, R.A. and Glezer, A.M. (1999) Size effect in nanocrystalline materials. I. Structure characteristics, thermodynamics, phase equilibrium, and transport phenomena, *Physics of Metals and Metallography* **88** (1), 45-66 (Engl. transl.).
19. Andrievski, R.A. and Glezer, A.M. (2000) Size effect in nanocrystalline materi-materials. II. Mechanical and physical properties, *Physics of Metals and Metallography* **89** (1), 83-102 (Engl. transl.).
20. Weertman, J.R., Farkas, D., Hemker, K., Kung, H., Mayo, M., Mitra, R., and Van Swygenhoven, H. (1999) Structure and mechanical behavior of bulk nano-

30

crystalline materials, *MRS Bulletin* **24** (2), 44-50.

21. Andrievski, R.A.(1997) Physical-mechanical properties of nanostructured TiN, *Nanostruct. Mater.* **9**, 607-610.
22. Andrievski, R.A. (1997) Nanocrystalline borides and related compounds, *J. Solid State Chemistry* **133**, 249-253.
23. Ma, K.J., Bloyce, A., Andrievski, R.A., and Kalinnikov, G.V. (1997) Microstructural response of mono- and multilayer hard coatings during indentation microhardness testing, *Surf. Coating Technol.* **94-95**, 322-327.
24. Andrievski, R.A., Urbanovich, V.S., Ogino, Y., and Yamasaki, T. (1999) Consolidation processes in nanostructured high melting point compound-based materials, in *The Sixth Conference and Exhibition of the European Ceramic Society*, Institute of Materials, Cambridge, pp. 389-390.
25. Andrievski, R.A., Kalinnikov, G.V., and Shtansky,D.V.(2000) Features of fracture surface and grain boundary structure of boride/nitride materials, in S.Komarneni, J.C. Parker, and H. Hahn (eds.), *Nanophase and Nanocomposite Materials III*, MRS, Warrendale.
26. Andrievski, R.A., Kalinnikov, G.V., and Shtansky, D.V. (2000) High-resolution scanning and transmission electron microscopy of nanostructured boride/ nitride films, *Phys. Sol. State* **42**, 741-746 (Engl. transl.).
27. Andrievski, R.A., Kalinnikov, G.V., Jauberteau, J., and Bates, J. (2000) Some peculiarities of fracture of nanocrystalline nitride and boride films, *J.Mater.Sci* **35**, to be published.
28. Gryaznov, V.G., Polonsky, I.A., Romanov, A.E., Trusov, L.I. (1991) Size effect of dislocation stability in nanocrystals, *Phys. Rev.* **B44**, 42-46.
29. Oden, M., Liungcrantz, H., Hultman, H. (1997) Characterization of the induced plastic zone in a single crystal TiN (001) film, *J. Mater. Sci.* **12**, 2134-2138.
30. Ma, K.J. and Bloyce, A. (1995) Observation of deformation and failure mechanisms in TiN coatings after hardness indentation and scratch testing, *Surface Engineering* **11**, 71-74.
31. Bendavid, A., Martin, P.J., Netterfield, R.P., and Kinder, T.J. (1994) The properties of TiN films deposited by filtered arc evaporation, *Surf. Coating Technol.* **70**, 97-103.
32. Knotek, O., Breidenbach, R., Jungblut, F., and Lofler, F. (1990) Superhard Ti-B-C-N coatings, *Surf. Coating Technol.* **43/44**, 107-115.
33. Knotek, O. and Loffler, F. (1992) Superhard physical vapor deposition coatings in the Ti-B-C-N system, *J. Hard. Mater.* **3**, 29-38.
34. Mitterer, C., Rauter, M., and Rodhammer, P.(1990) Sputter deposition of ultrahard coatings within the system Ti-B-C-N, *Surf. Coating Technol.* **41**, 351-363.
35. Hammer, P., Steiner, A., Villa, R., Baker, M., Gibson, P.N., Haupt, J., and Gissler, W. (1994) Titanium boron nitride coatings of very high hardness, *Surf. Coating Technol.* **68/69**, 194-198.
36. Andrievski, R.A., Kalinnikov, G.V., Potafeev, A.F., Ponamarev, A.M., and Sharivker, S.Y.(1995) Effect of deposition regimes and substrate type on the microhardness value of TiB₂ films prepared by magnetron sputtering, *Inorg. Mater.* **31**, 1536-1540 (Engl. transl.).
37. Kelosoglu,E. and Mitterer, C.(1998) Structure and properties of TiB₂ based co-

atings prepared by unbalanced DC magnetron sputtering, *Surf. Coating Technol.* **98**, 1483-1489.

38. Shtansky, D.V., Levashov, E.A., Sheveiko, A.N., and Moore, J.J. (1999) Optimization of PVD parameters for the deposition of ultrahard Ti-B-Si-N coating *J. Mater. Synthesis & Processing* **7**, 187-193.

39. Badzian, A., Badzian, T., Drawl, W.D., and Roy, R. (1998) Syntesis and properties of B-C-Si and Si-N-C hard materials, *Diamond and Related Compounds* **7**, 1519-1524 (cited from [9]).

40. Veprek, S., Nesladek, P., Niederhofer, A., Glatz, F., Jilek, M., and Sima, M. (1998) Hardness of TiN/BN composites, *Surf. Coat. Technol.* **108/109**, 138-143 (cited from [9]).

41. Hegemann, D., Riedel, R., and Oehr, C. (1999) PACVD-derived thin films in the system Si-B-C-N, *Chem. Vap. Deposition* **5**, 61-65.

42. Veprek, S. (1992) Deposition of B_4C coatings, *Plasma Chem. Plasma Process* **12**, 219-222 (cited from [9]).

43. Veprek, S., Niederhofer, A., Moto, K., Nesladek, P., Mannling,H., and Bolom, T. (2000) Nanocomposites nc - TiN/a - Si_3N_4/a - & - nc - $TiSi_2$ with hardness exceeding 100 GPa and high fracture toughness, in S. Komarneni, J.C. Parker, and H. Hahn (eds.), *Nanophase and Nanocomposite Materials III*, MRS, Warrendale.

44. Helmerson, U., Todorova, S., Barnett, S.A., Sundgren, J.-E., Market, L.C., and Greene, J.E. (1987) Growth of single-crystal TiN/VN strained-layer superlattices with extremely high mechanical hardness, *J. Appl. Phys.* **62**, 481-484.

45. Shinn, M., Hultman, L., and Barnett, S.A. (1992) Growth , structure, and microhardness of epitaxial TiN/NbN superlattices, *J. Mater. Research* **7**, 901-911.

46. Andrievski, R.A., Anisimova, I.A., and Anisimov, V.P. (1992) Structure and hardness of multilayer TiN, NbN, ZrN, and CrN films, *Physica i Chimiya Obrabotki Materialov* No2, 99-103 (in Russian).

47. Palatnik, L.S., Il'inskiy, A.I., and Sapelkin, N.P. (1966) Hardness of deposited multilayer Cu/Fe films, *Sov. Phys.-Solid State* **8**, 2515-2517 (Engl. Transl.).

48. Il'inskiy, A.I. (1986) *Structure and Strength of Multilayer and Dispersereinforced Films*, Metallurgia, Moscow (in Russian).

49. Movchan, B.A., Demchishin, A.V., Badilenko, G.F., Bunshah, R.F., Sans, C., Deshpandey, C., and Doerr, H.J. (1982) Structure-properties relationships in microlaminate TiC/TiB_2 condensates, *Thin Solid Films* **97**, 215-219.

50. Holleck, H. (1986) Material selection for hard coatings, *J. Vac. Sci. Technol.* **A4**, 2661-2668.

51. Holleck, H. and Lahres, M. (1991) Two-phase TiC/TiB_2 coatings, *Mater. Sci. Technol.* **A140**, 609-615.

52. Barnett, S.A. (1993) Deposition and mechanical properties of superlatties thin films, in M.H. Francombe and J.L. Vossen (eds.), *Physics of Thin Films: Mechanic and Dielectric Properties*, vol.17, Academic Press, San Diego, pp.2-75.

53. Jonsson, B. and Hogmark, S. (1984) Hardness measurements of thin films, *Thin Solid Films* **114**, 257-269.

54. Clemens, B.M., Kung, H., and Barnett, S.A. (1999) Structure and strength of multilayers, *MRS Bulletin* **24** (2), 20-26.

55. Andrievski, R.A. (1998) The-state-of-the-art of nanostructured high melting

32

point compound-based materials, in G.-M. Chow and N.I. Noskova (eds.), *Nanostructured Materials. Science & Technology*, Kluwer Academic Publishers, Dordrecht, pp. 263-282.

56. Zgalat-Lozynsky, O.B., Ragulya, A.V. and Herrmann, M. (2000) Rate-controlled sintering of nanostructured titanium nitride powders, poster presentation P-23 on the NATO ASI *"Functional Gradient Materials and Surface Layers Prepared by Fine Particle Technology"* (June 18-28, 2000, Kiev, Ukraine).

57. Desmaison, J. and Desmaison, M. (1999) Boride/nitride composites: synthesis and properties, in Gogotsi, Y.G. and Andrievski, R. (eds.), *Materials Science of Carbides, Nitrides and Borides*, Kluwer Academic Publishers, Dordrecht, pp. 267-284.

58. Andrievski, R.A., Urbanovich, V.S., Kobelev, N.P., and Torbov, V.I. (1997) High-temperature consolidation and physical-mechanical properties of nanocrystalline TiN, *Reports of Russian Academy of Sciences* **356**, 39-41(Engl. transl.).

59. Andrievski, R.A. (1994) Preparation and properties of nanocrystalline high-melting compounds, *Russ. Chem. Rev.* **63**, 431-448 (Engl. transl.).

60. Vassen, R., Buchkremer, H., and Stover, D. (1999) Nanophase ceramics, in *The Sixth Conference and Exhibition of the European Ceramic Society*, The Institute of Materials, Cambridge, pp. 391-392.

61. Gadzyra, N.F., Gnesin, G.G., and Mikhailik, A.A. (1999) Sintering of SiC-C solution at high pressure, *Inorg. Mater.* **35**, 1237-1242. (Engl. Transl.)

62. Kovtun, V.I., Kurdumov, A.V., Zeliavskiy, V.B., Ostrovskaya, N.F., and Trefilov, V.I. (1992) Sintering of BN in shock waves, *Powder Metallurgy* No12, 38-44 (in Russian).

63. Kondo, K. and Sawai, S. (1990) Shock compaction of diamond powder, *J. Amer. Cer. Soc.* **73**, 1983-1985.

64. Andrievski, R.A., Anisimova, I.P., and Anisimov, V.P. (1991) Structural and microhardness of TiN compositional and alloyed films, *Thin Solid Films* **205**, 171-175.

65. Engstrom, C., Birch, J., Hultman, L., Lavoie, C., Cabral, C., Jordan-Sweet, J.L., Carlsson, J.R.A. (1999) Interdiffusion studies of single crystal TiN/NbN superlattice thin films, *J. Vac. Sci Technol.* **A17**, 2920-2927.

66. Oblezov, A.E., Kalinnikov, G.V., and Andrievski, R.A. (2000) Thermal stability of nanostructured boride/nitride films, poster presentation P-13 on the NATO ASI *"Functional Gradient Materials and Surface Layers Prepared by Fine Particle Technology"* (June 18-28, 2000, Kiev, Ukraine).

NANOSTRUCTURED METAL FILMS:
POLYOL SYNTHESIS, CHARACTERIZATION AND PROPERTIES

G.M. Chow
Department of Materials Science
National University of Singapore
Kent Ridge, Singapore 117543, Republic of Singapore

Email: mascgm@nus.edu.sg

1. Abstract

Nanostructured metal films have been deposited using the polyol method. It involved the reduction of constituent metal precursors in refluxing ethylene glycol (EG). Selected examples of our work on the polyol deposition of nanostructured metal films, characterization and properties are presented.

2. Introduction

Nanostructured materials may exhibit unique properties that arise from the size effect and the interphase/surface effect. Nanostructured films and coatings find many advanced applications [1]. Solution chemistry processing is suited for depositing films not only on planar but also non-planar and hidden surfaces. Metal film can be synthesized using aqueous electroless and electrodeposition. If the substrate or deposited material is susceptible to aqueous degradation, non-aqueous methods can be used. We have used a non-aqueous, electroless polyol process for deposition of nanostructured metal films. This method has been previously used for making micron-size and nanostructured powders [2, 3, 4, 5]. It involves dissolution of solid precursor, reduction of metal ions and precipitation of metal powders in refluxing ethylene glycol. In this process ethylene glycol acts as both solvent and reducing agent. Film deposition can occur on suitable conductor and insulator substrates in this simple single-step process, without any pre-deposition surface treatment such as the catalyzation of insulator surface as required in traditional electroless plating. In this paper, selected examples of our work on the polyol deposition of nanostructured metal films are discussed.

33

M.-I. Baraton and I. Uvarova (eds.),
Functional Gradient Materials and Surface Layers Prepared by Fine Particles Technology, 33–43.

3. Experimental Methods

Constituent metal acetates were suspended in ethylene glycol and the mixture was heated to the refluxing temperature of EG at 194 °C. The substrate was vertically suspended in the mixture and the deposition time (t) was counted from after the refluxing temperature was reached. After the reaction, the substrate was rinsed with methanol or acetone and air-dried. The deposited films were studied using x-ray scattering, anomalous x-ray scattering, extended x-ray absorption fine structure spectroscopy, transmission electron microscopy, atomic force microscopy, scanning electron microscopy, vibrating sample magnetometry, microhardness and dynamic microscratch tests. Details of experimental methods can be found in the references. The synchrotron experiments were performed in Pohang Light Source, Korea, and National Synchrotron Light Source, Brookhaven National Laboratory, USA.

4. Results and Discussions

4.1 Cu FILMS ON AlN

The fabrication of circuit and interconnects requires surface metallization of the ceramic substrates. Aluminum nitride (AlN) is a potential substitute candidate of alumina because of its high thermal conductivity and electrical resistivity, low dielectric constant and coefficient of thermal expansion. Aqueous electroless metallization of AlN requires special caution to prevent hydrolysis of the surface that may deteriorate the surface properties.

The non-aqueous, electroless, polyol method has been used to deposit Cu metal films on AlN [6]. The orientation of the AlN substrate, vertically (V) and horizontally (H) suspended, was investigated. The Cu film thickness was in the range of 2-4 μm and increased with deposition time from 15-60 min. Polycrystalline Cu peaks were detected for all samples when metallization time was ≥ 30 min. Cu_2O was also found in the H-films. The lattice parameters of Cu increased by 0.028% and 0.017% for the H-films and V-films, respectively, compared to the bulk value of Cu (3.615 Å). The SEM results showed that Cu particles were equiaxed for V-samples and irregularly-shaped for H-samples. The sheet resistance of Cu was lower for the V-films. It was suggested that oxide first formed in solution and subsequently settled and incorporated in the Cu films.

A combined study of grazing incidence asymmetric Bragg (GIAB) scattering and small angle x-ray scattering (SAXS) was carried out. The GIAB method revealed texturing and variation of chemical composition as a function of depth in the sample by changing the x-ray incidence angle. The SAXS provided information on particle size and/or column shape up to 10 nm range and a measure of agglomeration in the films. The GIAB results of the Cu_2O peak of H-samples showed that there was a relatively higher oxide concentration with respect to the metal near the surface. The GIAB-SAXS results further showed that 4 nm structures existed in the surface region of the V-films, whereas Cu_2O existed as much larger structures from the surface into the interior of the H-films. A subsequent depth study of these films using different x-ray energies with grazing incidence showed that the first 20-60 nm of the Cu film was textured, with strains ranging from +0.1 to -0.6 % [7]. The azimuthal ordering in the film plane depended on film metallization time and substrate orientation.

4.2 Ni FILMS ON Cu

Nanostructured Ni films were polyol deposited on Cu substrate [8]. The film thickness ranged from 0.9-1.3 μm. Long deposition time, for example, 3 h, resulted in a decrease in film thickness (0.9 μm), which was attributed to the corrosive attack of ethylene glycol on the deposited Ni films. It is known for other systems involving chemical oxidation of ethylene glycol at elevated temperatures, oxalic acid was identified as a product. This and other organic acids that were formed as intermediates are corrosive to metals [9]. The effect of Pt as nucleating aid was also investigated by adding a small quantity of hexachloroplatinic acid to the reaction mixture. It was found that Pt precursor only promoted rapid precipitation of Ni powders but it did not enhance film deposition. The XRD results showed that Ni films were polycrystalline fcc. Figure 1 shows the average crystallite size and strain (estimated from XRD line broadening) of Ni films. The crystallite size reached a maximum of about 138 nm for $t = 2$ h, but it significantly decreased to 40 nm at $t = 3$h. The decrease in crystallite size and film thickness at longer t resulted from the corrosive attack of the solvent on deposited films, leading to dissolution and refinement of deposited film particles. The micro-strain of these films changed from tensile to compressive with increasing t.

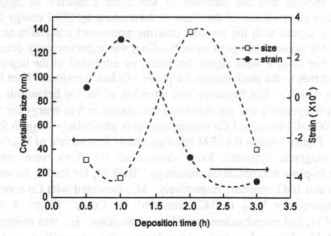

Figure 1. Dependence of crystallite size and strain of Ni films on deposition time

The XRD results of Ni films deposited using Pt catalysts showed only Ni with (200) texture, without any evidence of Pt as a separate phase, as Ni and Pt are miscible below 200 °C. The average Ni crystallite size increased with decreasing [Pt]/[Ni] precursor molar ratio from 33 to about 70 nm. The Ni films were magnetically saturated. Both Ni and Pt-catalyzed Ni films showed in-plane magnetization anisotropy. The saturation magnetization (M_s) of bulk fcc Ni is about 484 emu/cm^3. In-plane M_s ($M_{s//}$) increased with t and reached 405 emu/cm^3 for the Ni film deposited at 3 h. This could be due to increased film density at longer t. $M_{s//}$ increased whereas squareness ratio decreased with decreasing [Pt]/[Ni]. Pt had a predominant effect on $M_{s//}$ and a lower [Pt]/[Ni] ratio led to a higher M_s as expected. Lower M_s may be related to alloying of Pt with Ni. For Ni films deposited with or without Pt as catalysts, coercivity (H_c) was higher in the perpendicular ($\perp$) than the parallel ($//$) direction to the film plane.

4.3 Ni-Co FILMS ON Cu

4.3.1. Ni_xCo_{100-x} films

Nickel-cobalt finds applications in decoration as bright coatings, engineered hard coatings and magnetic films. Ni and Co has mutual equilibrium solubility. Ni_xCo_{100-x} films were polyol deposited on Cu substrates for 1 h [10]. The film thickness was in the range of 0.6-1.6 µm. The XRD results showed that there existed a single set of diffraction peaks. When x ≤ 30 %, only (111) reflections were observed. For x ≥ 50, higher fcc reflections were also detected. The average crystallite size increased with Ni concentration from 15 to 64 nm. As Ni was more easily reduced than Co, a higher Ni precursor concentration favored the formation of larger Ni-rich crystallites. The SEM results showed that the particles of the films consisted of agglomerated crystallites. The compositions of the films as determined by SEM energy dispersive analysis closely agreed with the nominal (starting precursor) compositions, with the exception that the nominal composition of $Ni_{10}Co_{90}$ was experimentally determined to be $Ni_{20}Co_{80}$. For Co_{100} a weak signal that could be attributed to the hcp phase was detected in addition to the predominant fcc phase. Cobalt transforms from hcp to fcc phase at about 420 °C. The formation and retention of fcc Co below this allotropic transformation temperature was possible since the change in free energy for the fcc → hcp is about 100 cal/g-atom, and the transformation is particularly sluggish for fine Co particles [11]. Figure 2 shows the SEM micrographs of film surface of $Ni_{50}Co_{50}$.

The magnetic hysteresis loops showed all the films were magnetically saturated with in-plane magnetization anisotropy. Bulk M_s for fcc Ni, fcc and hcp Co are 484, 1538 and 1442 emu/cm^3, respectively. $M_{s//}$ increased with Co concentration, from 236 emu/cm^3 for Ni_{100} to 1421 emu/cm^3 for Co_{100}. Figure 3 shows the dependence of H_c and microhardness on film composition. $H_{c\perp}$ was evidently larger than $H_{c//}$ and $H_{c\perp}$ followed a cyclic dependence with composition. A possible explanation of the higher $H_{c\perp}$ in these films is that the films consisted of two components: continuous film and a small fraction of magnetically isolated particles with some form of anisotropy (such as shape) in the perpendicular direction [12]. Calculations based on these assumptions showed that the hysteresis loops with in-plane magnetization anisotropy and higher perpendicular coercivity can be obtained. The magnetic properties of polyol deposited Ni-Co films compared well with that of electrodeposited films. The Vickers microhardness (HV) of the Cu substrate, bulk Ni and bulk Co samples were found to be 75, 144 and 288, respectively. The deposited films provided increased surface hardness for the underlying soft Cu substrate as expected. The hardness did not depend on film thickness. The $Ni_{50}Co_{50}$ sample showed the highest $H_{c\perp}$ and microhardness. The films showed good adhesion properties as determined by microscratch test that involved the deformation of the coating-substrate interface by straining the substrate. The critical load for film delamination is expected to increase with the film thickness for a given composition and substrate material. The critical load for both partial and extensive delamination of the Ni_xCo_{100-x} films did not systematically vary with thickness systematically. However the critical load increased with increasing Ni concentration (Fig. 4).

Figure 2. SEM micrograph of $Ni_{50}Co_{50}$ film

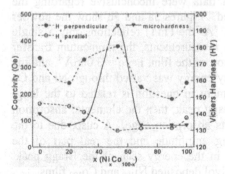

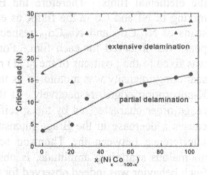

Figure 3. Dependence of coercivity and microhardness on composition of Ni_xCo_{100-x} films

Figure 4. Dependence of film delamination on composition of Ni_xCo_{100-x} films

4.3.2.   *Anomalous scattering study of Ni-Co films*
Conventional XRD has been well established to study the formation of solid solutions
[13]. For materials with large crystallite size, the appearance of a single set of
diffraction peaks and the disappearance of elemental peaks are commonly accepted as
evidence of formation of an alloy.   In a random, substitutional solid solution, the
lattice parameter changes with composition and the variation can be qualitatively
followed using Vergard's law. However, when the crystallite size is reduced to below
a critical length scale, a nanostructured solid solution cannot be unequivocally
differentiated from a nanocomposite using conventional x-ray diffraction [14]. For a
nanocomposite where the two phases have close lattice parameters and x-ray structural
coherence, the Bragg peak of one phase has some degree of overlap with that of the
other phase.   Because of the effect of size broadening and the contribution to
diffraction amplitude by structural coherence of the two phases, a single peak appears
for a particular Bragg reflection when the size is below a certain limit.   Yet, this
single peak has an average lattice spacing that bears no correspondence in real space,
and it can be easily mistaken as the evidence of formation of a solid solution.
Therefore conventional XRD, used for investigating the long range structure, should
not be used alone to study the structure of nanostructured alloys and composites made
of structurally coherent and immiscible materials. For example, characterization
techniques such as extended x-ray absorption fine structure (EXAFS) and solid state
nuclear magnetic resonance (NMR) have been used to investigate the averaged local
atomic environment of Co-Cu, in order to identify if the nanostructured material
existed as solid solution or composite [5]. Both techniques rely on the use of suitable
reference materials as standards. Although EXAFS has been commonly accepted as a
very useful tool to study nanostructured metastable alloys, it is not suitable for
investigating materials consisted of elements (such as Ni and Co) with very close
lattice parameters, backscattering amplitudes and phase shifts.

The structure of polyol deposited Ni-Co films have been investigated using
anomalous x-ray scattering (AXS) and EXAFS [15]. The EXAFS results of the Ni
edge and Co edge of $Ni_{90}Co_{10}$ and $Ni_{50}Co_{50}$ did not show noticeable difference from
the elemental films.   Therefore the EXAFS data were inconclusive regarding the
mixing of Ni and Co in the films, as expected. Figures 5a and 5b show the powder
scans of $Ni_{50}Co_{50}$ and $Ni_{90}Co_{10}$, respectively. At the (111) reflection, only a single
peak was observed for each film. For AXS measurements, the momentum transfer
was fixed to the positions of the (111) reflection of the film, i.e. $q = 3.086 Å^{-1}$, and the
scattering intensity was monitored as the x-ray energy was varied through Ni and Co
K-absorption edges, respectively. If the element in question is related to the long-
range order characterized by the specified Bragg peak, then the elemental absorption
causes a decrease in the Bragg intensity at its absorption edge. A cusp, due to the
interference between the Thomson scattering amplitude and the real part of the
anomalous scattering amplitude, is observed in the energy scan of the Bragg peak.
Such behavior was indeed observed for the polyol deposited Ni_{100} and Co_{100} films.

Figures 6a and 6b show the AXS measurements of the (111) peaks of $Ni_{50}Co_{50}$
and $Ni_{90}Co_{10}$, around the K-absorption edges of Ni and Co, respectively. Both Ni and
Co were related to this Bragg peak for $Ni_{50}Co_{50}$.   Due to their very close lattice

parameters, it was not possible to determine if the peak with both elements represented an alloy or a composite. For $Ni_{90}Co_{10}$, only Ni was found to be associated with the Bragg peak, whereas Co absorption was clearly absent. However, the existence of Co in the $Ni_{90}Co_{10}$ film was confirmed by EXAFS and energy-dispersive x-ray analysis. Therefore, within the detection limit of AXS, Co was not associated with the (111) Bragg peak of this film. Only Ni solely contributed to this long range order. The detection limit of long range order using XRD is roughly 3 nm. The absence of Co in the (111) Bragg peak of $Ni_{90}Co_{10}$ suggested that Co did not alloy with Ni in this atomic arrangement. Using anomalous x-ray scattering, it was shown that nanostructured NiCo films did not necessarily form solid solution as expected from their phase diagram or suggested by the results of conventional x-ray diffraction.

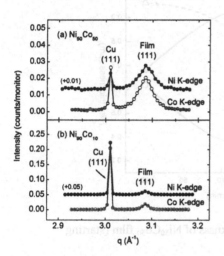

Figure 5. Powder diffraction of a)$Ni_{50}Co_{50}$, b)$Ni_{90}Co_{10}$

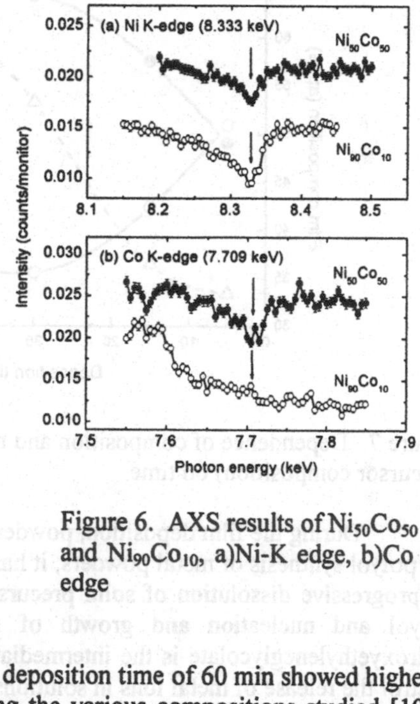

Figure 6. AXS results of $Ni_{50}Co_{50}$ and $Ni_{90}Co_{10}$, a)Ni-K edge, b)Co K-edge

4.3.3. $Ni_{50}Co_{50}$ films on Cu

The polyol synthesized $Ni_{50}Co_{50}$ films using deposition time of 60 min showed highest perpendicular coercivity and hardness among the various compositions studied [10]. The dependence of critical loads for partial and extensive delamination of the Ni_xCo_{100-x} films on compositions as investigated by microscratch testing also showed a change of slope at the composition of $Ni_{50}Co_{50}$ [12]. The growth process of $Ni_{50}Co_{50}$ films and properties have been studied as a function of deposition time, t, to gain further insight on the process-properties relationships [16].

Figure 7 shows the dependence of composition (as determined by EDX) and thickness of deposited films on t. Samples deposited using 5 and 30 min were denoted as 5NC and 30NC, respectively. The atomic ratio of Ni and Co concentrations in sample 5NC remained almost the same as the nominal precursor composition. However, with increasing t, Co concentration increased to a maximum of about 64 at.

%. At $t = 60$ min, the stoichiometry of Ni and Co was restored to nearly the same as nominal compositions. The polyol chemistry changed within $5 < t < 60$ min, as shown by the stoichiometric deviation in favor of Co precipitation. The melting temperatures of Ni and Co are 1455° C and 1495 °C, respectively. The deposition temperature was about 1/8 of the elemental melting temperatures, which perhaps would not be sufficiently high enough for the occurrence of through-thickness diffusion to achieve compositional homogenization. Therefore, the thicker films deposited at longer t could be conceivably compositionally graded through the film thickness.

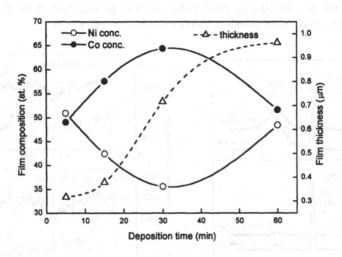

Figure 7. Dependence of composition and thickness of $Ni_{50}Co_{50}$ film (starting precursor composition) on time

During the film deposition, powder precipitation also occurred in solution. In the polyol synthesis of metal powders, it has been suggested that the reaction proceeds by progressive dissolution of solid precursors, reduction of dissolved species by the polyol and nucleation and growth of metal particles in solution [17]. Metal hydroxyethyleneglycolate is the intermediate solid phase that serves as a reservoir to control the release of metal ions in solutions. The competition of powder precipitation with film deposition is undesirable and should be avoided. During the reflux, it is possible that the solution chemistry and thus the vapor pressure related to a particular element varied with time. Such variation would cause a stoichiometric deviation of composition in solution and change the flux of the elemental atoms in question to the substrate as a function of t, resulting in the time (or film thickness)-dependence of composition.

The crystallite size of deposited films was between 9 to 11 nm and it did not vary significantly with film thickness. This showed that grain growth did not occur at the deposition temperature with increasing t. The lattice strain of the films decreased with thickness, which may be attributed to that the thicker films deposited in longer t

underwent annealing to relieve the stress as compared to the thinner films deposited in shorter t. Another possible stress relief mechanism is the formation of misfit dislocations in thicker films to reduce the strain. The SEM results showed that the apparent film density increased with increasing t. The films showed in-plane magnetization anisotropy with higher perpendicular coercivity. Details of growth mechanism and factors contributing to higher perpendicular coercivity were discussed [16].

4.3.4. *Electrodeposition of $Ni_{50}Co_{50}$ films on Cu*

The polyol process involves the reduction of metal precursors in refluxing ethylene glycol, resulting in powder precipitation in solution and/or film deposition on substrates. However, these competing reactions affect the chemistry of the solution and the deposition chemistry at the fluid/solid interface, thus rendering the control of film properties difficult. The powder precipitation during film deposition is undesirable as it represents a waste of reactants, and should be avoided in order to optimize film properties. The nucleation and growth of particles in precipitation is controlled by the degree of supersaturation, which strongly depends on temperature. On the other hand, the electric field selectively deposits metal film on the conductor surface based on the relative standard potentials, and is thus expected to minimize the free precipitation of metal powders in solution.

The effects of processing temperature, applied voltage and the addition of sodium acetate as a supporting electrolyte in the polyol electrodeposition of $Ni_{50}Co_{50}$ (nominal precursor composition) films on Cu were investigated [18]. It was found that the deposition chemistry depended on the above factors, which influenced film thickness, compositional stoichiometry, microstructures and local atomic environment. Increasing reaction temperature favored powder precipitation in the solution. The applied electric field aided film deposition at low temperatures without concurrent powder precipitation. Electrodeposition of films competed with powder precipitation at higher temperatures, yet the applied electric field could not prevent undesirable powder precipitation. At the refluxing temperature, the films were not stoichiometric as compared to nominal starting compositions. These tended to be Co-rich compared to the powders precipitated under same conditions. The presence of sodium acetate favored the deposition of Co at high temperatures even in the absence of applied electric field.

4.4 Fe-Ni FILMS ON Cu

Fe-Ni thin films are useful in magnetic recording heads due to their low coercivity, high permeability and near zero magnetostriction. Solution synthesis of Fe-Ni films has conventionally been carried out using electrodeposition. Fe-Ni powders were synthesized using the polyol method [17, 19]. Due to the low yield of Fe disproportionation, the Fe concentration in the powders could not exceed 30 at. %, even by increasing the ratio of Fe precursor in the reaction.

The electroless polyol deposition Fe-Ni film on Cu has been recently reported [20]. Varying the position of the substrate could increase the Fe concentration in deposited Fe-Ni films. When the substrate was placed above the refluxing solution, Fe concentration in the film could be increased to 90 at. %. The mechanical and magnetic properties depended on the crystallinity of the films, which apparently depended on the substrate position during deposition.

5. Summary

Nanostructured metal films have been successfully deposited using the simple, inexpensive polyol process. Selected examples of our work on polyol synthesis, characterization and properties of Cu, Ni, Ni-Co and Fe-Ni films were overviewed. The chemistry varied with deposition time in the current processing methodology where the precursor concentrations were initially fixed and depleted with increasing reaction time. The competing powder precipitation further complicated film deposition chemistry. Further work is needed to understand the deposition chemistry for both time-dependent and steady state precursor concentrations. The polyol method is an attractive method for synthesizing nanostructured metal films for both conventional and advanced applications.

Acknowledgment

The support of this work by the Academic Research Fund, the National University of Singapore, and the Office of Naval Research, USA, is gratefully acknowledged. The author thanks S.H. Lawrence, L.K. Kurihara, L.J. Martinez-Miranda, J. Zhang, Y.Y. Li and H. Yin for their valuable contributions to this work.

References

1. Chow, G.M., Ovid'ko, I.A., Tsakalakos, T. (2000) *Nanostructured Films and Coatings*, Kluwer Academic Publishers, Dordrecht.
2. Fiévet, F., Lagier, J.P., and Figlarz, M. (1989) Preparing monodisperse metal powders in micrometer and submicrometer sizes by the polyol process, *Mater. Res. Soc. Bull.* December, p.29-34.
3. Viau, G., Ravel, F., Acher, O., Fiévet -Vincent, F., and Fiévet, F. (1994) Preparation and microwave characterization of spherical and monodisperse $Co_{20}Ni_{80}$ particles, *J. Appl. Phys.* 76, 6570-6572.
4. Kurihara, L.K, Chow, G.M., and Schoen, P.E. (1995) Nanocrystalline metallic powders and films produced by the polyol method, *Nanostruc. Mater.* 5, 607-613.
5. Chow, G.M., Kurihara, L.K., Kemner, K.M., Schoen, P.E., Elam, W.T., Ervin, A., Keller, S., Zhang, Y.D., Budnick, J., and Ambrose, T. (1995) Structural, morphological and magnetic study of nanocrystalline cobalt-copper powders synthesized by the polyol process, *J. Mater. Res.* 10, 1546-1554.
6. Chow, G.M., Kurihara, L.K., Ma, D., Feng, C.R., Schoen, P.E., and Martinez-Miranda, L.J. (1997) Alternative approach to electroless Cu metallization of AlN by a nonaqueous polyol process, *Appl. Phys. Lett.* 70, 2315-2317.
7. Martinez-Miranda, L.J., Li, Y., Chow, G.M., Kurihara, L.K. (1999) A depth study of the structure and strain distribution in chemically grown Cu films on AlN, *Nanostruc. Mater.* 12, 653-656.
8. Zhang, J., Chow, G.M., Lawrence, S.H. and Feng, C.R. (2000) Nanostructured Ni films by polyol electroless deposition, *Mater. Phys. Mech.* 1, 11-14.

9. Jordan, C.B. (1986) *Engine Coolant Testing*, Second Symposium, (American Society for Testing of Metals) p. 249.

10. Chow, G.M., Ding. J., Zhang, J., Lee, K.Y. and Surani, D. (1999) Magnetic and hardness properties of nanostructured Ni-Co films deposited by a non-aqueous electroless method, *Appl. Phys. Lett.* **74**, 1889-1891.

11. Centre D'Information Du Cobalt, ed. (1960), Cobalt Monograph, Belgium, p.p. 75-77.

12. Chow, G.M., Zhang, J., Li, Y.Y., Ding, J., and Goh, W.C. (2000) Electroless polyol synthesis and properties of nanostructured Ni_xCo_{100-x} films, *Mater. Sci. Eng. A. in press*.

13. Cullity, B.D. (1978) *Elements of x-ray diffraction*, 2nd edition, Addison-Wesley Publishing, USA, pp. 375-376.

14. Michaelsen, C. (1995) On the structure and homogeneity of solid solutions: the limits of conventional x-ray diffraction, *Phil. Mag. A.* **72**, 813-828.

15. Chow, G.M., Goh, W.C., Hwu, Y.K., Cho, T.S., Je, J.H., Lee, H.H., Kang, H.C., Noh, D.Y., Lin, C.K., and Chang, W.D. (1999) Structure determination of nanostructured Ni-Co films by anomalous x-ray scattering, *Appl. Phys. Lett.* **75**, 2503-2505.

16. Zhang, J. and Chow, G.M. (2000) Electroless polyol deposition and magnetic properties of nanostructured $Ni_{50}Co_{50}$ films, *J. Appl. Phys. in press*.

17. Viau G., Fiévet-Vincent, F., and Fiévet, F. (1996) Nucleation and growth of bimetallic CoNi and FeNi monodisperse particles prepared in polyols, *Solid State Ionics* **84**, 259-270.

18. Li, Y.Y., Goh, W.C., Blackwood, D.J., Huang, Y.Z., and Chow, G.M. (2000) Polyol electrodeposition of nanostructured NiCo films, *submitted*.

19. Viau, G, Fiévet -Vincent, F., and Fiévet, F. (1996) Monodisperse iron-based particles: Precipitation in liquid polyols, *J. Mater. Chem.* **6**, 1047-1053.

20. Yin, H. and Chow, G.M. (2000) Anomalous Fe deposition of Fe-Ni films by polyol processing, *submitted*.

CHEMICAL PHENOMENA AT THE SURFACE OF NANOPARTICLES

Marie-Isabelle BARATON
University of Limoges
Faculty of Sciences, SPCTS – UMR 6638 CNRS
F-87060 Limoges (France)
e-mail: baraton@unilim.fr

1.   Introduction

In principle, the precise knowledge of the composition and the structure of a material allows the determination of the bulk properties. In a reverse way, the reproducibility of the composition and structure of the material should ensure the reproducibility of the material properties. However, when the size of a solid is decreased down to the nanometer scale, the overall properties of the nanomaterial are no longer controlled by the bulk structure but by the surface properties. Surface characterization then becomes a necessary prerequisite for the control of nanopowders.

The aim of this work is to demonstrate that Fourier transform infrared (FTIR) spectrometry can be successfully applied to the study of the chemical reactions taking place at the surface of semiconductor nanomaterials and to the analysis of the effects of these surface reactions on the semiconductor electrical conductivity. These fundamental investigations constitute an important step toward improvements of resistive gas sensors fabricated from nanosized powders by screen-printing technology. Examples of titanium oxide and tin oxide nanopowders are discussed. Both materials are semiconductors and are widely used in the fabrication of gas sensors. Chemical modifications of the surface species are monitored *in situ* by FTIR spectrometry and are proved to affect the gas sensing properties. The results presented in this chapter concern nanopowders but the study of thin layers can be similarly performed by diffuse reflectance infrared spectroscopy (DRIFTS). Previous studies have demonstrated that comparable results were obtained on nanopowders by infrared transmission analysis and on thin layers by DRIFTS measurements [1].

2.   Definition of a Surface and its Importance in Nanopowders

Fundamental questions arising about surfaces can be formulated as follows:
   1) How thick should a surface layer be considered ?
   2) What is the chemical composition of a surface ?

45

M.-I. Baraton and I. Uvarova (eds.),
Functional Gradient Materials and Surface Layers Prepared by Fine Particles Technology, 45–60.
© 2001 *Kluwer Academic Publishers. Printed in the Netherlands.*

3) What is the best technique to analyse the surface chemical composition ?

There is no one and only answer to the first question. Indeed, the depth of a surface layer is defined by the phenomenon to be studied [2]. For example, adsorption only concerns the adsorbed molecules or atoms and the very first atomic layer of the solid. On the contrary, changes in the electrical properties of a semiconductor usually affect over one hundred atomic layers.

It is well-known that any surface is more or less contaminated by the ambient environment and therefore foreign atoms are always present in the first atomic layers of materials. Therefore, the chemical composition of a surface has to be specifically studied. This is particularly critical in the case of nanoparticles which can actually be considered as surfaces in three dimensions [3]. The choice of the technique for such studies is not straightforward.

Indeed, the choice of a technique for surface analysis depends on the phenomenon or property to be studied. The thickness of the surface layer that will be perturbed by the considered phenomenon should be compared to the depth that the envisaged method can sample. In addition, it must be kept in mind that the depth of sampling of any technique does not exclusively depend on the material and on the energy of the probing particles (either electrons, or ions, or photons). If only one phenomenon has to be studied, it is quite possible to find the relevant technique for surface analysis with the appropriate depth of sampling. However, when two phenomena are strongly related, although involving different surface thicknesses, the choice of a technique for the simultaneous study of the two phenomena becomes extremely limited.

In addition, the tremendous problem with surfaces is that their composition, structure and reactivity are dependent not only on the synthesis conditions but also on the surrounding environment of the material. Indeed, a surface is in constant evolution to balance the energy at the interface. Current techniques for surface analysis can easily detect contamination in the order of 1% of a monolayer [2,3]. So, we can consider that a surface is clean if its contamination is below this detection limit. But to maintain a clean surface for about one hour a ultra high vacuum of 10^{-9} Torr is required [3]. Most of the time, surfaces are exposed to gases, liquids or solids and therefore contaminated by atoms or molecules from the ambient atmosphere. Therefore, one can never define the surface composition of any material without describing its surrounding environment. The consequence is that the surface properties may depend on the environmental conditions. Obviously, this consideration applies to nanoparticles.

In nanoparticles, the role played by the surface becomes critical. For example, in nanocrystals considered as 2 nm diameter spheres, 50% of the atoms are located on the surface [4]. The high reactivity of the nanoparticle surface leads to a high level of surface contamination. In addition, the forces between nanoparticles are responsible for a strong agglomeration, thus preventing the dispersion of nanoparticles in liquids and polymer matrices as well as their consolidation with a reduced porosity. The remedy is the control of the surface chemistry either during the synthesis process or by surface chemical modification afterwards.

Therefore, a control of the nanoparticles properties cannot be achieved without the control of their surface. This is an important drawback which can be turned into an advantage, since controlled modifications of the surface chemical composition can be used to tailor the nanoparticle properties.

3. Presentation of a Versatile Technique for Surface Analysis

Although Fourier transform infrared spectrometry is not usually considered as a sensitive technique for surface analysis, it can be a high-performance tool for the surface analysis of nanosized powders, under particular conditions [5]. This chapter does not intend to extensively describe the technique, but a complete review can be found in [5-7]. Only a brief summary is given here as a reminder. Moreover, this technique can be successfully used to follow the variations of the electrical conductivity of a semiconductor material when modifying its environment.

3.1. FOURIER TRANSFORM INFRARED SURFACE SPECTROMETRY

As explained above, once a clean surface is in contact with the atmosphere it becomes covered with various adsorbed species. Indeed, the existence of a surface means that bonds are broken and that the atoms at the very surface are in low coordination number and therefore highly reactive. Molecules surrounding a clean surface will therefore react with these unsaturated surface atoms to balance the forces at the interface. One of the most probable reaction occurring at a surface is the dissociation of water to form hydroxyl groups. These hydroxyl groups have a very important role in the surface chemistry. Depending on the nature and coordination of the surface atom to which they are bonded, these hydroxyl groups can be proton donor (Brönsted acid site) or proton acceptor (Brönsted base site). Once the surface is hydrolyzed, it is extremely difficult to remove these surface OH groups. Even though a heat treatment (up to 1000°C) under dynamic vacuum may eventually lead to a complete dehydroxylation of the surface, usually it simultaneously causes a surface reconstruction and the structure of the original clean surface is definitely lost. The study of these surface hydroxyl groups, and particularly that of their vibrational frequencies, brings information on the coordination number and chemical state of the surface atoms to which the OH groups are bonded. Other molecules can also adsorb on and eventually react with a clean surface, such as CO and CO_2 which may lead to surface carbonates formation. Some of these adsorbed species can be more or less easily removed from the surface by a heat treatment under dynamic vacuum, referred to as *activation*. It is obvious that the desorption of these species by activation leaves the surface in a non-equilibrium state. This non-equilibrium state of the surface can only exist under dynamic vacuum and when any molecule is in contact with this activated surface, it rapidly adsorbs and possibly reacts to restore the equilibrium. When these molecules are relevantly chosen, they can selectively adsorb on electron acceptor surface sites (Lewis acid sites) or electron donor surface sites (Lewis base sites). They may additionally interact with the OH surface groups.

To characterize a surface, Fourier transform infrared spectrometry is employed in two distinct and complementary ways. The first way is the use of FTIR spectrometry to determine the chemical nature of the surface chemical species which can be for example hydroxyl, carbonate, nitrate groups. The second way is to study the infrared spectrum of probe-molecules purposely adsorbed on an activated surface. The vibrational frequencies of these adsorbed molecules are perturbed with respect to those of the gas phase. The perturbations depend not only on the nature of the surface atoms to which the probe-molecules are bonded, but also on the coordination number of these atoms within the surface.

Two main types of adsorption processes have been more or less arbitrarily discriminated: *physisorption*, which causes only slight modifications in the infrared spectrum of the adsorbed molecule compared to that of the gas phase, and *chemisorption* which may lead to very strong perturbations of the spectrum, particularly in case of dissociative reactions. Obviously, both physisorption and chemisorption can simultaneously occur. Discrimination of physisorbed and different types of chemisorbed species can be obtained by thermal desorption. Details on the characterization procedure along with examples of surface characterization can be found in [5-11].

It must be emphasized, that in this case, surface characterization concerns only the *very first atomic layer* of the material.

3.2. ELECTRICAL CONDUCTIVITY VARIATIONS OF SEMICONDUCTORS

When the nanosized powder to be analyzed is a semiconductor material, FTIR spectrometry can bring additional information. Indeed, according to the Drude-Zener theory, the energy absorbed by a semiconductor sample over the total infrared spectrum is partly due to the free carriers [12,13]. Therefore, an increase of the free carrier density leads to an increase of the overall infrared absorption. At the upper limit, a conductor material, such as a metal, is opaque to the infrared radiation.

Let us assume that the variations of the electrical conductivity are induced by adsorption of oxidizing or reducing gases on the semiconductor surface. Indeed, the adsorption of oxygen, for example, on a semiconductor, causes a decrease of the electron density in the conduction band due to the formation of ionosorbed oxygen species, such as O^- or O_2^-. As a consequence, the electrical conductivity of a n-type semiconductor, such as tin oxide or titanium oxide, decreases. On the contrary, when a reducing gas, such as CO, adsorbs, electrons are injected into the conduction band and the electrical conductivity increases. The effects of these adsorptions on the infrared spectrum of the n-type semiconductor are a decrease of the overall IR absorption under oxidizing gases and an increase of the overall IR absorption under reducing gases. In terms of infrared energy transmitted by the sample, an increase of the electrical conductivity leads to a decrease of the transmitted infrared energy whereas a decrease of the electrical conductivity leads to an increase of the transmitted infrared energy. It must be noted that the variation of the free carrier density is related to the thickness variation of the depletion layer which usually concerns *several nanometers* (around 3 nm in tin oxide [14]).

From the previous section (cf. section 3.1), it appears that FTIR spectrometry can be used to monitor *in situ* the chemical reactions taking place at the first atomic layer. In addition, the variation of the overall infrared absorption brings information on the variation of the free carrier density in a thickness of a few nanometers. In other words, FTIR spectrometry allows the simultaneous study of two strongly correlated phenomena involving different thicknesses: i) the surface chemical reactions and ii) their consequences on the electrical conductivity. The correlation between these two phenomena constitutes the fundamental mechanism of the gas detection by semiconductor-based sensors.

4.   Chemical Surface Characterization of Semiconductor Nanopowders

As explained above and detailed in [5,6], the first step in the standard procedure for surface characterization of nanosized powders consists in heating the nanopowder under dynamic vacuum to eliminate physisorbed and weakly chemisorbed species (activation). However, this heat-treatment cannot be directly applied to semiconductor materials, such as tin oxide. Indeed, when a semiconductor sample is activated, oxygen atoms desorb from both surface and bulk. This leads to a strong reduction of the sample and, in the case of a n-type semiconductor, the electrical conductivity increases. This conductivity increase resulting from an increase of the free carrier density leads to a loss of transparency of the sample to the infrared radiation (cf. section 3.2). As a consequence, after a standard activation a tin oxide (SnO_2) nanopowder, for example, becomes totally opaque to the infrared beam. To avoid the desorption of oxygen, the tin oxide sample is therefore activated under few mbar of oxygen.

Although titanium oxide (titania, TiO_2) is also a n-type semiconductor material, the desorption of oxygen atoms under activation is not as drastic as in the case of tin oxide. The activation of the titania nanopowder can be performed according to the standard procedure even though an increase of the sample absorption over the total infrared range is observed, corresponding to the expected electrical conductivity increase.

4.1. TITANIUM OXIDE

A commercial titania nanopowder (Degussa P25) with a particle size of 21 nm has been used for this study. The infrared spectrum of the nanopowder at room temperature and under vacuum is presented in Figure 1a along with the spectrum of the same sample recorded after activation at 450°C (Figure 1b). The disappearing of the very intense band centered around 3320 cm^{-1} indicates the elimination of water molecules bonded to the surface OH groups [6].

At the end of the activation treatment (Figure 1b), several bands in the 3780-3600 cm^{-1} range are observed. All of them are assigned to ν(OH) stretching vibrations [15,16]. The multiplicity of these bands indicates that several types of OH groups are present on the titania surface. According to the literature [16-22], the number and the vibrational frequencies of these OH groups depend on the extent of the dehydroxylation and on the possible presence of impurities, which makes the comparison between different samples and between the results from different works extremely difficult. Moreover, this particular commercial sample is a mixture of anatase and rutile phases and the OH groups on the anatase surface are not exactly the same as those on the rutile surface [21]. It must also be taken into account in the band assignments that a hydroxyl group can be bonded to one titanium atom, or linked to two titanium atoms, or bridged to three titanium atoms. As a consequence, all these OH surface groups have a different reactivity and it has been proved that some of them are Brönsted acid sites whereas some other ones are Brönsted base sites [23].

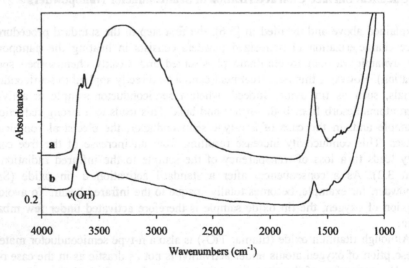

Figure 1. Infrared spectra of the TiO$_2$ nanopowder:
a) at room temperature and under vacuum; b) after activation at 450°C.
(The spectra have been shifted for clarity sake).

4.2. TIN OXIDE

The tin oxide nanopowder investigated in this study has been synthesized by laser evaporation of commercial micronsized SnO$_2$ powder [24]. The average particle size is 15 nm and the nanopowder is crystallized in the rutile form.

As previously explained, the tin oxide nanopowder is activated under a low oxygen pressure (50 mbar). Figure 2 compares the spectrum recorded at room temperature and under vacuum with the spectrum recorded after activation at 400°C under oxygen. Like in the case of titania, a decrease of the broad band centered at 3200 cm^{-1} is observed although weaker. It corresponds to the elimination of water molecules adsorbed on the surface. In addition, the multiplicity of the ν(OH) bands, observed in the 3800-3000 cm^{-1} region after activation, indicates a large number of different types of OH groups [25]. They originate from the diversity of the coordination types of the tin atoms within the surface. Indeed, it is known that Sn^{2+} and Sn^{4+} along with oxygen vacancies can be present on the surface [26]. The bands in 1500-1000 cm^{-1} region are assigned to the corresponding δ(OH) bending vibrations of these OH groups.

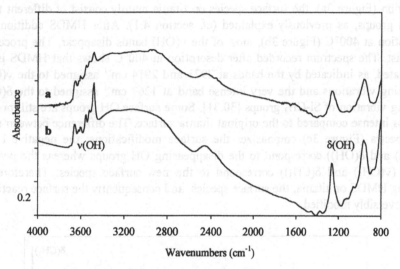

Figure 2. Infrared spectra of the SnO$_2$ nanopowder:
a) at room temperature and under vacuum; b) after activation at 400°C under 50 mbar O$_2$.
(The spectra have been shifted for clarity sake).

4.3. SURFACE MODIFICATION OF TITANIA

It is well-known that any modification of the surface groups affects both the surface reactivity and, in the case of a semiconductor, the work function [27,28]. It is therefore expected that the gas sensing properties of a semiconductor, which are related to both the reactivity and the electrical properties, will be strongly perturbed if the chemical surface species are modified [29,30]. In particular, selectivity and sensitivity of gas sensors can, in principle, be improved by modifying the surface chemistry of the nanopowder used to fabricate the gas sensor by screen-printing technology [30].

Surface modifications of the commercial titania nanopowder have been monitored *in situ* by FTIR spectrometry. Hexamethyldisilazane (HMDS) is widely used for surface modifications because it exhibits a reasonably high vapor pressure at room temperature, it is easy to handle and environmentally friendly. It usually reacts with surface hydroxyl groups leading to ammonia formation according to the following reaction [30-32]:

$$2 \text{ Ti-OH} + (CH_3)_3Si\text{-}(NH)\text{-}Si(CH_3)_3 \rightarrow 2 \text{ Ti-O-Si}(CH_3)_3 + NH_3$$

To graft HMDS on titania, the titania nanopowder is first activated at 400°C. Then, after cooling at room temperature under dynamic vacuum, 7 mbar of HMDS vapor is introduced in the cell. After a few minutes contact, HMDS is thermally desorbed up to 400°C. All the steps of the grafting process are followed *in situ* by recording the infrared spectra [33].

Figure 3 shows the evolution of the titania surface under grafting. After activation (Figure 3a), the surface species on titania mainly consist of different types of OH groups, as previously explained (cf. section 4.1). After HMDS addition and desorption at 400°C (Figure 3b), most of the $\nu(OH)$ bands disappear. The process is very fast. The spectrum recorded after desorption at 400°C shows that HMDS is not eliminated, as indicated by the bands at 2973 and 2914 cm^{-1} assigned to the $\nu(CH_3)$ stretching vibrations and the very intense band at 1267 cm^{-1} assigned to the $\delta(CH_3)$ bending vibration of Si-CH$_3$ groups [30,31]. Some surface OH groups are still present but less intense compared to the original titania surface. The difference between these two spectra (Figure 3c) emphasizes the surface modifications: the negative bands ($\nu(OH)$ and $\delta(OH)$) correspond to the disappearing OH groups whereas the positive bands ($\nu(CH_3)$ and $\delta(CH_3)$) correspond to the new surface species. Therefore, by grafting HMDS on titania, the surface species, and consequently the surface reactivity, are irreversibly modified.

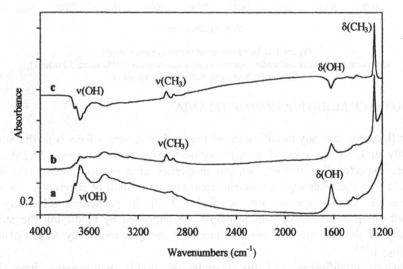

Figure 3. Infrared spectra of the TiO$_2$ nanopowder:
a) after activation at 400°C; b) after HMDS grafting; c) difference spectrum b-a
(The spectra have been shifted for clarity sake).

5. Application of FTIR Spectrometry to the Study of Gas Sensing Properties

As explained in section 3.2, the variation of the infrared energy transmitted by a semiconductor sample when its gaseous environment is changing, is related to the variation of the electrical conductivity. In this section, results obtained on tin oxide and titanium oxide nanopowders are discussed

5.1. TIN OXIDE

Tin oxide is one of the most popular material used for the fabrication of semiconductor-based sensors for CO detection. The chemical reactions leading to changes in the electrical conductivity are usually proposed as follows, depending on the temperature [34-37]:

$$CO + O^- \rightarrow CO_2 + e^-$$
$$CO + O^{2-} \rightarrow CO_2 + 2e^-$$

On the activated and oxidized surface of a tin oxide nanopowder (cf. section 4.2), CO is adsorbed at 350°C in presence of oxygen. The comparison of the infrared spectra recorded before and after CO addition (Figures 4a,b) clearly indicates the formation of CO_2 (band centered at 2348 cm^{-1}) along with carbonate groups (bands in the 1500 cm^{-1} region) [36-38]. These carbonates can be formed first by CO adsorption and then transformation into CO_2 or the first step can be the oxidation of CO into CO_2 which then adsorbs as carbonates according these sequences:

$$CO + O^{2-} \rightarrow CO_3^{2-} \quad \text{and} \quad CO_3^{2-} \rightarrow CO_2 + O^- + e^-$$
$$\text{or} \quad CO_3^{2-} \rightarrow CO_2 + \frac{1}{2}O_2 + 2e^-$$
$$CO + O^{2-} \rightarrow CO_2 + 2e^- \quad \text{and} \quad CO_2 + O^{2-} \rightarrow CO_3^{2-}$$

It must be noted that the formation of carbonate groups (CO_3^{2-}) according to whatever sequence does not participate in the variation of the electrical conductivity because the involved electrons stay localized. Only the formation of CO_2 leads to the release of free carriers.

   In addition to the formation of these new species, an increase of the overall absorption of the sample is observed. Indeed, the baseline of the SnO_2 spectrum shifts toward higher absorbance values under CO addition (Figures 4a,b), due to the increase of the free carrier density.

   Figure 5 shows the variation of the infrared energy *transmitted* by the tin oxide sample versus gas (O_2 and CO) exposures. This curve can be compared to the sensor response curve. A decrease of the electrical conductivity (corresponding to an increase of the transmitted infrared energy) is observed when oxygen is adsorbed whereas an increase of the electrical conductivity (corresponding to a decrease of the transmitted infrared energy) is caused by CO adsorption. When CO is adsorbed in absence of oxygen, a strong reduction of the SnO_2 sample is observed and the oxidation state is

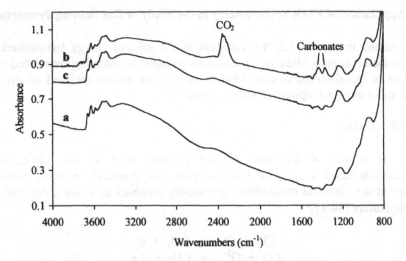

Figure 4. Infrared spectra of the SnO₂ nanopowder at 350°C:
a) under oxygen; b) after CO addition in presence of oxygen; c) after evacuation.
(The spectra have NOT been shifted).

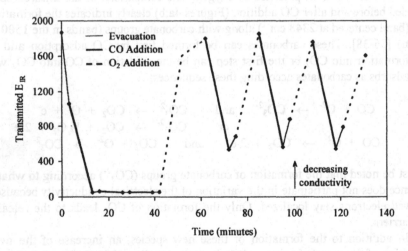

Figure 5. Variations of the infrared energy transmitted by the SnO₂ nanopowder at 350°C versus gas exposures.

not restored by evacuation. In fact, the sample is so reduced that the addition of a second and third CO dose in absence of oxygen is no longer detected by the material due to a saturation. The reproducibility of the sensor response is ensured by the presence of oxygen allowing a complete recovery of the SnO₂ oxidation state after elimination of both CO_2 and carbonate groups by evacuation (Figure 4c).

5.2. TITANIUM OXIDE

An important drawback of the semiconductor-based sensors is their sensitivity to humidity causing adverse effect on the sensor response. The grafting of HMDS on the titania surface, described in section 4.3, by modifying the surface hydroxyl groups, is expected to have an influence on the humidity effects. In this section, the effect of humidity on the sensitivity toward CO is compared for the activated titania ("clean" surface) and the HMDS-grafted titania.

5.2.1. *Response of the Clean TiO₂ Surface Toward Dry CO*

The sensitivity of the titania surface activated at 400°C was checked toward dry CO. A first dose of CO was added for 10 minutes, and then evacuated. Other CO doses were subsequently added. The spectrum of the titania sample recorded after 10 minutes under CO is presented in Figure 6. To emphasize the modifications caused by CO addition, the spectrum of the titania sample before CO addition has been substracted. The formation of CO_2 is observed at 2348 cm^{-1}, responsible for the expected increase of the electrical conductivity through release of electrons in the conduction band. This increase of electrical conductivity translates into a decrease of the infrared energy transmitted by the titania sample, as previously explained (cf. section 3.2). The evolution of the transmitted infrared energy is shown in Figure 7. After evacuation of the first CO dose, the transmitted infrared energy does not get back to its original intensity. Each CO dose causes a decrease of the transmitted infrared energy, whereas each evacuation leads to an increase of the IR energy. But, the sample is steadily reduced by the subsequent CO additions, which is quite expected due to the absence of an oxidizing environment. Simultaneously, a perturbation of the OH surface groups by CO addition is observed on the infrared spectra (Figure 6).

5.2.2. *Response of the Clean TiO₂ Surface Toward Dry and Wet CO*

The second series of experiments on titania consisted in the introduction of four dry CO doses followed by four "wet CO" doses, that is a mixture of CO and water vapor (10%), and then followed by four new dry CO doses. The response toward the first four CO doses is obviously similar to that just described (Figures 8a, 9). The addition of wet CO causes a strong decrease of the electrical conductivity although without any reproducibility. In parallel, the infrared spectra show the intensity increase of the ν(OH) absorption range (Figure 8b). The overall effect of the four wet CO doses is an oxidation. When dry CO is added again (Figures 8c, 9), the energy evolution is similar to that observed during the addition of the first dry CO doses. Moreover, the baseline drift shows the same downward trend, thus indicating an overall reducing effect.

5.2.3. *Response of the HMDS-Grafted Titania toward Dry and Wet CO*

The same sequence of dry and wet CO doses was adsorbed on the HMDS-grafted titania (cf. section 4.3). As in the previous case, the evolution of the transmitted infrared energy versus gas exposures shows the reducing effect of the dry CO doses (Figures 10a, 11). Concomitantly, no perturbation is observed on the IR spectrum and

56

the formation of CO_2 is hardly visible (Figure 10a). The oxidation state of the sample was not restored by evacuation (Figure 11). But, unlike the previous case, the "addition-evacuation" sequences of dry CO doses led to an oxidizing effect. This has been explained by a rearrangement of the surface under CO because, in addition to the modification of the surface species, the HMDS grafting has a reducing effect, partly because of NH_3 formation (cf. section 4.3) [30,31,39].

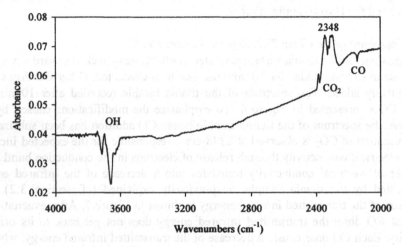

Figure 6. Infrared spectrum of the TiO_2 nanopowder recorded at 400°C after CO addition. The spectrum of TiO_2 recorded before CO addition has been substracted.

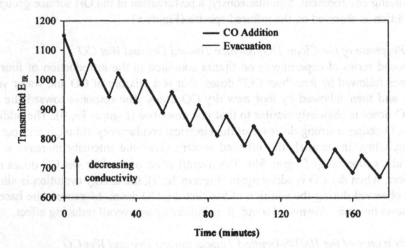

Figure 7. Evolution of the infrared energy transmitted by the "clean" TiO_2 nanopowder versus gas exposures.

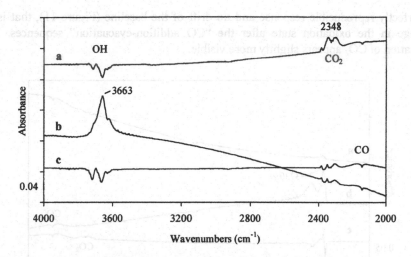

Figure 8. Evolution of the infrared spectrum of the "clean" TiO$_2$ nanopowder during the additions of :
a) a dry CO dose; b) a (CO+H$_2$O) dose; c) a new dry CO dose (cf. text).
(The difference spectra have been shifted for clarity sake).

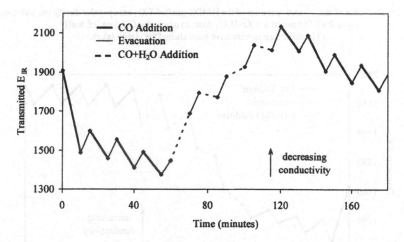

Figure 9. Evolution of the infrared energy transmitted by the "clean" TiO$_2$ nanopowder versus gas exposures.

The addition of the first two wet CO doses amplified the oxidizing effect (Figure 11). At this step, a new band appears at 3730 cm^{-1} (Figure 10b). This band has been assigned to the formation of new Si-OH surface groups by reaction of water with the Si(CH$_3$)$_3$ grafted groups [31,39]. Then, for the subsequent two wet CO doses, the reducing action becomes preponderant and no further changes are noted in the IR spectrum.

When dry CO is added again, a reversible shift of the 3730 cm^{-1} band is slightly visible (Figure 10c). But, surprisingly, the transmitted infrared energy evolution shows

58

a perfectly reproducible response and no drift of the baseline (Figure 11), that is no change in the oxidation state after the "CO addition-evacuation" sequences. The formation of CO_2 appears slightly more visible.

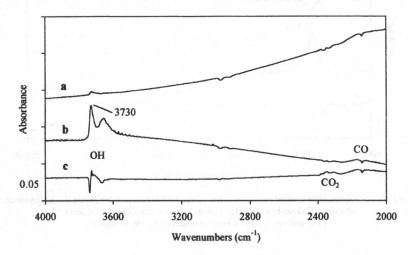

Figure 10. Evolution of the infrared spectrum of the HMDS-grafted TiO_2 nanopowder during the additions of : a) a dry CO dose; b) a $(CO+H_2O)$ dose; c) a new dry CO dose (cf. text). (The difference spectra have been shifted for clarity sake).

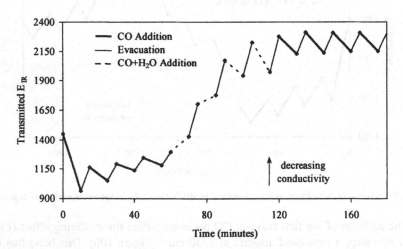

Figure 11. Evolution of the infrared energy transmitted by the HMDS-grafted TiO_2 nanopowder versus gas exposures.

Although the reason for this apparent stability of the response is not totally understood yet, these results clearly demonstrate that, by changing the chemical species on a nanopowder surface, the overall properties of the material are affected like, in this particular case, the response of a semiconductor to gases. Therefore, a specific property of a nanopowder can be modified on purpose by controlling the surface composition and the surface chemistry. The obvious counterpart is that a contamination of the surface may adversely affect the desired property.

6. Conclusion

In conclusion, surface studies, and particularly those of nanosized powders, are generally difficult to perform mainly because the requirement of adequacy between the property to be analyzed and the investigation tool is not easily satisfied.

Fourier transform infrared spectrometry can be successfully applied to the investigation of nanosized powders. The first atomic layer can be characterized as far as the chemical composition is concerned. But, an important achievement of FTIR spectrometry, when applied to semiconductor materials, is to allow the analysis of the chemical reactions and interactions taking place at the very first atomic layer simultaneously with the study of the electrical conductivity variations under different environments.

The results obtained in the field of gas sensors clearly demonstrate that a careful control of the surface composition and of the surface reactivity is essential for reproducible and optimized overall properties of nanopowders.

7. References

1. Baraton, M.-I. (2000) Surface Characterization of Nanostructured Coatings: Study of Nanocrystalline SnO$_2$ Gas Sensors, in G.M. Chow *et al.* (eds.), *Nanostructured Films and Coatings*, NATO-ASI Series, Kluwer Academic Publishers, Dordrecht, pp. 187-201.
2. Somorjai, G.A. (1990) *Introduction to Surface Chemistry and Catalysis*, Wiley, New York.
3. Somorjai, G.A. (1998) From Surface Materials to Surface Analysis, *MRS Bulletin* 23(5), 11-29.
4. Veprek, S. (1997) Electronic and Mechanical Properties of Nanocrystalline Composites when Approaching Molecular Size, *Thin Solid Films* 297, 145-153.
5. Baraton, M.-I. (1999) FTIR Surface Spectrometries of Nanosized Particles, in H.S. Nalwa (ed.), *Handbook of Nanostructured Materials and Nanotechnology*, Academic Press, San Diego, pp. 89-153.
6. Baraton, M.-I. (1998) The Surface Characterization of Nanosized Powders: Relevance of the FTIR Surface Spectrometry, in G.M. Chow *et al.* (eds.), *Nanostructured Materials*, NATO-ASI Series, Kluwer Academic Publishers, Dordrecht, pp. 303-317.
7. Baraton, M.-I. (1994) IR and Raman Characterization of Nanophase Ceramic Materials, *J. High Temp. Chem. Processes* 3, 545-554.
8. Boehm, H.-P. and Knözinger, H. (1983) Nature and Estimation of Functional Groups on Solid Surfaces, in J.R.A. Anderson and M. Boudart (eds.), *Catalysis* Vol. 4, Springer-Verlag, Berlin, pp. 39-207.
9. Hair, M.L. (1967) *Infrared Spectroscopy in Surface Chemistry*, M. Dekker, New York.
10. Knözinger, H. (1976) Specific Poisoning and Characterization of Catalytically Active Oxide Surfaces, *Advances in Catalysis* 25, 184-201.
11. Davydov, A.A. (1984) *Infrared Spectroscopy of Adsorbed Species on the Surface of Transition Metal Oxides*, John Wiley & Sons, New York.
12. Harrick, N.J. (1962) Optical Spectrum of the Semiconductor Surface States from Frustrated Total Internal Reflection, *Phys. Rev.* 125(4), 1165-1170.
13. Baraton M.-I., Merhari L., Chancel F., and Tribout J. (1997) Chemical characterization by FT-IR spectrometry and modification of the very first atomic layer of a TiO$_2$ nanosized powder, *MRS Symp. Proc. Vol. 448*, MRS Publisher, Warrendale, pp. 81-86.

60

14. Ogawa, H., Nishikawa, M., and Abe A. (1982) Hall Measurement Studies and an Electical Conduction Model of Tin Oxide Ultrafine Particle Films, *J. Appl. Phys.* **53**, 4448-4455.
15. Primet, M., Pichat, P., and Mathieu, M.-V. (1971) Infrared Study of the Surface of Titanium Dioxides. I. Hydroxyl Groups, *J. Phys. Chem.* **75**(9), 1216-1220.
16. Morrow, B.A. (1990) Surface Groups on Oxides, in J.L.G. Fierro (ed.), *Spectroscopic Characterization of Heterogeneous Catalysis* (Part A), Elsevier, Amsterdam, pp. A161-A224.
17. Busca, G., Saussey, H., Saur, O., Lavalley, J.-C., and Lorenzelli, V. (1985) FT-IR Characterization of the Surface Acidity of Different Titanium Dioxide Anatase Preparations, *Applied Catal.* **14**, 245-260.
18. Morterra, C. (1988) An Infrared Spectroscopic Study of Anatase Properties, *J. Chem. Soc., Faraday Trans. I* **84**(5), 1617-1637.
19. Ho, S.-W. (1996) Surface Hydroxyls and Chemisorbed Hydrogen on Titania and Titania Supported Cobalt, *J. Chinese Chem. Soc.* **43**, 155-163.
20. Tsyganenko, A.A. and Filimonov, V.N. (1972) Infrared Spectra of Surface Hydroxyl Groups and Crystalline Structure of Oxides, *Spectr. Letters* **5**(12), 477-487.
21. Primet, M., Pichat, P., and Mathieu, M.-V. (1968) Etude par Spectrométrie Infrarouge des Groupes Hydroxyles de l'Anatase et du Rutile, *C. R. Acad. Sc. Paris* **267B**, 799-802.
22. Yates, D.J.C. (1961) Infrared Studies of the Surface Hydroxyl Groups on Titanium Dioxide, and of the Chemisorption of Carbon Monoxide and Carbon Dioxide, *J. Phys. Chem.* **65**, 746-753.
23. Primet, M., Pichat, P., and Mathieu, M.-V. (1971) Infrared Study of the Surface of Titanium Dioxides. II. Acidic and Basic Properties, *J. Phys. Chem.* **75**(9), 1221-1226.
24. Riehemann, W. (1998) Synthesis of Nanoscaled Powders by Laser-Evaporation of Materials, in K.E. Gonsalves, M.-I. Baraton *et al.* (eds.), *MRS Symp. Proc. Vol. 501*, MRS Pub., Warrendale, USA, pp. 3-13.
25. Thornton, E.W. and Harrison, P.G. (1975) Tin Oxide Surfaces. I. Surface Hydroxyl Groups and the Chemisorption of Carbon Dioxide and Carbon Monoxide on Tin(IV) Oxide, *J. Chem. Soc., Faraday Trans. I* **71**, 461-472.
26. Cox, D.F., Fryberger, T.B., and Semancik S. (1988) Oxygen Vacancies and Defect Electronic States on the SnO$_2$(110)-1x1 Surface, *Phys. Rev. B* **38**(3), 2072-2083.
27. Many, A., Goldstein, Y., and Grover, N.B. (1965) *Semiconductor Surfaces*, North-Holland Publishing Co, Amsterdam.
28. Morrison, S.R. (1990) *The Chemical Physics of Surfaces*, Plenum Press, New York.
29. Bruening, M., Cohen, R., Guillemoles, J.F., Moav, T., Libman, J., Shanzer, A., and Cahen, D. (1997) Simultaneous Control of Surface Potential and Wetting of Solids with Chemisorbed Multifunctional Ligands, *J. Amer. Chem. Soc.* **119**, 5720-5728.
30. Chancel, F., Tribout, J., and Baraton, M.-I. (1998) Effect of Surface Modification on the Electrical Properties of TiO$_2$ and SnO$_2$ Nanopowders, in K.E. Gonsalves, M.-I. Baraton *et al.* (eds.), *MRS Symp. Proc. Vol. 501*, MRS Pub., Warrendale, USA, pp. 89-94.
31. Chancel, F., Tribout, J., and Baraton, M.-I. (1997) Modification of the Surface Properties of a Titania Nanopowder by Grafting: A Fourier Transform Infrared Analysis, *Key Eng. Mater.* **136**, 236-239.
32. Hertl, W. and Hair, M.L. (1971) Reaction of Hexamethyldisilazane with Silica, *J. Phys. Chem.* **75**(14), 2181-2185.
33. Baraton, M.-I., Chancel, F., and Merhari, L. (1997) In Situ Determination of the Grafting Sites on Nanosized Ceramic Powders by FTIR Spectrometry, *Nanostructured Materials* **9**, 319-322.
34. Henrich, V.E. and Cox, P.A. (1994) *The Surface Science of Metal Oxides*, Cambridge University Press, Cambridge.
35. Clifford, P.K. (1981) *Mechanisms of Gas Detection by Metal Oxide Surfaces*, Ph.D. Thesis, Carnegie Mellon Univ., Pittsburg.
36. Harrison, P.G. and Willett, M.J. (1988) The Mechanism of Operation of Tin(IV) Oxide Carbon Monoxide Sensors, *Nature* **332**(6162), 337-339.
37. Willett, M.J. (1991) Spectroscopy of Surface Reactions, in Mosley, P.T., Norris, J.W.O., and Williams D.E. (eds.), *Techniques and Mechanisms in Gas Sensing*, Adams Hilger, Bristol, pp. 61-107.
38. Busca, G. and Lorenzelli, V. (1982) Infrared Spectroscopic Identification of Species Arising from Reactive Adsorption of Carbon Oxides on Metal Oxide Surfaces, *Mater. Chem.* **7**, 89-126.
39. Baraton, M.-I. and Merhari, L. (1998) Surface Properties Control of Semiconducting Metal Oxides Nanoparticles, *NanoStruct. Mater.* **10**(5), 699-713.

AMPHIPHILIC TEMPLATES IN THE SYNTHESIS OF NANOSTRUCTURED COMPOSITES – FROM PARTICLES TO EXTENDED STRUCTURES

Sichu Li, Limin Liu, Blake Simmons, Glen Irvin, Christy Ford, Vijay John*,
Gary L. McPherson
Tulane University
Chemical Engineering Department, New Orleans, LA 70118

Arijit Bose, Paul Johnson
University of Rhode Island
Chemical Engineering Department, Kingston, RI 02881

Weilie Zhou, Charles O'Connor
University of New Orleans
Advanced Materials Research Institute, New Orleans, LA 70122

1. Abstract

Surfactant self-assembly is used to develop templates for materials synthesis. The reverse micelle microstructure of the anionic, AOT, serves as a template for the enzymatic synthesis of polymer microspheres and the encapsulation of nanoparticles within these microspheres. Transformation from reverse micelles to a rigid organohydrogel structure is conducted through the addition of the zwitterionic phospholipid, lecithin. The gel serves as a template for the extended synthesis of silica networks and provides opportunities to synthesize polymer-inorganic structured nanocomposites.

2. Introduction

The tendency of amphiphilic molecules to self-assemble to well-defined microstructures has significant implications in the synthesis of composite materials. Surfactants are examples of such amphiphilic species that aggregate to a variety of microstructures such as micelles and inverse micelles, bilayer vesicles, and lyotropic liquid crystals. These self-assembling microstructures have tremendous potential as scaffolds and templates in the synthesis of nanostructured materials. There are several aspects to such synthesis. First, the geometry of the self-assembled structure may serve to template the geometry of the resulting material.[1] Second, the ability of amphiphiles

M.-I. Baraton and I. Uvarova (eds.),
Functional Gradient Materials and Surface Layers Prepared by Fine Particles Technology, 61–67.
© 2001 *Kluwer Academic Publishers. Printed in the Netherlands.*

to tremendously increase oil-water contact areas indicates that aqueous phase materials synthesis may be combined with organic phase materials synthesis. This aspect in principle, allows the synthesis of polymer-ceramic nanocomposites with unique structure-function properties. In recent years, there has been significant interest in using self-assembled surfactant microstructures as templating media for the synthesis of novel materials. Mesoporous solids[2], porous polymers[3], biomimetic ceramics[4], etc. are just some of the fascinating examples of materials that have been synthesized in surfactant systems. The literature in this area is vast, and there is tremendous potential for use in the development of materials for structural and device applications.

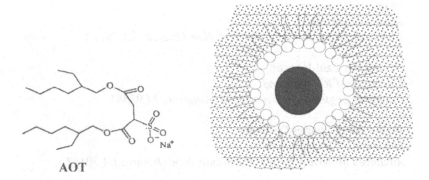

AOT

Figure 1: Structure of AOT (left) and of a reverse micelle (right). The presence of inorganic nanoparticles is also shown.

Our research is directed towards a systematic understanding of the role of surfactant templating, using a model twin-tailed surfactant AOT (bis 2-ethylhexyl sodium sulfosccinate). Due to its structure, AOT has a packing parameter $P=v/al$ greater than unity, where v is the hydrocarbon chain volume, a, the headgroup area, and l, the effective hydrocarbon chain length.[5] The surfactant therefore tends to form water-in-oil microemulsions conventionally termed as inverse micelles or reverse micelles, at relatively low concentrations (above the cmc). Figure 1 illustrates the structure of the surfactant and its assembly into reverse micelles containing stabilized water microdroplets in an organic solvent. The reverse micellar environment has been used extensively in the synthesis of inorganic nanoparticles, since the aqueous droplets serve as compartmentalized microreactors for such synthesis.[6] The water pools serve as microreactors for the synthesis of the inorganic materials, and restrict the particle growth to the nanoscale. This system has been successfully used in the synthesis of semiconductor quantum dots and superparamagnetic ferrites with novel optical and magnetic properties.[6,7]

In this paper, we describe how the reverse micellar solution can be exploited for the synthesis of polymer-nanoparticle composites. We then illustrate methods to convert the reverse micellar solution into a rigid organohydrogel by the addition of a specific dopant to the system. The transformation of a liquid micellar solution to a gel state leads to a rigid system with a very high oil-water interfacial area. In specific instances, these gels can sustain equal volumes of a microaqueous phase and a hydrocarbon phase thus allowing combinations of aqueous phase synthesis and organic phase synthesis.

3. Nanocomposite Particle Synthesis in Reverse Micelles.

We describe an approach to the synthesis of polyphenolics coupled to semiconductor and magnetic particle synthesis. Polyphenol synthesis is catalyzed by an oxidative enzyme (horseradish peroxidase), present in the water pools of the micelles. Figure 2 illustrates the reaction mechanism, and the system. The reaction proceeds through formation of the phenoxy radical which migrates to ortho positions on the aromatic ring, following which chain-monomer and chain-chain coupling occurs (Figure 2a). The monomer (4-ethylphenol as an example) when added to the solution, partitions to the oil-water interface as a consequence of its own amphiphilicity, and hydrogen bonds to the AOT carbonyl and sulfonate head groups (Figure 2b).[8] The reaction therefore proceeds at the oil-water interface.

The enzymatic polymerization of phenols and aromatic amines is very feasible in the reverse micellar environment.[9] The polymer that is formed precipitates out of solution in the form of connected microspheres (Figure 2c).

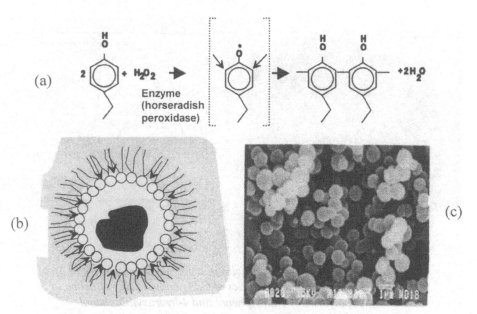

Figure 2: (a) reaction mechanism (b) monomer partitioning to the oil-water interface (c) morphology of precipitated polymer.

In our recent work, we have shown that the precipitating polymer encapsulates intramicellar solutes such as inorganic nanoclusters or proteins.[10,11] Of particular interest is the encapsulation of magnetic ferrites that are first synthesized in the micelles. On conducting polyphenol synthesis subsequently, we observe that the ferrite particles are largely taken up into the polymer matrix, a ship-in-a-bottle approach to microencapsulation. Figure 3a illustrates a cross-section transmission electron micrograph of one of the polymer particles showing the presence of ferrite nanoparticles in the matrix. A similar approach is used in the attachment of CdS nanoparticles.[10] In this case however, we use 4-hydroxythiophenol as a comonomer to synthesize copolymers of 4-ethylphenol and 4-hydroxythiophenol. The thiol groups bind oxidatively to the semiconductor nanoparticles, and the particle is covalently attached to the polymer. The polymer can therefore be dissolved in a solvent with retention of particle attachment. Figure 3b illustrates the fluorescence characteristics of these polymers that are doped with CdS nanoparticles. As the hydroxythiophenol content of the polymer increases, the polymer is able to bind increasing levels of CdS.

Thus, we have developed a rather general method of polymer-nanoparticle composite preparation. The templating aspect of the oil-water interface in reversed micelles to conduct interfacial polymerization leads to encapsulation of intramicellar solutes. Both the peroxidase enzyme and other enzymes solubilized in the micelles can also be incorporated into the polymer leading to additional interesting aspects in the preparation of polymer-enzyme composites.[12]

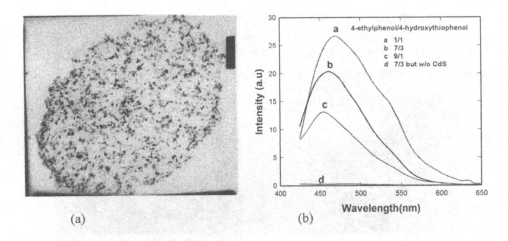

(a)   (b)

Figure 3: (a) *Ferrite nanoparticles in a polymer microsphere (0.4 μm)*
(b) *fluorescence characteristics of CdS nanoparticles attached to
copolymers of 4-ethylphenol and 4-hydroxythiophenol*

4. Synthesis of Composites with Extended Nanostructures

In our current work, our objective is to prepare extended nanostructures using the templating aspect of surfactants that self-assemble to gels. We have found that the

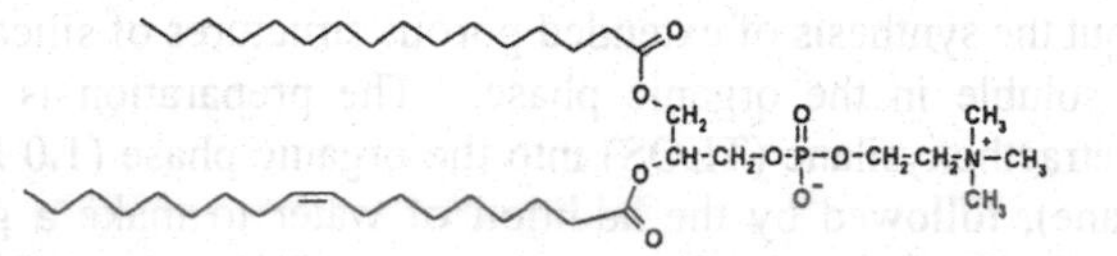

Figure 4: Structure of phosphatidylcholine (lecithin)

addition of lecithin (phosphatidylcholine) (Figure 4) to AOT water-in-oil microemulsions can result in the formation of a rigid gel when additional water is added to the system.[13] In other words, the gel phase can be sustained with equal volume fractions of water and the organic phase, implying the presence of spatially immobilized and extended hydrophilic and hydrophobic microstructures. The gel is neither an organogel nor a hydrogel, but an immobilized bicontinuous structure of water and hydrocarbon networks. A typical gel has an AOT/lecithin/isooctane/water composition of 15/14/29/42 (wt%).

The fact that the gel can sustain equal volumes of an aqueous phase and an organic hydrocarbon phase implies that this might be a medium in which aqueous phase synthesis can be combined with organic phase synthesis, leading to structured composite materials with novel application possibilities. For example, the organic phase could be used in the synthesis of hydrophobic polymers while the aqueous phase can be used in the synthesis of hydrophilic polymers or inorganic materials. Thus, there is the possibility of obtaining structured polymer/polymer nanocomposites or

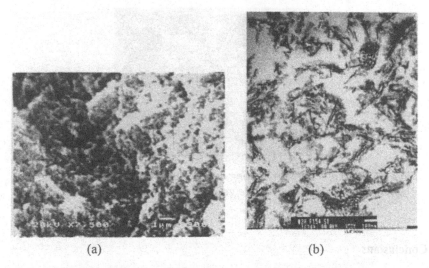

(a)   (b)

Figure 5: SEM (a) and TEM (b) of reticulated silica formed in the gel system. The TEM is at a magnification of 50,000.

polymer/ceramic nanocomposites. The oil-water interface in these systems may also be exploited in interfacial polymerization. And if synthesis is carried out in simply one of the microphases (organic or aqueous), there is the possibility of generating materials with structured porosities. We report two examples of such structured synthesis.

66

We have carried out the synthesis of extended porous structures of silica using a precursor that is soluble in the organic phase. The preparation is started with the incorporation of tetraethoxysilane (TEOS) into the organic phase (1.0 M AOT + 0.4 M lecithin in isooctane), followed by the addition of water to make a gel. The ratio of TEOS to isooctane is 1.3:1, and the final water content W_0 (water/AOT molar ratio) is 140. The gel is then kept at 60 to 70°C for 4 to 5 days to allow the reaction to proceed. After completion of the reaction, the surfactants are removed by washing with isooctane.

Results of the synthesis are shown in the scanning (SEM) and transmission (TEM) electron micrographs of Figure 5a and 5b, respectively. While the precursor (TEOS) is soluble in the organic phase, the hydrolysis products deposit in the aqueous microphase. The reticulated, extended network structure of silica is clearly seen. Such structure evolution may have relevance to biological siliceous structures. The highly open structures may serve to host cells and enzymes, thus making these materials useful in applications related to biomaterials development. As an example of such extended nanocomposites, we carried out the combined synthesis of inorganic silica and polyethylphenol (through peroxidase catalysis). Figure 6 illustrates the SEM of enzymatically synthesized polyethylphenol coating extended silica structures. A significant amount of the porososity of the silica microstructure is retained. The interesting aspect of this synthesis is the demonstration of coupling inorganic materials synthesis with polymer synthesis in microstructured media.

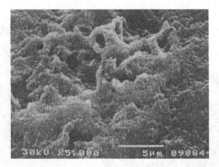

Figure 6: SEM of a polymer coating on macroporous silica. The polymer was synthesized enzymatically in the gel phase together with silica synthesis.

5. Conclusions

Thus, these examples in nanocomposite synthesis add to the class of novel materials synthesized in self-assembling media, either based on surfactant aggregation, block copolymer assembly, or latex templating.[14] Amphiphile self-assembly can be used as building blocks for the structured 3-dimensional synthesis of nanocomposite materials. The approach complements traditional deposition techniques where functionally gradient materials are built by layer-by-layer deposition using gas or liquid phase precursors. While the scaffolding afforded by surfactant self-assembly can be exploited for 3-dimensional structures, there are many aspects to be considered. These include the stability of the scaffold under various pH and ionic strength conditions, and the specific microstructure of the self-assembled system. While there are many emerging examples of templates that lead to network structures, rod-like structures, connected particulates, etc., it is still difficult to design the template to generate a desired morphology. Nevertheless, as we learn more about surfactant self-assembly, it may be possible to understand patterns of morphology generation that will eventually lead to the ability to design specific microstructures.

6. Acknowledgments

Support from DARPA (Grant MDA972-97-1-0003) and NSF (Grant 9909912) is gratefully acknowledged. We also gratefully acknowledge collaboration from Dr. Joseph Akkara and Dr. David Kaplan.

7. References

1. Braun, P.V., Osenar, P., Tohver, V., Kennedy, S. B. and Stupp, S.I. (1999), *J. Am. Chem. Soc.*, **121**, 7302.
2. Kresge, C.T., Leonowicz, M.E., Roth, W.J., Vartuli, J.C. and Beck, J.S. (1994), *Nature* **368**, 321.
3. Schmulhl, N., Davis, E., and Cheung, H.M. (1998), *Langmuir* **14**, 757.
4. Walsh, D., Hopwood, J.D., and Mann, S. (1994), *Science*, **264**, 1576.
5. Israelachvili, J. (1992), *Intermolecular and Surface Forces, Academic Press*, New York.
6. Pileni, M.P. (1997), *Langmuir*, **13**, 3266.
7. Lopez-Quitela, M.A., and Rivas, J. (1993), *J. Colloid Interface Sci.*, **158**, 446.
8. Rao, A.M., John, V.T., Gonzalez, R.D., Akkara, J.A. and Kaplan, D.L. (1993), *Biotechnol. Bioeng.*, **41**, 531.
9. Karayigitoglu, C., Kommareddi, N., John, V.T., McPherson, G., Akkara, A., and Kaplan,D. (1995), *Materials Science and Engineering C: Biomimetic Materials, Sensors and Systems*, **2**, 165.
10. Premachandran, R., Banerjee, S., John, V.T., Mcpherson, G., Akkara, J., and Kaplan, D.L. (1997), *Chemistry of Materials* **9**, 1342.
11. Kommareddi, N.S., Tata, M., John, V.T., McPherson, G.L., Herman, M., Lee, Y.S., O'Connor, C.J., Akkara, J.A., and Kaplan, D.L. (1996), *Chemistry of Materials* **8**, 801.
12. Banerjee, S., Premachandran, R., Wu, K., John, V.T., McPherson, G., Akkara, J., Kaplan, D. (1998) in R.A. Gross, D. Kaplan, G. Swift (eds.) *Enzymes in Polymer Synthesis. ACS Symposium Series*, **684**, 125.
13. Li, S., Irvin, G., Simmons, B., John, V.T., McPherson, G.L., Zhou, W. (2000), *Colloids and Surfaces*, in press.
14. Zhao, D., Yang, P., Chmelka, B.F., and Stucky, G.D., (1999), *Chem. Mater.* **11**, 1174.

EFFECT OF WATER-SOLUBLE ALUMOXYCARBOXYLATES ON THE PROPERTIES OF ALUMINA MOLDED BY DIE PRESSING

M. SZAFRAN, E. ZYGADŁO-MONIKOWSKA, G. ROKICKI,
Z. FLORJAŃCZYK, E. ROGALSKA-JOŃSKA, P. WIŚNIEWSKI
Warsaw University of Technology, Faculty of Chemistry,
ul. Noakowskiego 3, 00-664 Warsaw, Poland

1. Introduction

Studies on obtaining new ceramic materials are at present one of most dynamically developing fields of modern material science. Ceramic materials owe their progress to low thermal conductivity, resistance to corrosive action of many media at elevated temperatures, mechanical strength, etc.

During the molding from non-plastic materials (Al_2O_3, Si_3N, AlN, etc.), various types of organic additives such as dispersants, binders, plasticizers, surfactants and adhesion promoters (coupling agents) are added in order to control the rheological properties, thickening and uniformity of thickening of the powders, as well as high strength of green ceramic samples after powder compaction.

The mechanical strength of green ceramic bodies depends mainly on the polymeric binder-to-ceramic and polymer-to polymer interactions. The presence in the polymer of polar groups, such as carboxylic, hydroxyl, ester, etc. leads to the formation of strong interactions with the surface of ceramic powders and the possibility of formation of hydrogen bonds leads to an increase in the cohesion forces.

The achievement of good wetting of the ceramic powder by the polymer is an essential parameter in obtaining high mechanical strength and elasticity of the ceramic material after drying. Therefore, external non-fugitive plasticizers should be added before drying. The plasticizer also acts as a lubricant during molding (e.g. pressing) of ceramic samples. An appropriate wettability of the ceramic powder achieved by adding a polymeric binder to the composites is one of the parameters of essential importance for obtaining high strength before and after sintering. This is facilitated by coupling agents used as adhesion promoters. Silane compounds are the most known adhesion promoters produced on an industrial scale [1-3]. These compounds contain Si–OR bonds which undergo hydrolysis with the formation of SiOH groups reacting with active groups present on the surface of the ceramic grain. They are mainly used in non-aqueous media.

In this paper the alumoxycarboxylates elaborated by us have been evaluated as coupling agents in the technology of obtaining ceramic materials based on Al_2O_3. Alumoxycarboxylates, due to solubility in water, can be applied in modern processing techniques, in which casting slips containing water are used. A favorable effect of them on the grain surface – polymeric binder interaction has been found, and thus on the

69

M.-I. Baraton and I. Uvarova (eds.),
Functional Gradient Materials and Surface Layers Prepared by Fine Particles Technology, 69–74.
© 2001 *Kluwer Academic Publishers. Printed in the Netherlands.*

thickening ability during pressing and on the properties of the sintered samples. An additional advantage of the aluminum carboxylates used is the fact that they themselves are precursors of a ceramic material and during sintering they undergo transformation to aluminum oxide.

2. Experimental

2.1. SYNTHESIS OF ORGANOALUMINUM COMPOUNDS AND THEIR CHARACTERIZATION

The carboxylates were obtained in two steps. In the first step diethylaluminum alkoxide was obtained according to the following reaction scheme (eqn. 1).

$$(C_2H_5)_3Al + CH_3(OCH_2CH_2)_nOH \longrightarrow (C_2H_5)_2AlO(CH_2CH_2O)nCH_3 + C_2H_6 \uparrow \quad (1)$$

The reactions were carried out in an inert gas atmosphere, in a 100 mL reaction vessel equipped with a stirrer, reflux condenser and dropping funnel. Triethylaluminum (0.0746 mol) in the form of a 25 wt. % solution in toluene was introduced to the reaction vessel, to which an equimolar amount of poly(oxyethylene) glycol methyl monoester containing a various number of ethylene oxide monomeric units (n = 1, 3 or 7) was slowly added dropwise. Ethane evolves during the reaction, which is carried out for 3 hours until the ethane evolution is completed.

In the second reaction step diethylaluminum carboxylate is obtained from diethylaluminum alkoxide and carboxylic acid anhydride (eqn. 2):

$$CH_3(OCH_2CH_2)_nOAl(C_2H_5)_2 + H_2C{-}CH_2 \xrightarrow{100°C} CH_3(OCH_2CH_2)_nOCCH_2CH_2COAl(C_2H_5)_2 \quad (2)$$

I

The reactions were carried out in a 100 mL autoclave under an inert gas atmosphere using a toluene solution of the aluminum alkoxide obtained in the previous step and an equimolar amount of succinic anhydride. The reaction was performed at 100°C for ca. 12 h, to complete decay of the anhydride carbonyl groups bands observed in IR spectra ($v_{C=O}$ 1865 and 1790 cm^{-1}).

Studies on the aluminum content indicate a good correlation between the theoretical (Al_T) values calculated for compound **I** and the experimental (Al_E) ones: $I_{n=1}$, $Al_T = 10.38$, $Al_E = 10.89$; $I_{n=3}$, $Al_T = 7.76$, $Al_E = 8.43$; $I_{n=7}$, $Al_T = 5.15$, $Al_E = 5.28$. The ^{1}H NMR spectra of the products are in agreement with those expected for compounds **I**. The ^{27}Al NMR studies indicate that compounds **I** can occur in associated forms, in which the aluminum atoms may have the degree of coordination 4, 5 and 6. Cryoscopic molecular weight determinations carried out in benzene show that the average degree of association is close to 2 (between 1.70 and 2.15).

The alumoxycarboxylates **II** were prepared by slow addition of water to toluene solutions of compound **I** at room temperature (eqn. 3). After ethane evolution an emulsion was obtained, which was left at room temperature until toluene was completely evaporated [4].

$$m\,[(C_2H_5)_2AlOCCH_2CH_2C(OCH_2CH_2)_nOCH_3]_2 \xrightarrow[-2m\,C_2H_6]{m\,H_2O} \begin{array}{c} -Al-O- \\ | \\ OCCH_2CH_2C(OCH_2CH_2)_nOCH_3 \end{array} \quad (3)$$

$$\overset{\|}{O} \quad \overset{\|}{O} \qquad\qquad \overset{\|}{O} \quad \overset{\|}{O}$$

II

DSC and DTG analysis indicate that compounds **II** are completely amorphous up to ~200 °C. The aluminum content determined in products $II_{n=1}$, $II_{n=3}$ and $II_{n=7}$ was equal to 13.84, 8.67 and 4.92 %, respectively, and the theoretical values calculated for monomeric units were equal to 12.38, 8.22 and 5.6 %, respectively. In IR spectra of the products strong bands characteristic for stretching vibrations of the CO bonds in the carboxylic anion occur at 1605, 1589 and 1465 cm^{-1}, in the ester group at 1735 cm^{-1}, and in the oxyethylene groups at ~1100 cm^{-1}. The band of AlO vibrations is present at 991 cm^{-1}.

NMR spectra were recorded on a Varian 300 MHz spectrometer in C_6D_6 at 35°C.

The average molecular weights of the diethylaluminum alkoxides (n = 1, 2, 7) and diethylaluminum carboxylates ($I_{n=1, 3, 7}$) were determined cryometrically in benzene. DSC studies were performed on a Perkin-Elmer apparatus at a heating rate of 20°C/min in the temperature range from −100 to 200°C. FT-IR spectra were measured on a Bio-Rad 165 spectrophotometer using KBr pellets. DTG, DTA and TG analyses were carried out by means of a Derivatograph Q1500D MOM Budapest apparatus.

DTA, DTG and TG studies have been carried out for a selected alumoxycarboxylate comprising 7 oxyethylene units ($II_{n=7}$). The decomposition was performed in an oxidizing atmosphere. A number of transformations connected with the oxidation of the organic substituent and gradual mass loss of the sample were observed when heating the sample in the atmosphere of air. At ca. 500°C complete transformation to the inorganic compound Al_2O_3 occurs, which is accompanied by a 89 % mass loss. The measurement performed in an atmosphere of air indicates that in the first step decarboxylation of the compound occurs at ca. 250°C, similarly as in the previous case when at 500°C the alumoxycarboxylate underwent transformation to Al_2O_3. The alumoxycarboxylates were used as coupling agents in the form of aqueous solutions of 8 wt. % concentration during the pressing of Al_2O_3 with the addition of the emulsion of acrylate-styrene copolymers with a built-in amphiphilic macromonomer.

2.2. MATERIALS AND METHODS OF STUDY

The poly(acrylate-styrene) emulsions comprising butyl acrylate, styrene, acrylic acid and the amphiphilic macromonomer L25 were obtained according to the method described in [5].

Al_2O_3 ZS 402/M (Martinswerk) of average grain size ca. 1 μm and specific surface measured by the BET method of 6.41 m^2/g was used.

The thickening ability of the ceramic powder involving the additives studied was determined according to the modified method elaborated by Shtern et al. [6]. It consists in the determination of the average pressure P_{av} necessary to press in the same matrix profiles of identical apparent density but of two different heights (5 and 15 cm). The smaller the P_{av} value, the greater the thickening ability of the granulated product. The bending strength of the sintered samples was determined using an Instron 1115 testing machine at a rate of crosshead movement of 0.02 mm/min (ring – bowl).

2.2.1. *Preparation of the granulated product, molding of ceramic samples and their sintering*

A casting slip was prepared from alumina of solid phase concentration of 70 mass %, with the addition of a deflocculant (ammonium salt of poly(acrylic acid) - Dispex A-40 of Applied Colloids) (0.25% with respect to the solid phase), polymeric binder (0.5 and 2.0 wt. %) and alumoxycarboxylate coupling agent. A granulated product was obtained from the thus prepared casting slip and the fraction of which of the 0.2-0.5 mm diameter was employed for further studies.

In order to determine the effect of the additives on the apparent density, open porosity, water absorbability and mechanical strength after sintering as well as the Weibull modulus of samples pressed uniaxially, cylindrical profiles of 20 mm in diameter and ca. 2.5 mm in height have been prepared. The samples were pressed uniaxially on a hydraulic press under a pressure of 50 MPa.

The ceramic samples obtained were dried at 105°c for 24 h, and then sintered in a Carbolite furnace type HTC 18/8; heating rate up to 500°C: 3°C/min, heating rate in the 500-1650°C range: 5°C/min; heating at 1650°C: 1 h; cooling rate: 5°C/min.

3. Results and Discussion

It was found that addition of water-soluble alumoxycarboxylates to a slip casting consisting of alumina powder and poly(acrylate-styrene) binder leads to improvement of mechanical strength and uniformity of sintered ceramic bodies obtained by die-pressing method. The alumoxycarboxylate concentration did not exceeded 0.005 wt. % with respect to the ceramic powder. The thickening ability, measured by the value of average pressure P_{av} necessary to press in the same matrix profiles of identical apparent density (2,70 g/cm^3) but of two different heights (5 and 15 mm) increased ca. four times, and when 2 % of the binder was added – nearly two times (Table 1).

The use of alumoxycarboxylates resulted also in an increase in density of the molded samples by ca. 5 %, increase in density after sintering by ca. 3 % (from ca. 0.94 to ca. 0.97), increase in bending strength by ca. 20 % and increase in the Weibull modulus from 12.7 to 17, i.e. by over 30 % (Table 2).

Such an action of water-soluble alumoxycarboxylates is caused probably by the fact that they facilitate the cleavage of the aluminum oxide agglomerates due to their structure similar to that of aluminum oxide, and on the other side due to the polymeric binder. The alumoxane interacts with the Al_2O_3 particles and the poly(oxyethylene) fragment facilitates the wettability by the polymeric binder. It should be stressed that the ceramic powder can be covered by the obtained alumoxycarboxylate adhesion promoters in the same technological process of preparation of a slip casting.

TABLE 1. Effect of the alumoxycarboxylate coupling agent on the thickening ability of aluminum oxide molded uniaxially with the use of the poly(acrylate-styrene) binders.

Binder	Binder content	P_{av}[a] (MPa)	
	(wt. %)	Without coupling agent	With coupling agent[d]
PB-1[b]	0.5	122	54
	2.0	57	58
PB-2[c]	0.5	106	29.5
	2.0	41	39

a) Average pressure necessary to press in the same matrix profiles of identical apparent density (2.7 g/cm³) but of two different heights (5 and 15 mm)

$$P_{av} = \frac{3P_{3h} + P_h}{3 + 1}$$

b) Copolymer consisting of styrene (28.5 %), butyl acrylate (65 %), acrylic acid (1.5 %) and amphiphilic macromonomer L25 (5 %)

c) Copolymer consisting of styrene (28.5 %), butyl acrylate (65 %), acrylic acid (5 %) and amphiphilic macromonomer L25 (1.5 %)

d) Alumoxycarboxylate $II_{n=7}$ content: 0.005 wt. %

TABLE 2. Effect of the coupling agent on the properties of samples of aluminum oxide sintered at 1650 °C for 1 h involving a poly(acrylate-styrene) binders (0.5 wt. %).

Ceramic binder	d_{vo}[b]	d_v[c]	d_w[d]	P_o[e]	δ_{zg}[f]	Weibull
	(g/cm³)	(g/cm³)		(%)	(Mpa)	modulus[g]
PB-1[a]						
without coupling agent	2.40	3.69	0.941	4.1	270±46	12.7
with coupling agent[h]	2.52	3.81	0.970	1.5	320±34	17.1
PB-2[a]						
without coupling agent	2.41	3.66	0.933	4.6	219±17	10.1
with coupling agent[h]	2.52	3.81	0.972	1.5	323±36	15.5

a) Composition of the poly(acrylate-styrene) binders see Table 1
b) Density of green samples (before sintering)
c) Density of samples after sintering at 1650 °C for 1 h
d) Relative density of samples after sintering at 1650 °C for 1 h
e) Open porosity of samples after sintering at 1650 °C for 1 h
f) Bending strength of profiles after sintering at 1650 °C for 1 h
g) Weibull modulus of profiles after sintering at 1650 °c for 1 h
h) Alumoxycarboxylate $II_{n=7}$ content 0.005 wt. %

74

Alumoxycarboxylate can be added simultaneously with other components of the ceramic casting slip, which is undoubtedly an advantage of the elaborated coupling agents.

An additional advantage of the alumoxycarboxylate coupling agents is the fact that during sintering they are an additional source of aluminum oxide (sintering in an oxidizing atmosphere) or aluminum nitride (sintering in a nitrogen atmosphere). The aluminum oxide or nitride formed during sintering fills the pores of the ceramic sinter, and thus contributes to an increase in its density and mechanical strength.

4. Conclusion

The water-soluble alumoxycarboxylates cause an improvement of the rheological properties of the granulated products, which leads to an increase in their density before sintering, and thus an increase in their density after sintering. The increase in the Weibull modulus of samples molded with these alumoxycarboxylates indicates a much greater uniformity.

This work was financially supported by the Polish State Committee for Scientific Research within the grant no. 3 T09B 033 17.

5. References

1. Plueddemann E. P., Silane Coupling Agents, Plenum Press, New York 1982.
2. Eaborn, C., Bott, R.W., Organometallic Compounds of the Group IV Elements (Ed. Mac Diarmid, A.G.) Marcel Dekker, New York 1968.
3. Eaborn, C., Organosilicon Compounds, Butterworths Scientific Publications, London 1960.
4. Florjańczyk, Z., Zygadło-Monikowska E., Rogalska E., *Proc. 33rd Intersociety Energy Conversion Engineering Conference*, 1999, Vancouver, BC
5. Szafran, M., Florjańczyk, Z., Rokicki, G., Zygadło-Monikowska, E. and Langwald, N., Water based polymeric binders in ceramic processing 9th Cimtec – World Ceramic Congress, Ceramics: Getting into 2000s, Part B, p. 515, 1999, P. Vincenzini (Editor), TECHNA Srl, 1999.
6. Mironiec S.W., Swistun J., Serdiuk G.G. and Shtern M.B. „Opredielenie uplotniaemosti, bokowo dawlenia i vnieshniego trenija metaliceskih poroshkov", *Poroshkovaya metallurgia* 12, 1999.

NEW WATER THINNABLE POLYMERIC BINDERS IN DIE PRESSING OF ALUMINA POWDERS

M. SZAFRAN, G. ROKICKI, P. WIŚNIEWSKI
Warsaw University of Technology, Faculty of Chemistry,
ul. Noakowskiego 3, 00-664 Warsaw, Poland

1. Introduction

From the phenomenological point of view classic press molding of ceramic bodies from loose ceramic powders is a process of dense packing of fine grains in rigid form. The non-uniformity in the thickening of the samples pressed along the direction of the pressure applied is the greatest shortcoming of the classic die pressing. It results from the non-uniform distribution of the pressing forces in the sample, which is the effect of friction forces between the ceramic powder particles, and also between the ceramic powder particles and walls of the matrix.

The shortcomings of classical press molding can be minimized by selecting appropriate materials for the matrices for pressing, appropriate pressing parameters (pressing pressure, pressing rate, etc.), and by optimizing the rheological properties of the ceramic powders formed involving polymeric binders. The type and amount of organic additives used in the forming of ceramic products has, to a great degree, a decisive effect on the ceramic properties of the end products. This concerns especially the bodies obtained from non-plastic starting materials, i.e. those which contain a small amount of clay-like minerals or without these minerals. The role of substances enabling the forming and sintering of ceramic products obtained form non-plastic starting materials are played mainly by polymers of various molecular weight [1,2]. Water-soluble polymeric binders, such as poly(vinyl alcohol), methylcellulose, carboxymethyl-cellulose and poly(oxyethylene) glycol, are most often used in the press molding of ceramic powders. A disadvantage of use of such binders is the relatively small mechanical strength of samples obtained, which precludes the processing of green profiles. In recent years various types of polymers are used in the form of aqueous emulsions. Poly(vinyl acetate) and polyacrylic emulsions are used most often [3,4]. There is growing interest in emulsion polymers characterized by a large interface surface (above 5 $m^2 \cdot mL^{-1}$) as well as high surface functionality. These types of systems find specific application, mainly in microencapsulation [5]. The synthesis of such polymers requires the incorporation of a comonomer introducing the required functional groups modifying the structure of the polymer chain. The emulsions thus obtained are characterized by a smaller content of the external emulsifier, since the incorporated comonomer plays a stabilizing role.

75

M.-I. Baraton and I. Uvarova (eds.),
Functional Gradient Materials and Surface Layers Prepared by Fine Particles Technology, 75–80.
© 2001 *Kluwer Academic Publishers. Printed in the Netherlands.*

Studies on the application of water-thinnable poly(acrylate-styrene) binders containing built-in amphiphilic macromonomers playing the role of surface active agents as well as internal plasticizers for forming aluminum oxide by means of uniaxial pressing are presented in the paper.

2. Experimental Part

2.1. MATERIALS

2.1.1. *Preparation of poly(acrylate-styrene) emulsions containing amphiphilic macromonomers*

Molecules terminated with β-hydroxyacrylate group resulting from the reaction of acrylic acid with an epoxide group of the fatty alcohol glycidyl ether (eqn. 1) according to the method described in [6] were used as amphiphilic macromonomers.

$$H_2C = CH - C\underset{OH}{\overset{O}{\diagup}} + R-(-OCH_2CH_2-)_n-O-CH_2-CH-CH_2 \xrightarrow{cat.}$$

$$R-(-OCH_2CH_2-)_n-O-CH_2-\underset{OH}{CH}-CH_2-O-\underset{O}{C}-CH = CH_2 \qquad (1)$$

The copolymerization of styrene with butyl acrylate, acrylic acid and an amphiphilic macromonomer was carried out under the atmosphere of deoxygenated nitrogen in a glass reaction vessel equipped with a high-speed mechanical stirrer. Redistilled water, initiator ($NH_4S_2O_8$, 1 mass % with respect to the monomers) and emulsifier were introduced to the reaction vessel. The monomers (30 mass %) were added after dissolution of the components added earlier. The reaction vessel content was stirred and heated maintaining the temperature at 70±1°C. The polymerization was performed to 99.5 % monomer conversion. All the copolymers were characterized by high molecular weight of 400,000 – 600,000 g/mol.

Aqueous solutions of commercial poly(vinyl alcohol) of molecular weight 30,000 g/mol and degree of hydrolysis of 88 %, and that of methylcellulose of viscosity of 400 mPa·s (of a 2 % solution) were also used.

2.1.2. Alumina

Al_2O_3 ZS 402/M of Martinswerk of average grain size 1.5 μm and specific surface measured by the BET method equal to 6.41 m^2/g was used for the studies.

2.2. METHODS OF STUDY

The glass transition temperature (T_g) of the binders was determined by the DSC method by means of the Pyris 1 apparatus of Perkin-Elmer. The molecular weight of the synthesized binders was determined by the GPC method on a gel chromatograph LC-10AD of Shimadzu. Microscopic studies of both green and sintered tapes were carried out on a scanning electronic microscope LEO 1530. The bending strength of the sintered samples was determined using an Instron 1115 testing machine at a rate of crosshead movement of 0.02 mm/min (ring – bowl). The thickening ability of the ceramic powder involving the binders studied was determined according to the modified method elaborated by Shtern et al. [5]. It consists in the determination of the average pressure P_a necessary to press in the same matrix profiles of identical apparent density but of two different heights (5 and 15 cm). The smaller is the P_a value, the greater is the thickening ability of the granulated product.

2.2.1. *Preparation of the granulated product, molding of ceramic samples and their sintering*

A casting slip was prepared from alumina of solid phase concentration of 70 mass %, with the addition of a deflocculant (ammonium salt of poly(acrylic acid) - Dispex A-40 of Applied Colloids) (0.25% with respect to the solid phase) and polymeric binder (0.5-2.0 mass %). A granulated product was obtained from the thus prepared casting slip and the fraction of which of the 0.2-0.5 mm diameter was applied for further studies.

In order to determine the effect of the binder on the apparent density, open porosity, water absorbability and mechanical strength after sintering as well as the Weibull modulus of samples pressed uniaxially, cylindrical profiles of 20 mm in diameter and ca. 2.5 mm in height have been prepared. The samples were pressed uniaxially on a hydraulic press under a pressure of 50 MPa.

The ceramic samples obtained were dried at 105°C for 24 h, and then sintered in a Carbolite furnace type HTC 18/8; heating rate up to 500°C: 3°C/min, heating rate in the 500-1650°C range: 5°C/min; heating at 1650°C: 1 h; cooling rate: 5°C/min.

3. Results and Discussion

Water-thinnable poly(acrylate-styrene) and two types of water-soluble polymers: poly(vinyl alcohol) (PVA) and methylcellulose (MeCellul) were chosen for studies on the application of polymeric binders in the die pressing of ceramic powders. The chemical composition of poly(acrylate-styrene) binders is presented in Table 1.

TABLE 1. Chemical composition of emulsion poly(acrylate-styrene) binders applied in the uniaxial pressing.

Symbol of the poly(acrylate-styrene) binder	T_g °C	Chemical composition (mass %)			
		Butyl acrylate	Styrene	Acrylic acid	Macromonomer L25
PB-1	-7.6	65	28.5	1.5	5.0
PB-2	1.8	65	28.5	5.0	1.5
PB-3	1.7	65	28.5	6.5	0
					Macromonomer L18
PB-4	-13.2	62	28.5	1.5	8.0
PB-5	2.1	60	28.5	1.5	10.0

L 18

$$CH_3\left[CH_2\right]_n\left[O-CH_2-CH_2\right]_m OCH_2-CHCH_2-OC-CH{=}CH_2$$
OH
n=15-21, m=18

L 25

$$CH_3\left[CH_2\right]_n\left[O-CH_2-CH_2\right]_m OCH_2-CHCH_2-OC-CH{=}CH_2$$
OH
n=15-21, m=25

78

The effect of the chemical structure of poly(acrylate-styrene) binders on the thickening ability of alumina powders during die pressing with their involvement and properties of the molded samples before and after sintering are presented in Table 2.

TABLE 2. Effect of the chemical structure of poly(acrylate-styrene) binders on the properties of the molded samples with Al_2O_3 formed by uniaxial pressing (pressing pressure 50 MPa; sintering at 1650°C for 1 h).

Symbol of binder[a]	P_h[b] MPa	P_{3h}[c] MPa	P_a[d] MPa	d_w[e]	P_a[f] %	δ_g[g] MPa	Weibull modulus[h]
PB-1	132±7	92±3	102	0.934	4.6±1.1	219±17	15.5
PB-2	117±4	72±2	83	0.965	1.6±0.9	314±30	12.7
PB-3	73±3	30±5	41	0.972	0.5±0.1	272±30	12.1
PB-4	49±3	32±4	37	0.933	6.3±2.3	276±64	5.5
PB-5	43±2	37±3	39	0.920	7.6±1.7	319±66	8.1
PVA	108±8	76±6	84	0.955	5.5±1.3	234±46	5.3
MeCellul	128±4	80±6	92	0.965	2.3±0.3	218±49	7.5

a) Composition of the poly(acrylate-styrene) binders is presented in Table 1
b) Pressure necessary to obtain a ceramic sample of 5 mm in height and 2.7 g/cm³ in density
c) Pressure necessary to obtain a ceramic sample of 15 mm in height and 2.7 g/cm³ in density
d) Average pressure of press molding of ceramic samples of 2.7 g/cm³ density

$$P_a = \frac{3P_{3h} + P_h}{3+1}$$

e) Apparent density of samples after sintering at 1650°C for 1 h
f) Open porosity of samples after sintering at 1650°C for 1 h
g) Bending strength of samples after sintering at 1650°C for 1 h
h) Weibull modulus of samples after sintering at 1650°C for 1 h

The thickening abilities of alumina powders pressed depends on the chemical structure of the poly(acrylate-styrene) binder as well as on the type and amount of the amphiphilic macromonomers built-in. The built-in acrylic acid and macromonomer caused an increase in the wettability of the ceramic powders by such binders, which led to a more uniform covering of the ceramic powder with the binder applied and increase in their ability to press molding. The highest thickening was observed for binders containing the high amount of amphiphilic macromonomer (PB-4 and PB-5) or acrylic acid (PB-3).

The apparent density is one of the most often applied parameters for the evaluation of the properties of sintered bodies. The greater is the apparent density of the pressed samples before sintering, the greater is their density after sintering. The greatest density values for the pressed samples were achieved when acrylate-styrene copolymers containing a built-in acrylic acid (PB-3) were used (Table 1).

Besides density, as a parameter to macroscopically characterize the pressed ceramic sample, its uniformity is an extremely important parameter. The greater uniformity of density, the less defects in the microstructure of the sintered sample, and thus the greater mechanical strength and also Weibull modulus. As far as the mechanical strengths of the pressed and sintered samples involving the emulsion of poly(acrylate-styrene) and water-soluble binders were similar, the Weibull modulus values of them differed considerably. The Weibull modulus of sinters with the use of water-soluble binders (PVA and MeCellul) did not exceed 7, so for sinters pressed involving poly(acrylate-styrene) emulsions it was nearly twice as high as that of water-soluble ones. At an appropriately chosen ratio of styrene, butyl acrylate, acrylic acid and amphiphilic macromonomer of long poly(oxyethylene) chain (PB-1, PB-2 and PB-3) the Weibull modulus reached the values of 12-15.5. Such high values of the Weibull modulus indicate a relatively high uniformity of ceramic sinters obtained using emulsion acrylic binders. Also the smallest open porosity (0.5 and 1.6 %) of sinters was observed for ceramic samples prepared using poly(acrylate-styrene) emulsion containing acrylic acid (PB-3) or macromonomer with long hydrophilic poly(oxyethylene) fragment (PB-2).

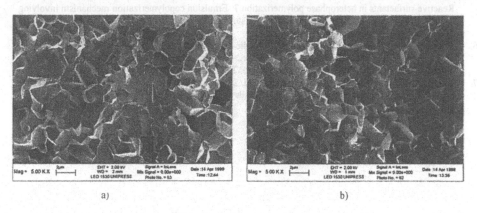

a)   b)

Figure 1. The fractures of the alumina powder molded by die pressing using PB-4 (a) and PB-1 (b) poly(acrylate-styrene) binders and sintered at 1650 °C/1 h.

The samples obtained involving binders PB-1 and PB-2 were very uniform, while those obtained involving binders PB-4 and PB-5, besides small pores had bigger ones. Pictures of such samples taken in SEM are presented in Figure 1.

4. Conclusions

The studies on the application of water-thinnable poly(acrylate-styrene) binders in the forming of aluminum oxide by the uniaxial pressing method showed that by building-in of various functional groups to the polymer molecule, by changing the length of the hydrophilic and hydrophobic part of the amphiphilic side chain it is possible to achieve their affinity to the ceramic powders used, and thus to affect directly the electric double layer on the ceramic grain – liquid phase interface. This affects the rheological properties of the casting slips and that of the granulated products obtained from these casting slips. This has a direct effect on the thickening ability of the ceramic powders pressed, which leads to an increase in density and thickening uniformity of the samples before sintering, and also in an increase in density, mechanical strength and Weibull modulus of ceramic sinters.

80

This work was financially supported by the Warsaw University of Technology within the grant no. 504/G/1020/0132

5. References

1. Bortzmeyer, D. Die Pressing and Isostatic Pressing, *Materials Science and Technology*, Part I, Processing of Ceramics, Winheim, 1996.
2. Wu, X.L.K. and McAnany, W.J. (1995) Acrylic Binder for Green Machining, *Am. Ceram. Soc. Bull.* **74**, 61.
3. Mistler, R.E. (1990) Tape casting: The Basic Process for Meeting the Needs of the Electronics Industry, *Am. Ceram. Soc. Bull.* **69**, 1022.
4. Riedel, G. and Krieger, S. (1996), Acrylate als Binder für trockengepresste Si_3N_4 – Strukturkeramik, *Keramische Zeitschrift* **48**, 193.
5. Schoonbrood, H.A.S., Unzue, M.J., Beck, O.J., Asua, J.M., Goni, A.M. and Sherrington D.C. (1997) Reactive surfactants in heterophase polymerization.7. Emulsion copolymerization mechanism involving three anionic polymerizable surfactants (surfmers) with styrene-butyl acrylate-acrylic acid *Macromolecules* **30**, 6024.
6. Szafran, M., Florjańczyk, Z., Rokicki, G., Zygadło-Monikowska, E. and Langwald, N., Water based polymeric binders in ceramic processing 9th Cimtec – World Ceramic Congress, Ceramics: Getting into 2000s, Part B, p. 515, 1999, P. Vincenzini (Editor), TECHNA Srl, 1999.
7. Mironiec, S.W., Swistun, J., Serdiuk, G.G. and Shtern, M.B. (1990) Opredielenie uplotniaemosti, bokovo davlenia i vnieshniego treniya metalicheskih poroshkov, *Poroshkovaya metallurgia* 12.

ELECTRICAL CONDUCTION IN IRON OXIDE POWDER COATED WITH POLYMETHYL METHACRYLATE

Vladislav Skorokhod, Richard P. N. Veregin and Michael S. Hawkins
Xerox Research Centre of Canada
2660 Speakman Drive, Mississauga, Ontario L5K 2L1, Canada

Abstract Current-voltage and transient behaviours of iron oxide powder coated with polymethyl methacrylate (i.e., a model xerographic carrier) was studied by measuring current through powder piles sandwiched between two planar electrodes. A space charge limited process was identified as the major factor controlling transverse DC conduction in the coatings. Current-voltage hysteresis and the characteristic transient behaviour observed in the present structure indicate the presence of intrinsic long-lifetime traps in polymethyl methacrylate, that may play a significant role in its triboelectric response.

Introduction

Many xerographic printers and copiers involve a two-component development process, where toner particles are triboelectrically charged by vigorous mixing with xerographic carrier (Figure 1 a). The mixture (developer) is afterwards transported towards the development system, where, under high electric field, toner particles are transferred from the carrier onto photoreceptor to develop the latent image (Figure 1 b). Usually, xerographic carrier consists of coarse ferromagnetic powder (core) with an average particle diameter of 15 – 150 μm, coated with triboelectrically active polymer such as acrylics, amines, etc.

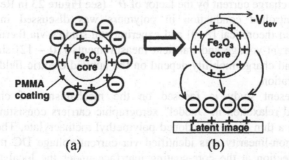

(a)   (b)

Figure 1. Xerographic development process: (a) toner and carrier coating are triboelectrically charged against each other by vigorous mixing; (b) charged toner particles are transferred onto the latent image under a development voltage V_{dev}, while space charge of an opposite sign is left behind in the carrier coating.

81

M.-I. Baraton and I. Uvarova (eds.),
Functional Gradient Materials and Surface Layers Prepared by Fine Particles Technology, 81–86.
© 2001 *Kluwer Academic Publishers. Printed in the Netherlands.*

Charge transport mechanisms in carrier coatings influence the entire xerographic process by controlling triboelectric charge level and its dissipation rate (Figure 1 b). So far, electrophysical studies in xerographic developers have been limited to the measurements of low-voltage conductivity as a function of toner concentration [1 – 3]. Understanding the processes of local charge transport, accumulation and relaxation in carrier coatings at the carrier-carrier and carrier-toner contact point, can provide additional insight into the triboelectric performance of xerographic materials. Although the best approach to localized electrophysical analysis would be current-voltage and transient measurements in an individual carrier bead, its realization could be associated with some technical difficulties and would require expensive precision equipment. A simpler and more practical experimental approach is suggested in the present study, where electrical measurements were carried out on uncompacted powder piles sandwiched between two parallel planar electrodes.

Electrical conductivity of packed powders (both coated and uncoated) is known to be a function of packed density and individual Ohmic conductivity of the core and coating [4 – 5]. However, in the present situation, the electrophysical response of xerographic carrier is complicated, to a great extent, by the current-voltage non-linearity of the both phases, when the carrier is subjected to development fields of over 100 kV/cm.

One of the mechanisms typically causing current-voltage non-linearity in thin insulating films is space charge injection at a conductor-insulator Ohmic contact, where the work function for electrons in the conductor is lower than that in the insulator. Space charge conduction in planar structures is described in monographs [6, 7]. Typically, space charge limited current-voltage dependence is given as:

$$J \propto \frac{\theta}{\delta^3} V^2 \tag{1}$$

where J is the current density, δ is the film thickness, and θ is the free-to-trapped charge ratio. It should be pointed out that, at low voltage, when the concentration of injected charge is lower than that of intrinsic charge, Eq. (1) is not valid. In such a case, the current-voltage dependence is linear (Ohmic). The presence of empty electron or hole levels usually referred to as traps, reduces space charge currents due to a reduction in charge carrier mobility. Trap saturation at high injection levels leads to an instantaneous increase in space charge current by the factor of θ^{-1} (see Figure 23 in Ref. 7).

Trap-controlled conduction in polymers was discussed in a number of xerography-related theoretical [8, 9] and experimental studies via thermally stimulated currents, and current-voltage and transient measurements [10 – 12] showing that trap depth, lifetime and charge mobility depend on the applied electric field, injection level and trap concentration.

The present study is focused on the mechanisms of charge transport, accumulation and relaxation in "model" xerographic carriers consisting of iron oxide core coated with a thin layer of undoped polymethyl methacrylate. The mechanism of current-voltage non-linearity was identified via current-voltage DC measurements as space charge injection at the core-coating interface under the local transverse inter-particle electric field. Current transient measurements indicated the presence of long-life unsaturated intrinsic traps in polymethyl methacrylate.

Experimental

Iron oxide powder (carrier core) consisting of spherical polycrystalline aggregates with diameters ranging from 15 to 65 µm was coated with 1, 1.25 and 1.5 wt. % polymethyl methacrylate by using the available coating process [13]. Preliminary SEM studies showed that, even at 1 wt. % coating, the iron oxide particles were nearly entirely covered with the polymer.

Current-voltage and transient dependencies were measured by using the two-probe method. A 3-mm thick layer of coated powder was spread between two circular planar stainless-steel electrodes. A load of 4 kg was applied to the upper electrode to insure good inter-particle and particle-electrode contact. The applied load did not cause powder compaction, nor it changed powder apparent density. Close packing was deliberately avoided, since it could modify powder surface properties and mechanically damage the polymer coating. Transients were measured by applying 10-second voltage pulses, followed by short-circuiting the voltage source and measuring the current by a Keithley electrometer. All measurements were conducted at the room temperature and ambient humidity.

Results and Discussions

Figure 2 shows the DC current-voltage dependencies measured in the iron oxide powder coated with 0, 1, 1.25 and 1.5 wt. % polymethyl methacrylate. For uncoated iron oxide powder (the filled circles in Figure 2), current increases linearly with voltage up to 100 V, with a constant conductance of 10^{-4} S. The further increase in voltage results in a rapid non-linear increase in current. The high-voltage portion of the current-voltage curve for the uncoated powder can be best fitted with the Poole-Frenkel-type dependence $I/V \propto \exp(\gamma V^{1/2})$, as shown in Figure 3.

The data in Figure 2 indicate that the presence of polymer coating substantially modifies the DC current-voltage behaviour of powder piles. For 1 wt. % coating, the current voltage dependence is linear only at voltages below 12 V, and then becomes quadratic between 12 and 100 V. As indicated above, the quadratic behaviour is typical for the space charge injection process [7]. In the range of 12 – 100 V, the current-voltage behaviour of the core powder remains linear, with a conductance of $10^3 - 10^4$ times higher than that of the powder containing 1 wt. % polymer coating. Therefore, the observed non-linearity is due to space charge injection at the core-coating interface that could be expected to occur in the vicinity of the interparticle contact points.

For the powders coated with 1.25 and 1.5 wt. % of polymer, linear current-voltage behaviour persists at even higher voltages, up to 60 – 100 V, followed by a rapid increase in current. Similarly to the powder containing 1 wt. % coating, the departure from current-voltage linearity can be ascribed to the onset of space charge limited currents. However, the expected quadratic current-voltage dependency is most likely obscured by the non-linearity of the core powder at voltages of 100 V and higher.

The voltages corresponding to the onset of space charge process (V_{SC}) were determined from the data in Figure 2 as the voltage of the departure from current-voltage linearity. Figure 4 shows that $(V_{SC})^{1/2}$ linearly increases with coating weight, due to the expected increase in coating thickness. The relationship between V_{SC} and the film thickness δ is defined in planar thin films as $(V_{SC})^{1/2} \propto \delta$ [7]. Although, in the present coated powder, this relationship can be significantly modified due to a number of factors such as particle roughness, coating non-uniformity and the presence of uncoated patches or loose coating material, V_{SC} measurement can be potentially used as a method to determine coating thickness.

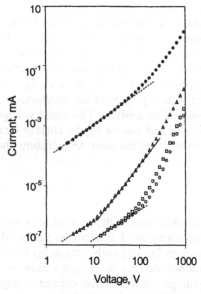

Figure 2. Current-voltage dependencies of the piles of iron oxide powder coated with 0 (●), 1 (Δ), 1.25 (□), and 1.5 (O) wt. % polymethyl methacrylate. Slopes of 1 and 2 are shown with the dashed and solid lines, respectively.

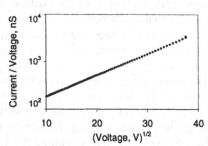

Figure 3. Poole-Frenkel plot of the current-voltage dependence of uncoated iron oxide.

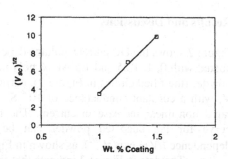

Figure 4. Change of the space charge onset voltage square root $(V_{SC})^{1/2}$ with coating weight.

Further DC current-voltage analysis shows that the current-voltage dependence demonstrates a significant hysteresis between 0 and 100 V, when voltage is ramped from 0 to 2000 V within a total time of 5 minutes, and then lowered to 0 V within 1 minute (Figure 5). The low-voltage Ohmic behaviour observed at the increasing voltage cycle was not detected when the voltage rapidly decreases, and the current-voltage behaviour remains quadratic even at very low voltages (< 12 V), with current being an order of magnitude higher than during the increasing voltage cycle. The observed current hysteresis can be explained from the viewpoint of the expected presence of initially unsaturated intrinsic traps in the coating polymer, such that, at high voltage, part of the injected charge becomes trapped, thus saturating the trap sites. Provided that trap lifetime is sufficiently long, the traps remain saturated when the voltage is decreased which leads to the increase in current. Similarly, current-voltage hysteresis was observed in the specimens containing 1.25 and 1.5 wt. % polymer (Figure 6).

The presence of trap-controlled conduction in polymethyl methacrylate coatings was verified via transient current measurements (Figures 7 and 8). It should be pointed out that the capacitance of powder piles, measured under standard conditions at a frequency of 1 kHz and amplitude of 1 V, was in the range of 60 – 70 pF. Thus, in the absence of a non-liner process such as trap-controlled space charge limited conduction, 100 ms would be sufficient to restore zero current through the powder piles (with an electrometer internal resistance of 100 MΩ). Preliminary transient studies of the uncoated core powder showed that zero current restores rapidly, within 50 – 100 ms, regardless of the amplitude of the applied voltage pulse.

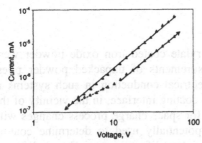

Figure 5. Current-voltage dependence of iron oxide powder coated with 1 wt. % polymethyl methacrylate, at increasing (Δ) and decreasing (▲) voltage.

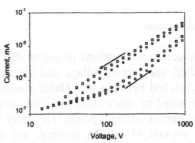

Figure 6. Current *versus* increasing and decreasing voltage in iron oxide powder coated with 1.25 (□), and 1.5 (O) wt. % polymethyl metacrylate.

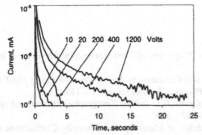

Figure 7. Transients of the iron oxide powder with 1 wt. % coating at various pulse amplitudes.

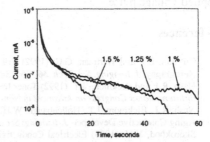

Figure 8. Effect of coating weight on the transients at a pulse amplitude of 800 V.

The data in Figure 7 indicate that the length of the transient process in the powder with 1 wt. % coating polymer depends, to a great extent, on the voltage pulse amplitude. Time of current decay from 10^{-5} to 10^{-7} mA changes from 1.8 s at a pulse amplitude of 10 V to 25 s at that of 1200 V. It can be observed from Figure 7 that the most significant change in the transient behaviour occurs at pulse amplitudes between 200 and 400 V. It is believed that major part of the trap saturation process occurs within this voltage region. Thus, transient current is reduced by the presence of unsaturated traps when the applied voltage is 200 V or below. At pulse amplitudes of over 400 V, the transient process becomes trap free, leading to an increase in transient current.

In addition to trap saturation, the change of transient behaviour with voltage is also associated with the change of injection level and the ratio of free-to-trapped charge that, in turn, effects charge mobility [8, 9]. Rigorous analysis of the transient behaviour in the present system can be greatly complicated by the uncertainty of the local transverse electric field applied to the coating at inter-particle contacts, due to the non-uniformity of coating thickness.

Figure 8 shows that increase in coating weight results in a decrease in transient time. This effect is most likely associated with the fact that increasing coating weight and, consequently, coating thickness leads to a decrease in effective transverse electric field in the coating at the inter-particle contact. In addition, increase in coating thickness in the presence of unsaturated traps can effect charge mobility [9] that also contributes to the decrease in transient current.

Conclusions

Conduction mechanisms in polymethyl methacrylate coated iron oxide powder were studied via current-voltage and transient measurements of unpacked powder piles, sandwiched between two planar electrodes. Electrical conduction in such systems is dominated by space charge injection at the core-coating interface, in the vicinity of the inter-particle contact points. The onset voltage for space charge process changes with the amount of coating material, and can be potentially used to determine coating thickness. The presence of intrinsic unsaturated long-life traps was detected in the undoped polymethyl methacrylate via DC current-voltage hysteresis measurements, and then confirmed by transient measurements. The length of the transient process and transient current magnitude depend on coating material content and the amplitude of applied voltage pulse.

References

1. Gutman, E.J. and Hartman, G.C. (1992) Triboelectric Properties of Two-Component Developers for Xerography. *J. Imaging. Sci. Tech.* **36** 335 – 349.
2. Nash, R.J. and Bickmore, J.T. (1992) Toner Impaction and Conductivity Aging. *Proceedings of IS&T 8-th International Congress on Advances in Non-Impact Printing Technologies* 131 – 139.
3. Nash, R.J., Bickmore, J.T., Hollenbaugh, W.H. and Wohaska, C.L. (1996) Xerographic Response of and Aging Conductive Developer. *J. Imaging. Sci. Tech.* **40** 347 – 353.
4. Skorokhod, V.V. (1963) Electrical Conductivity, Modulus of Elasticity and Viscosity Coefficients of Porous Bodies. *Powder Metallurgy* 188 – 200.
5. Kendall, K. (1990) Electrical Conductivity of Ceramic Powders and Pigments. *Powder Technology* **62** 147 – 154.
6. Mott, N and Gurney, R.W. (1948) *Electronic Processes in Solids*, Oxford.
7. Simmons, J.G. (1971) *DC Conduction in Thin Films*, Mills and Boon, London.
8. Chen, I. and Tse, M.K. (1999) Electrical Characterization of Semi-insulating Devices for Electrophotography. *Proceedings of IS&T 15-th International Congress on Advances in Non-Impact Printing Technologies* 486 – 489.
9. Novikov, S.V. (1999) An Unusual Dependence of the Charge Carrier Mobility in Disordered Organic Materials on Trap Concentration: Real Phenomenon or Artifact? *J. Imaging. Sci. Tech.* **43** 444 – 449.
10. Gamoudi, M., Rosenberg, N., Guillaud, G., Maitrot, M and Mensard, G. (1974) Analysis of Deep and Shallow Trapping of Holes in Anthracene. *J. Phys. C: Solid State Phys.* **7** 1149 – 1159.
11. Kojima, K., Takai, Y and Ieda, M. (1982) Carrier Traps in Polyethylene Naphtalate (PEN): Photoeffect. *Japan. J. Appl. Phys.* **21** 1025 – 1027.
12. Murti, K., Murray, H. and Baillie, S. (1986) Charge Trapping in an Electron Transport Active Matrix. *J. Phys. D: Appl. Phys.* **19** 1265 – 1281.
13. Veregin, R., Allison, G., Kovacs, G. and Gerroir, P. US Pat. 6,051,354, *Coated Carrier*.

FROM NANOSIZE POWDERS TO A DIESEL SOOT CONVERTER

Z.S. RAK

Netherlands Energy Research Foundation ECN
Westerduinweg 3, 1755 ZG Petten, The Netherlands

Abstract

The reduction of diesel soot emissions in exhaust systems may be performed by trapping particulates by an appropriate filter and then cleaning the filter walls by burning off the soot. A novel design of an electrochemical reactor for filtering and continuous combustion of soot particles is presented in this paper. Such a reactor consists of a porous, oxygen ion conducting material covered by catalytically active, electron conductive electrodes, electrical connections and an external power supply. The manufacturing process of high porosity, ion and electron conducting ceramic monoliths (type of functionally graded material) from the nanosize powders by extrusion and coating techniques is described in detail. The optimisation of the various and sometimes conflicting properties of the developed porous material implies the formation of a specific microstructure, characterised by a very high porosity, between 75-80%, and low pressure drop, below 150 mbar. Pores in the developed material are characterised by a bimodal distribution, micrometric (100-300 μm) and nanometric (20-100 nm) in size. The electrode coating with a thickness of 6-8 μm adheres to the basic ceria-gadoline (CGO) material very well in spite of a big discrepancy in coefficients of thermal expansion between them. Commercial nanosize powders were used for manufacturing the basic monolithic structure and as well as the deposited functional coating.

1. INTRODUCTION

Nanostructured ceramics are materials with the characteristic length scale of which is of the order of a few nanometers, typically 1-10 nm. Nanostructured ceramics may be in or far away from thermodynamic equilibrium. The properties of nanoceramics deviate from those of single crystals or coarse-grained polycrystals and/or glasses with the same average composition. The motivation for manufacturing and extensively studying of nanocrystalline ceramics lies in their unique properties. The behaviour of a nanocrystalline ceramic material is dominated by events at the grain boundaries due to the fact that a larger fraction of atoms resides on the grain boundaries. Therefore the nanocrystal ceramics can deform plastically and extensively by grain boundary sliding. Such "superplastic deformation" is in contrast to the usual brittle behaviour

87

M.-I. Baraton and I. Uvarova (eds.),
Functional Gradient Materials and Surface Layers Prepared by Fine Particles Technology, 87–103.
© 2001 *Kluwer Academic Publishers. Printed in the Netherlands.*

characteristic of commercial ceramics. Also due to the many grain boundaries, nanocrystalline ceramics have also been used as solid state bonding agents to join together other larger grain ceramics materials at moderate temperatures. As a result of this, nanocrystalline ceramics are considered to have a low thermal conductivity. The explanation of this phenomenon is that the thermal conductivity is directly proportional to the mean free path length of phonons. Moreover, in nanocrystalline grain sizes, where the distance between grain boundaries approaches the nanometre scale, the grain boundaries can also contribute meaningfully to the scattering process. Another characteristic property of nanoceramics is their lower hardness in comparison to large grain ceramics at room temperature and tendency to crack less easily [1, 2]. The main advantages of nanoceramics compared to conventional microsize ceramics are: the possibility of mixing of different ceramic powders in a nanosize level, thus a better homogeneity of the final material, and sharply reduced sintering temperature to be applied to reach a full densification of the ceramic material because of the high free specific surface of the nanocrystalline powders. Due to their high specific surface, the nanocrystalline ceramics are characterised by extremely low bulk densities, a strong agglomeration, have high sintering activity and show high reactivity, specifically with non-oxide nano-phase powders. The costs of nanopowders are also still very high due to a batch production on an order for a specific customer. The production of big quantities of ultrafine powders can therefore result in a drastic cut of their prices. However the sinter activity can be used when the appropriate processing technologies are used. To reduce grain growth during the sintering of nanocrystalline powders, a range of possibilities such as an introduction of inclusions of foreign phases, are used [3, 4].

Up to now the main applications of nano-phase ceramics were in the preparation of coatings, as a binder, and only to a lesser extent in the production of bulk products because the production of bulk nanocrystalline ceramics from nanocrystalline powders are difficult to obtain. Due to their small size, nanocrystalline particles are quite susceptible to the formation of interparticle London - Van der Waals bonds, in either wet or dry state. There are also problems with compaction, which cannot be attributed to agglomeration. One is the large number of particle-particle point contacts per unit volume in a nanocrystalline powder, as compared with a powder of micrometre sized particles. Each of these point contacts represents a source of friction resistance to the compaction of the powder, and thus the total frictional resistance to compaction can be much higher than for a powder composed of larger particles. Consequently, for a given applied stress during compaction, there tends to be less particle-particle sliding and particle rearrangement than one might expect for conventional ceramics. A number of manufacturing techniques are used for compaction of ultrafine powders, such as a cold uniaxial pressing, cold isostatic pressing, centrifugication and osmotic consolidation. Pressure-supported sintering processes like hot pressing, hot isostatic pressing, plasma activering sintering, shock compaction, and sinter-forging are used as well for their consolidation. However, the success with sintering of nanosize powders has been mixed. Typically, it is difficult to obtain full density while retaining a nanocrystalline (<100 nm) grain size, if pressureless sintering is used. On the contrary, pressure assisted sintering techniques, especially hot pressing and sinter-forging, have produced fully dense grained ceramics, below 100 nm. The advantage of these hot deformation techniques is that the plastic deformation of the nanocrystalline ceramics can be exploited to close large pores which otherwise require extremely high sintering temperatures to close during pressureless sintering. With the hot consolidation techniques, grain sizes in the 15-34 nm range have been obtained for several dense ceramics [1-4].

The nanocrystalline ceramics have already found an extensive application in the so called "3-way" car catalytic converter, where they are applied as a coatings (mainly for gasoline-powered vehicles). The converter unit contains a cordierite honeycomb structure, which directs the exhaust gas through a large number of parallel channels. The channel walls are wash-coated with a thin, nanocrystalline layer of a γ-alumina active catalyst support containing dispersions of cerium oxide particles and of noble metal particles. The ceria has a multifunctional role: it acts as an "oxygen storage" component to buffer transient shifts in the fuel/air ratio, it acts to stabilise the transition alumina structure and noble metal dispersion, and it has a synergistic effect on the noble metal catalysts of reactions such as that between CO and NO/NO_2. All such catalysts rely for their efficiency on a nanoscale and uniform dispersion of the chemically active phases: ceria, noble metals, and other additives, e.g. to suppress formation of hydrogen sulphide [5, 6].

A second prospective application for nanocrystalline ceramics in the bulk form and also as a coating could be in cleaning combustion off-gases of particles, consisting of the soot and soluble organic fraction (SOF), generally named particulate material (PM) emitted from diesel engines. A concept for an electrochemical reactor acting as a trap for the continuous removal of soot particles from diesel exhaust gas has been already developed [7]. In this concept, the materials commonly used in the manufacturing process of Solid Oxide Fuel Cells (SOFC) have been employed: a nanosize ceria-gadoline (CGO) powder for a monolith production and nanosize perovskite-oxides, e.g. lanthanum strontium iron cobalt perovskite-type oxides (LSFC), as a coating [8-12]. Therefore, the problems associated with the processing of a nanosize CGO powder for manufacturing of a highly porous monolithic structure, coated with a nanosize perovskite-type electrode material are the main topic of this paper.

2. METHODS OF REMOVAL OF DIESEL PARTICULATES

It is expected that the market for diesel engines in Europe will grow significantly in the near future due to their superior thermal efficiency, durability and reliability and, as a consequence of this, their economy in comparison to the traditional gasoline-powered engines [12]. However, the stringent legislation in the EU (EURO 4 in 2004/5) and USA (FTP 75 and EPA regulations) poses some major challenges for the diesel passenger car which will be difficult to meet without secondary systems, see Table 1[13-15].

Diesel exhaust consists of gaseous, liquid, and solid emissions. Gaseous emissions consist of N_2, CO_2, CO, H_2, NO/NO_2, SO_2/SO_3, HC (C_2-C_{15}), oxygenates, and organic nitrogen and sulphur compounds. Liquid emissions include H_2O, H_2SO_4, HC (C_{15}-C_{40}), oxygenates, and polyaromatics. Solid emissions are made up of dry soot, metals, inorganic oxides, sulphates, and solid hydrocarbons [16]. The physical and chemical processes responsible for soot formation are well described in [17]. These particulates consist of soot nuclei (carbon) including inorganic material, adsorbed hydrocarbons (SOF), SO_3 and some water. The size of the individual soot spheres is about 25 nm, the size of total particulates about 200 nm. Polynuclear aromatic hydrocarbons (PAHs), which are partly present in the gas phase and are partly adsorbed onto the soot spheres, consist of a large number of compounds originating from lubricant oil, diesel fuel, and radical hydrocarbon degradation. Formation processes occur during combustion. PAHs are thought to have adverse health effects and some of these compounds have been found to be carcinogenic [13,17].

Table 1. European and USA emission limits for light duty diesel vehicles [13-16].

Year/Legislation/Country	Emissions, g/mile		
	HC + NOx	CO	Particulates (+ NO_x)*
1996, EURO II, EU			
- Direct Injection	0.90	1.00	0.10
- Indirect Injection	0.70	1.00	0.08
2000, EURO III, EU	0.56	0.64	0.50*
2005/8, EURO IV/V, EU	0.30	0.50	0.25*
1994, FTP 75, USA	0.40		0.08
2004, FTP test, USA	0.20		0.08

Several techniques are promising for reducing emissions from diesel engines, such as modified or alternative fuels, engine modifications and finally after-treatment technologies [13-17]. HC and CO can be relatively removed easily by utilisation of an oxidation catalyst. Diesel fuel typically contains 0.15-0.30 wt% sulphur. In the engine, 98% of this element is combusted to sulphur dioxide and the remainder to sulphates. However, the problem of sulphur contamination is expected be solved in the near future by the implementation of low sulphur fuels, with sulphur content below 0.05 wt% [18]. Emissions of CO and HC from diesel engines are low and already fulfil standards, and therefore recently the major attention has been focused on the reduction of particulate (soot + SOF) and NO_x emissions. The soot can be easy removed by filtering and oxidation. Particulate filters or traps with tiny pores that capture particles are commonly used for this purpose.

As trapped material accumulates, resistance to flow increases and smaller particles are removed until resistance grows so much that regeneration is needed. In a diesel soot trap, resistance equates to a pressure drop across the bed that increases back pressure and fuel use. When a sensor indicates too great a resistance to flow, the filter is regenerated by heating it to burn off the collected soot. Filter efficiency for collection of soot lies, in general, between 60 to 90 %. Soot traps can be made from a ceramic or wire filter medium; the ceramic version (mainly from cordierite material) is favoured because it withstands high temperatures and has a higher trapping efficiency. However, regeneration ads significant cost and complexity, and it is a major obstacle to the widespread use of soot filters.

There are many systems under investigation to fulfil the requirements of future standards: particulate traps (wall-flow monoliths) made from cordierite, particulate traps made from SiC, ceramic foams, candle filters, metal wool filters and wire mesh filters [13,14, 16, 18-20]. All filters differ by type of filtration and particulate collection efficiency. Also, electrostatic precipitation or agglomeration and corona discharge have been tested. The latter two have hardly been applied due to the high voltages that would be required on board. There are also a number of non-catalytic and catalytic regeneration technologies for particulate traps to prevent the trapping becoming clogged with collected particulates what, in the long term, would result in engine malfunctioning due to increased back-pressure. The most common non-catalytic regeneration techniques are: burner installed up-stream of the particulate trap, installed electrical heaters, intake or exhaust gas throttling, etc. Catalytic regeneration of particulate traps is a logical choice for oxidation of carbonaceous materials. Many catalysts have been tested for this purpose: noble metals, transition metal based catalysts (Cu, Cr, Mn, Fe), perovskite type catalysts, as well as techniques such as injection of reagents before filter for decreasing the soot ignition temperature, use of microwaves for catalytic oxidation of soot, and many others [13, 16, 20]. All of them have their pro's and cons. However, some of these after treatment techniques (i.e. particulate removal using traps combined with catalytic coatings or with oxidising agent) are clearly still too fundamental to be applied soon. Some of the other techniques (trap plus a burner or electrically assisted regeneration, a trap combined with fuel additives, flow-through oxidation catalysts) can possibly play a significant role in emission reduction of particulates in the near future. Actually the regeneration is done by heating the filter electrically or by a burner to the ignition temperature of the soot. In general, such systems are too expensive and complicated in operation to gain general acceptance. Continuous combustion of the collected soot is hard to accomplish at the fairly low temperatures (300°C or lower) encountered in diesel engine exhausts. Several attempts have been made, but still no system is generally useful under all conditions. A completely new approach is the concept of the electrochemical reactor for filtering and continuous combustion of soot particles [7]. The concept is based, in general, on achievements in the field of SOFC [11,21,22]. The electrochemical reactor consists of a highly porous, oxygen ion conducting electrolyte (ceria based) covered by catalytically active, electron conductive perovskite-based electrodes [Figure 1]. The porous reactor structure acts as a mechanical filter, trapping the soot particles from the exhaust gas. By polarising the reactor with an external power supply, the combustion process of the collected soot particles can be forced to take place at a low temperature. The lower temperature limit is defined by the ionic conductivity of the electrolyte material. Ceria-gadoline mixture (90/10 by weight) was selected for this purpose which allowed the temperature of operation to go down to 250°C. By construction of a multilayered system, the voltage over the reactor and the filtering properties can be adjusted to the desired levels. The voltage required for this purpose should be between 24 and 60 V. The HC and CO compounds could also be partly oxidised in the electrochemical reactor [7].

The processes which go on in the reactor can be classified as follows:

1. Ionisation of oxygen from off-gases to oxygen ions:

$$2O_2 + 4e^- \rightarrow 2O^{2-} \qquad (1)$$

2. Migration of oxygen ions inside the monolithic structure to a soot particle,
3. Oxidation of the soot particle to carbon dioxide,

$$C + 2O^{2-} \rightarrow CO_2 + 4e^- \qquad (2)$$

4. Removal of CO_2 with off-gases.

The multichannel chemical reactor was constructed from single stacking tubes. The performed measurements showed an efficiency of above 90% for soot removal at low flow and 75 % for high flow (in a temperature range of 250-300°C). The backpressure was in a range of 60-160 mbar depending on the flow rate of off-gases (from 20 to 60 l/min). The fuel penalty from placing the reactor in the exhaust system was estimated to be 1% of the power output from the engine.

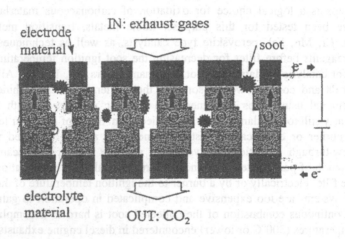

Figure 1 Principle of the electrochemical reactor [7].

3. EXPERIMENTAL PROCEDURE

3.1 MATERIALS

A cerium-gadolinium mixture (CGO) and lanthanum strontium iron cobalt oxides mixture (LSFC) of the perovskite type in the form of nanopowders were used as basic materials for preparation of the ceramic part of the diesel soot converter. Wax TP P was selected as a pore former and metylcellulose polymer (MC) as a temporary binder. The characteristics of all used materials are described in detail in this chapter.

Cerium-gadolinium oxides mixture

Cerium dioxide plays an important role in many catalytic processes. In recent years it has been increasingly used as a promoter in three-way catalysts (TWCs) for automotive emission control. CeO_2 has a fluorite-like cubic structure in which each cerium site is surrounded by eight oxygen sites in an fcc arrangement and each oxygen site has a tetrahedron of cerium sites. Under various redox conditions, the oxidation state of cerium may vary between 3+ and 4+. Owing to its nonstoichiometric behaviour, CeO_2 is shown to be a good promoter for oxygen storage while some noble metals (Pt, Pd, Rh, etc.) are used as major catalysts [22]. The ionic conductivity of CeO_2 can be increased by the addition of acceptor dopants. Doped ceria, which has poor sinterability and electronic conductivity in a reducing atmosphere, was considered to have potential as the solid electrolyte material of low-temperature SOFC's because its ionic conductivity at 800°C is close to or higher than that of YSZ at 1000°C [7, 8, 10, 23]. It is well understood that the oxygen vacancies, which are formed to charge, compensate the acceptors lead to higher ionic transport. In seeking a satisfactory strategy for the design of oxygen ion conducting solid electrolytes, it is possible to exploit the result that if the ionic radii of dopant and host cation are well matched, the strain energy term in the defect binding energy is minimised. The most detailed studies of this "size factor" in oxide electrolytes have been made in the system CeO_2-M_2O_3 (where M = Sc, Y, Gd, La). An experimental determination of the binding energy by Gerhard-Anderson and Nowick has revealed differences in the defect pair binding for various dopants; a minimum in binding energy is obtained when Gd_2O_3 is used as dopant [24]. It was also prove that the conductivity exhibits a maximum at the level of 10-20 mol% of acceptor impurities, its activation level increases with the concentration of dopant and depends upon their type [25]. Therefore the mixture of cerium-gadolinium oxides with a molar ratio of 90/10 from a commercial supplier was selected for experiments. Some parameters of the powder reported by the producer (*) and measured at ECN are presented in Table 2 [26].

Table 2. Selected properties of nanosize powders.

Powder/Parameter	Unit	CeO_2/Gd_2O_3 (90/10)	$La_{0.6}Sr_{0.4}Fe_{0.8}Co_{0.2}O_3$
* Content of CeO_2	%	89.00	-
* Content of Gd_2O_3	%	11.00	-
* Content of LSFC	%	-	95.00-99.00
* Content of Co_2O_3	mg/m^3	-	0.1
* Content of Fe_2O_3	mg/m^3	-	5.0
Particle size d_{50}	μm	0.48	0.37
Size of nanoparticles	nm	50-100	100-150
Specific surface area, (BET)	m^2/g	25.9	11.9
Coefficient of thermal expansion, α	1/°C x 10^{-6}	9.0 (20-900°C) 3.0 (20-400°C)	22.5 (20-900°C) 14.5 (20-400°C)
* Density	g/cm3	7.22	6.22-6.28

The powder was characterised by SEM. The material consists mainly of CGO agglomerates built from flakes (app. 1 um in length) formed from particles of size app. 50-100 nm. A typical shrinkage curve of the CGO taken in the dilatometric measurement is presented in Figure 2. The powder sinters at quite low temperatures, between 500 and 1100°C, due to the presence of nanosize particles.

94

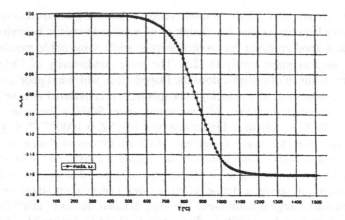

Figure 2 Dilatometric curve presenting the shrinkage behaviour of the GCO powder vs temperature of sintering

Electrode material

In place of expensive platinum, some perovskite-type electronic conductors could be used as electrode materials in the SOFC [7, 21,27]. Perovskite-type oxides (ABO_3 structure) are characterised by a mixed ionic-electronic conduction in solid electrolytes. Electronic conduction under a reducing atmosphere results from excess electrons generated by the reaction [7]:

$$O_o = \tfrac{1}{2}O_2(g) + V_o + 2e- \qquad (3)$$

Where, O_o and V_o represent oxygen ions at normal lattice sites and doubly ionized oxygen ion vacancies.

This same concept was also used in the preparation of electrode materials for the diesel soot convertor. As potential candidates for this application are considered strontium-doped $LaMnO_3$ ($La_{0.9}Sr_{0.1}MnO_3$), $La_{(1-y)}xSr\ _yFe_{(1-x)}$, and others compound from the perovskite-type family. The following perovskite–type oxides were tested for this purpose: in A sites: La_2O_3, SrO, K_2O, Li_2O, Cs_2O, and in B sites: Co_3O_4, Mn_2O_3, Fe_2O_3, Cu_3O_4, V_2O_5 [7, 9, 21, 27]. Another specific feature of these perovskite-type oxides is that they can catalyse the simultanous removal of NO and diesel soot particulates in the presence of oxygen [9]. Although the catalytic activity of perovskite-type oxides depends on both A-site and B-site cations, the substitution of potassium at A sites significantly promotes the oxidation of soot. Therefore the perovskite-type oxides are suggested for the catalytic diesel trap for simultaneous removal of soot particulates from off-gases. In the last case the catalyst/soot contact significantly affects the performance of the system. The perovskite-type compound with the formula $La_{0.6}Sr_{0.4}Fe_{0.8}Co_{0.2}O_3$ was obtained from a commercial supplier and used in the performed experiments. The as-supplied powder was first precalcinated at $650°C$ to remove all organic functional groups from the manufacturing process before it was characterised and used for a coating deposition. Some parameters of the powder are presented in Table 2. The powder consists of granulates with sizes of a few micrometres, built from agglomerates of nanocrystals, with sizes of approximately 100 nm.

Pore former

Wax coded TP P from a commercial supplier was used as a pore former. Wax TP P is the pore former commonly used for the production of porous ceramic products such as grinding wheels or catalyst supports. Four types of the pore former were used in the performed experiments: TP P 300, TP P 450, TP P 750 and TP P 1100. The first two types of pore former are characterised by a rectangular shape, the latter two by a ball shape. The particle size distribution of the pore formers according to the supplier was as follows:

* TP P 300   - below 350 μm,
* TP P 450   - 350-600 μm,
* TP P 750   - 600-1100 μm,
* TP P 1100   - 1000-1500 μm,

Other properties:
* Density   - app. 0.93 g/cm^3
*Drop point at   - at 105-115°C,
* Viscosity at 140°C - 100-350 cPs,

From the DTA, TG and DTG measurements it appeared that the pore former was completely removed from the ceramic body (by evaporation, pyrolysis and oxidation) above 600°C [26].

Temporary binder

The water soluble multicellulose type polymer grade A4M from a commercial supplier was selected for fabrication of ceramic monoliths by the extrusion technique. Cellulose is widely used as a binder for ceramic extrusions because it is non-ionic, water soluble and non-toxic. Additionally, metylocellulose polymers effectively reduce the surface tension of the aqueous system, resulting in improved wetting of the ceramic powder, provide water holding characteristics what translate to improved plasticity and workability, and reduce interparticle friction and erosion of metal components of extruders and dies. Thermal gelation is caused by the hydrophobic interaction between molecules containing methoxyl substitution. In a solution at low temperatures, molecules are hydrated and there is little polymer-polymer interaction other than entanglement. At higher temperatures, the molecules lose their water of hydration, which results in a decrease in viscosity. When a sufficient but not complete dehydration occurs, a polymer-polymer association takes place and the system approaches an infinite network structure, reflected by a sharp rise in viscosity. These network structures or gels are completely reversible. Thermal gelation also eliminates binder migration during drying [28]. The task of the polymer was, therefore, to maintain a good strength of the material in the green stage, improve the plastic properties of the paste for extrusion and reduce the friction between particles and particles and metal components of the mould/extruder. Metylocellulose burns out completely to carbon dioxide and water vapour at temperatures above 600°C.

Other additives

Alkali earth (Ca, Sr and Ba) salts in the form of acetates and a transition metal salt, Co-acetate, were used to study the effect of doping on the microstructure and electrical properties of ceramic materials in the system CeO_2-Gd_2O_3.

3.2 METHODS OF MANUFACTURING

The porosity of ceramic materials manufactured in a natural way is on the level of approximately 40 vol.%. To reach the porosity level above 70 vol.%, required in the case of catalytic diesel filters, the extrusion of ceramic plastic pastes incorporating a volatile/combustible additive was selected as a major shaping technique [29]. A number of paste compositions were prepared with a varied concentration and type of pore former. The aqueous solution of metylocellulose grade A4M was added as a temporary binder/lubricant to the composition of the paste to improve the rheological and lubrication properties of the paste and to maintain a good strength of the material in the green stage. All components of the paste were mixed together in a kneader type ZS 2L19 from Werner & Pfleiderer AG under vacuum at room temperature for 1 hour and then the paste was stored for 24 h in a plastic bag to improve its homogeneity.

The extrusion of tubes with external/internal diameter of 14.4/10.0 mm and later multichannel type of monoliths was performed. The monoliths with an external diameter of 30.9 and 19 round holes of 3.5 mm in diameter were made in the first phase of the project. Later the monoliths with an external diameter of 84.0 mm, wall thickness of 8 mm, and with square holes of sizes 8x8 mm^2 were extruded. The pressure of extrusion was 20-23 bar and the temperature was 20-22°C. The tubes were dried directly after the extrusion process using a rotating set-up of ECN design. The multichannels were dried in a microwave oven using 1 min heating/5 min cooling cycles to avoid rapid overheating of the samples which could result in melting of the pore former inside the green wet body followed by deformation of the samples. The sintering process was done in the temperature range from 800 to 1250°C with residence time ranging from one hour to four hours. The samples were sintered in the vertical position. The samples were extruded from the paste prepared from the as-delivered, green CGO powder and also from compositions prepared from the green powder, and precalcinated at 500, 700 and 900°C powders. This change in the manufacturing process was introduced to improve the extrusion properties of the plastic paste, to lower the drying shrinkage of the green product and to increase the drying rate of the multichannels without any deformation of the product.

In an attempt to increase the mechanical strength and, at the same time, decrease the backpressure of the porous material, acetates of Ca, Sr, Ba and Co (good solubility in water) were added to some samples. Additionally some monoliths sintered at 950°C for 2 h were impregnated with 5 wt% aqueous solutions of these acetates [8, 30]. The impregnated samples were sintered at 950°C for 1 h.

The selected samples in the tubular and multichannel forms were then coated with a 25 wt% water-based suspension of the LSFC powder with a suitable defloculant (Dolapix CE64). The coated samples were again sintered at 900°C for 1 h. Such prepared samples were used for characterisation and/or for preparation of the electrochemical reactor. Every second channel was closed alternately from each monolith side, silver electrical connections were joined to the electrode layer and connected to the external electric current source to complete the circuit.

3.3 CHARACTERISATION METHODS

The shrinkage of drying and sintering, open porosity, mechanical strength, backpressure, electrical resistance and microstructure were characterised on the dried/sintered samples. The open porosity of the sintered samples were measured using the water immersion technique, and also, for comparison purposes, this parameter was evaluated by mercury porosimetry and by optical microscopy. The microstructure was observed using optical microscopy and a scanning electron microscopy (SEM) on polished and fractured samples. The thickness of the LSFC coating was taken from the SEM pictures. The crushing strength was measured on the monolithic samples of diameter and height of app. 25 mm in the INSTROM machine with a cross head speed of 0.5 mm/min (in the axis parallel to the axis of the channels). The results are reported as an arithmetical average from 5 measurements or statistically evaluated from 10 samples. The backpressure parameter was measured on the tubular monoliths by an industrial partner. Electrical resistance measurements were carried out employing a d-c four-probe method. The details of measurements were given in [26].

4. RESULTS AND DISCUSSION

4.1 CGO-BASED POROUS MONOLITH

The composition of the plastic paste for extrusion consisted of:
- CGO powder in the range of 50 to 25 vol.%,
- Pore former in the range of 50 to 75 vol.%,
- MC A4M in the quantity of 4 wt% (calculated relative to the combined weight of CGO and pore former),
- Water in the quantity of 20-25 wt% (calculated relative to the combined weight of CGO and pore former).

The first experiments showed that it is necessary to introduce a maximum quantity of pore former (75 vol.%) to the paste to obtain the total porosity of the porous material above 70 vol.%. The second very important conclusion was that the pore former of a rectangular shape is only suitable for the manufacturing of a crack-free ceramic monolith. The fracture of the porous CGO monolith with a content of 70 vol.% of the pore former TP A 450 and TP P 750 proved the superiority of the former to the latter. In the latter case, lot of cracks were produced inside the porous body, which has a negative influence on the strength of the monolith. Therefore in the further research only the rectangular shape pore former TP P 450 was used. The third conclusion from this preliminary research was that the use of as-delivered fresh powder has a negative influence on the extrusion properties of the paste (non-equal rates of flow of the paste inside and outside of the monolith) and is a source of cracks during the drying process. To solve this problem the CGO powder was precalcinated at the temperatures of 500, 700 and 900°C for 2 h and the precalcinated powder substituted the fresh powder in the composition of the plastic paste partly or totally (100% precalcinated powder). The average grain size of fresh powder (soft agglomerates), $d_{0.5}$, was 4.681 µm, precalcinated at 700°C – 2.278 µm and at 900°C - 1.501 µm. This means that during the presintering step the size of particles becomes smaller and smaller. The SEM investigation allowed us to elucidate this phenomenon. The "soft" agglomerates in the fresh powder were totally destroyed during the extrusion process. Those agglomerates calcinated at 500°C or higher were much "harder" and managed to survive intact, partly or totally, in the extruded products. The performed tests with all presintered powders indicated that the temperature of precalcination 900°C was too high. The CGO material lost a lot of its reactivity and manufactured products were very weak, too weak even for the manual handling. Therefore 700°C was selected as a standard presintering temperature and a combination of the fresh powder with the presintered powder was used in further experiments to avoid the drying cracks and to ensure a good sinterability of the mixture. 950°C was selected as an optimal temperature for the sintering process with a residence time of 2-4 h. The detailed results of this investigation are reported in [25]. In such sintering conditions the growth of nanocrystals is very limited, the porosity is very high and the resulting strength is sufficient for this application. The standard composition is presented in Table 3 together with its further modifications and the properties of the produced monoliths.

Table 3. Standard composition of the paste for extrusion, its further modifications, and the properties of the manufactured monoliths sintered at 950°C for 2 h.

Composition	Standard M0	Modified M1 (infiltrated with 1wt% CoO)	Modified M2 (1 wt% CoO added to the paste)
CGO fresh, vol.%	0.0	10.0	10.0
CGO precalc. At 700°C, vol;.%	25.0	15.0	15.0
Pore former TP P450, vol.%	75.0	75.0	75.0
Properties of the CGO samples sintered at 950°C for 2h			
Drying shrinkage, %	8.3-14.5	-	-
Total (drying + sintering) shrinkage, %	21.0-23.5	25.1±0.4	27.9±1.4
Total porosity, %	76.5-77.0	68.4-69.7	60.0
Average pore size, um	149.0	-	-
Crushing strength, Mpa	2.24	6.8±3.9	16.9±0.6
Electrical resistance, kΩ/mm	111	23	11
Back pressure, kPa	832	104	5 596

The developed porous CGO material was, however, characterised by a low mechanical strength. Material with such low strength cannot survive for a long time (min. 100 000

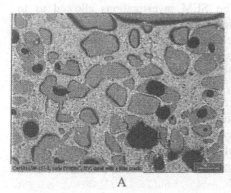

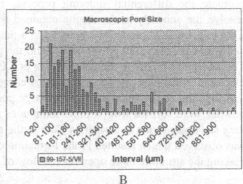

A   B

Figure 3 (A) Optical microscopy image of the polished cross-section of the porous CGO material sintered at 950°C for 2h. (B) Pore size distribution measured by the linear intercept technique for this same material. Total porosity (big pores only) – 53.8%, average pore diameter – 75 um.

km) in the real working conditions in diesel vehicles. Therefore it was necessary to improve the strength of material either by increasing the temperature of sintering or by addition of sintering additives. The second option was more economically attractive and therefore the alkali earth oxides (CaO, SrO, BaO) and a transition metal oxide (CoO) were introduced to the composition of the CGO material as dopants. The sintered CGO samples were either impregnated with a 5 wt% aqueous solution of the corresponding metal acetate or the metal acetate was introduced in the quantity of 1 wt% to the composition of the standard paste before the extrusion process.

The second parameter of the porous material, the back pressure, was also too high, 832 kPa, in comparison to the required value which was less than 150 kPa. The performed analysis of the macro- and mesopores size distribution by linear intercept of the polished CGO samples revealed that the majority of macropores are between 20 and 300 μm (Figure 3). The average pore size differs from sample to sample, depending strongly on the pore former used and the sintering temperature. The average macropore size for the standard composition sintered at 950°C for 2 h was 149.0 μm. The majority of pores, however, were closed. Thin walls were built from nanocrystalline CGO particles, closed packed, and with mesopores of sizes between 20-100 nm (mercury porosimetry measurement). To lower the backpressure parameter it was necessary to make the thin walls between the big pores more open. The following methods were considered for this purpose: precalcination of the CGO material at higher temperatures, introduction of a second fine pore former to the plastic paste or differential sintering shrinkage of the thick/thin parts of the ceramic body. The performed experiments with the standard composition, M0, doped with different metal cations showed the best improvement in the mechanical strength and decrease of the back pressure value for the composition with cobalt oxide additive [26]. This improvement, however, was combined with a strong decrease in the open porosity value, from 78 to 69%. The conclusion was obvious; cobalt oxide must be introduced into the porous CGO structure by the impregnation process but not by a simple addition of the CoO to the plastic paste before the extrusion process (see Table 3, data for compositions M2 and M3). For the composition M2 an additional heat treatment at 950°C for 1 h was also necessary to promote the differential sintering process. The SEM investigations allowed us to resolve the anomaly in the results obtained for the CoO doped samples. Addition of CoO to the CGO composition resulted in a much higher shrinkage of the ceramic body and lower porosity, but on the other hand the permeability of the system was growing. The addition of CoO to the composition of CGO material has also an influence on the electrical resistance of the system, lowering this value by a factor 4. The SEM investigation showed that the walls of large pores of samples made only from the CGO powder were formed from a thin, continuous film of CGO nanoparticles. Introducing very strong sintering aids such as CoO to the system, a higher shrinkage of the system was observed which resulted in the formation of cracks and holes in the thin walls thus making the structure more open for the flow of gases or liquids.

4.2 DEPOSITION OF LSFC COATING

Deposition of the electrode layer on the surface of the CGO monolith was made by a dip-coating technique using a 25wt% water-based suspension of the LSFC powder. The LSCF powder characteristics are given in Table 2. The CGO porous samples were coated once which resulted in a coating with the thickness of app. 6-8 μm. Coated samples were dried at ambient temperature and sintered at 900°C for 1h. To produce a thicker coating a number of dipping-sintering cycles must be repeated. The main concern was whether the LSFC coating adhered well to the CGO material due to a significant difference in their coefficients of thermal expansion, 22.5×10^{-6} 1/°C to 9.0×10^{-6} 1/°C, respectively, in the temperature range 20-900°C. The formed coating adhered very well to the surface of the porous CGO material creating a good packing layer on it (Figure 4). The heating/air cooling test between 20 and 600°C (typical temperature shock for an exhaust system of a diesel engine) repeated 10 times did not show any influence on the quality of the coating. The explanation of this good adhesion of the two materials can be found in their nanocrystalline structure. It is well known that nanocrystalline materials can deform plastically and extensively by grain boundary sliding showing the "superplastic deformation" not typical for brittle ceramic materials [1]. Such a phenomenon took place in the developed CGO/LSFC functionally graded composite.

4.3 SCALE-UP OF THE MONOLITHS

The mould for extrusion of experimental size monoliths was designed and manufactured at ECN. The monoliths with an external diameter of 84 mm were extruded and sintered at 950°C for 4 h. Sintered crack-free monoliths were infiltrated with aqueous Co-acetate solution and baked at 950°C for 1h, then coated with the LSFC suspension and again baked at 900°C for 1h. The total shrinkage of large size monoliths was slightly lower than reported in Table 3: 19.1 %. The other parameters were without any remarkable changes in comparison to the reported early results. The coated CGO monoliths of sizes: 80 mm in diameter and 160 mm in length were produced and supplied to a manufacturer of diesel exhaust systems to perform the field test on them in a 3.5 l diesel engine.

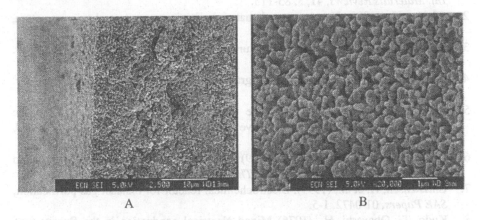

A   B

Figure 4 (A) SEM image of the fracture of the CGO material coated with the LSFC layer and sintered at 900°C for 1h. Thickness of the LSFC layer is the range 6-8 μm. (B) A top view of the coating; LSFC particles form a homogeneously packed porous layer with open pores of size 80-100 nm. The pores are sufficiently big for penetration of off-gases into the structure of the FGC material (working as a diesel soot trap and catalyst simultaneously).

5. Conclusions

- A ceramic catalytic monolith for an electrochemical reactor acting as a trap for the continuous removal of soot particles from diesel exhaust gas has been developed.
- The nanocrystalline powders of perovskite type, CeO_2-Gd_2O_3 oxides mixture (oxygen ion conductor) and strontium and cobalt doped lanthanium ferrite, $La_{1-y}Sr_yFe_{1-x}C_xO_z$ (electron conductor) were used for the manufacturing of the functionally graded composite-based component; a ceramic porous monolith with a diameter of 84 mm coated with a LSCF layer of thickness 6 to 8 μm.
- The CGO powder requires an additional operation, precalcination at 700°C, to improve the extrusion properties of the paste and ensure a crack-free drying process.
- The impregnation of sintered CGO monoliths with the transition metal oxide (CoO) improves the mechanical strength of the material and also lowers the backpressure parameter to the required value.
- The great mismatch in the thermal expansion coefficients of the two powders used for manufacturing the CGO monolith coated with the LSFC film did not have any negative influence on the adhesion of the coating to the bulk body, due to their inherent nanocrystalline character.
- The performed research proved that it is feasible to produce large ceramic components such as diesel filter soot traps using already available commercial nanocrystalline powders.

6. REFERENCES

1. Mayo, M.J., (1996) Processing of nanocrystalline ceramics from ultrafine powders, *Int. Materials Reviews,* **41**, 3, 85-115.
2. Gleiter, H., (2000) Nanostructured materials; basic concepts and microstructure, *Acta Materialia,* **48**, 1-13.
3. Sternitzke, M., (1997) Review: Structural Ceramic Nanocomposites, *J. of the Eur. Ceram. Soc.,* **17**, 1061-82.
4. Vassen, R., (1999) Densification and grain growth of nano-phase ceramics, *cfi/Ber. DKG,* **76**, 4, 19-22.
5. Marécot, P. et al., (1994) Influence of the redox properties of ceria on the preparation of three-way automotive Pt-Rh/alumina-ceria catalysts, *Applied. Catalysis B. Environmental,* 5, 57-69.
6. Djuricic, B., Pickering, St., (1999) Synthesis of alumina-ceria nano-nano composites for catalytic applications, *The Materials Challenge,* 12, 4-5.
7. Christensen, H. et al., (1999), Electrochemical reactor for exhaust gas purification, *SAE Papers,* 01-0472, 1-5.
8. Kudo, T., Obayashi, H., (1976) Mixed electrical conduction in the fluorite-type $Ce_{1-x} Gd_x O_{2-x/2}$, *J. of Electrochem.Soc.,***123**, 3, 415-419.
9. Christie, G.M., Van Berkel, F.P.F., (1996) Microstructure – ionic conductivity relationships in ceria-gadoline electrolytes, *Solid State Ionics,* **83**, 17-27.
10. Teraoka, Y. et al., (1995) Simultaneous removal of nitrogen oxides and diesel soot particulates catalyzed by perovskite-type ceramics, *Applied Catalysis B: Environmental,* 5, L181-L185.
11. Gauckler, L.J., Kleinlogel, C., (1999) Preparation and properties of nanosized ceria solid solutions for Solid Oxide Fuel Cells, *Proc. of 12th Int. Conf. On Solid State Ionics, Halkidiki,* Paper C-KE-01, 238-239.
12. Wanqin, J., et al., (2000) Fabrication of $La_{0.2}Sr_{0.8}Co_{0.8}Fe_{0.2}O_{3-\delta}$ mesoporous membranes on porous supports from polymeric precursors, *J. of Membrane Science,* **170**, 9-17.
13. Neeft, J.P.A., (1995) Catalytic oxidation of soot, *PhD Thesis,* TU Delft, The Netherlands.
14. Twigg, M.V., Wilkins, A.J.J., (1998) Autocatalysts – Past, Present, and Future, in *Structured Catalysts and Reactors,* Ed. by A. Cybulski, J.A. Moulijn, Marcel Dekker, Inc., 91-148.
15. Frost, J.C., Smedler, G., (1995) Control of NO_x emissions in diesel powered light vehicles, *Catal. Today,* **26**, 207-214.
16. Gulati, S.T., (1998) Ceramic catalyst supports and filters for diesel exhaust after treatment, in *Structured Catalysts and Reactors,* Ed. by A. Cybulski, J.A. Moulijn, Marcel Dekker Inc., 510-541.
17. Lox, E.S et al., (1991) Diesel emission control, in Catalysis and Automotive Pollution Control II, Ed. by A. Crucq, Elsevier Science Publ. B.V, Amsterdam, 291-320.

18. A status report (1992) Reducing truck diesel emissions:, *Automotive Eng.*,**100**, 2, 19-22.

19. Stobbe, Per, Høj, J.W., (2000) Diesel Exhaust gas filter, *European Patent no WO0014639A1* from January 13, 2000.

20. Global Developments in Diesel particulate Control, *SAE Papers P240*, Publ.: Soc. of Automotive Eng., Inc., February 1991.

21. Eichler, K., et al., (1993) The electrolyte of the SOFC as thick film, in *Third Euro-Ceramics*, Ed. P. Duran and J.F. Fernandez, 3, 1157-1162.

22. Zhang, Yu, et al., (1995) Nanophase catalytic oxides: I. Synthesis of doped cerium oxides as oxygen storage promoters, *Applied Catalysis B; Environment*, 6, 325-337.

23. Kim, S.G. et al., (1998) Fabrication of YSZ Films on Ceria Electrolyte by the Sol-Gel Method, in *Sol-Gel Processing of Advanced Materials*, Ed. by L.C. Klein et al, **81**, 359-364.

24. Zhen, Y.S. et al., (1988) Oxygen ion conduction in CeO_2 ceramics simultaneously doped with Gd_2O_3 and Y_2O_3, in *Science of Ceramics*, Ed. by D. Taylor, **14**, 1025-1030.

25. Kosacki, I. et al., (2000) Modeling and Characterization of Electrical Transport in Oxygen Conducting Solid Electrolytes, *J. of Electroceramics*, 4, 1, 243-249.

26. Rak, Z.S., (2000) CGO porous monoliths, *ECN report ECN-CX-00-009* from 20 January 2000, Petten, The Netherlands.

27. Iwahara, H., (1995) Technological challenges in the application of proton conducting ceramics, *Solid State Ionics*, 289-298.

28. Schueltz, L.E., (1986) Metylocellulose Polymers as Binders for Extrusion of Ceramics, *Am. Ceram. Soc. Bull.*, **73** 12, 1556-59.

29. Supelveda, C., (1997) Gelcasting of foams for porous ceramics, *Am. Ceram Soc. Bull.*, **76**, 10, 61-65.

30. Moure, C., et al., (1999) Microstructure and ionic conductivity of Ca-doped CeO_2-Gd_2O_3 solid electrolytes, *Proceed. of 6th ECS Conf. in British Transactions*, **60**, 2, 135-136.

NANOCRYSTALLINE NICKEL AND NICKEL-ALUMINA FILMS PRODUCED BY DC AND PULSED DC ELECTROPLATING

J. Steinbach and H. Ferkel
Institute for Materials Science and Materials Technology
Technical University Clausthal
Agricolastr. 6, 38678 Clausthal-Zellerfeld, Germany
E-Mail : jan.steinbach@tu-clausthal.de; hans.ferkel@tu-clausthal.de

1. Introduction

Al_2O_3 nanoparticle reinforced nickel can be produced by electrolytical co-deposition of nickel and alumina nanoparticles under direct current (DC) plating conditions from electrolytical nickel baths in which alumina nanopowders are dispersed [1-8]. It is also known that high current density favors the formation of fine grained nickel films due to an increasing nucleation rate of nickel nuclei on the growing film with increasing current density [e.g. 9]. Recent investigations also showed that when alumina nanoparticles are added to the electrolytical bath, besides incorporating in the growing nickel film these particles inhibit growth of nickel nuclei and therefore allow deposition of nanocrystalline Ni/Al_2O_3 composite films with an average nickel matrix grain size around 30 nm under DC conditions [6]. It was also found by transmission electron microscopy (TEM) investigations that the size distribution of the codeposited nanoparticles is nearly the same as that of the employed nanopowder prior deposition and therefore size selection during embedding of particles in the film is negligible [7]. This is of importance when uniform composite films are required. However, when these films are plated under pulsed direct current (PDC) conditions where the pulse length allows only limited growth of nickel nuclei, large nanoparticles or parts of particle agglomerates might not be sufficiently embedded by growing nuclei during one pulse to be trapped on (in) the film. Therefore these particles might escape from the surface before the next pulse occurs, which not necessarily promotes a continuous growth of the same nuclei. Furthermore under PDC conditions the ion concentrations in the cathode film are fluctuating, which will also affect the growing film in a different manner than in the case of DC plating. Under those conditions the homogeneity of the particle

M.-I. Baraton and I. Uvarova (eds.),
Functional Gradient Materials and Surface Layers Prepared by Fine Particles Technology, 105–110.
© 2001 *Kluwer Academic Publishers. Printed in the Netherlands.*

distribution and the particle size distribution in the PDC plated films might be different from those in the DC plated composites. This issue is the subject of current investigations in our laboratories and we present first results on the plating of Ni/Al₂O₃-composite films under PDC conditions.

2. Experimental Procedures

Nanocrystalline Ni/Al₂O₃ composite films were produced in a typical sulfamate bath of following composition: 485 g/l nickel-(II)-sulfamate (Ni(NH₂SO₃)·4 H₂O), 22 g/l boric acid (H₃BO₃), 8 g/l nickel chloride (NiCl₂ · 6 H₂O), 0.5 g/l surfactant, and 10 g/l alumina nanopowder. The commercially available nanopowder (NanoTek aluminium oxide 0100, Nanophase Technologies Corp.) was produced by physical vapor synthesis (PVS). Size and structure of the powder were determined by TEM (CM200, Philips) equipped with an energy dispersive X-ray (EDX) analytic unit and by X-ray diffraction (XRD) (D500, Siemens) using the $Cu_{K\alpha}$ radiation. Fig. 1 shows a TEM picture of the employed alumina nanopowder (left) and the corresponding particle size distribution (right). As can be seen the particles are of spherical shape and their size distribution can be characterized by a log-normal distribution with a median particle diameter d_m of 25 nm and a geometric standard deviation σ of 1.7. The particles are crystallized in the γ-phase.

The dispersion of the particles in the bath at 50°C was assisted by ultrasonic treatment during plating on flat and polished copper cathodes which had a surface area of 16 cm². The anode was made from pure nickel and the ratio of surface area between anode and cathode was 6:1. Under DC conditions the current density was 1.5 A/dm² at a voltage of 2-3 V. In the case of PDC deposition the duty cycle (ratio between pulse

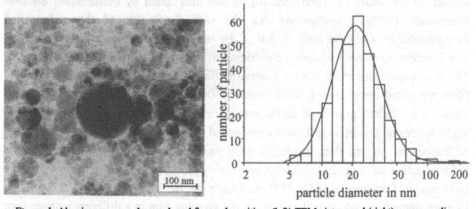

Figure 1. Alumina nanopowder employed for co-deposition. (left) TEM picture and (right) corresponding particle size distribution with logarithmic scale for the particle size, the smooth line describes a log-normal distribution with $d_m = 25$ nm and $\sigma = 1.7$.

length Δt_{on} and pulse pause Δt_{off}) was kept constant at 1:2.5. The pulse current density j_p was 10 A/dm^2 at a pulse potential between 8 and 10 V. Films formed at higher j_p exhibit a rough surface and dark gray nickelhydroxide deposits.

Specimens from the plated films were fabricated before and after heat treatment (850°C for 1 h) for TEM inspection by dimple grinding followed by ion etching.

3. Results and Discussion

Figure 2 compares TEM pictures of two specimens plated under DC and PDC (Δt_{on} = 200 ms) conditions in the same nickel sulfamate bath containing 10 g/l nanopowder at different magnifications. The left and right parts of the figure show the sample produced under DC and under PDC conditions, respectively. As can be seen the composite plated under DC condition has a nickel matrix grain size around 500 nm whereas the sample plated under PDC condition exhibits a much smaller matrix grain size around 30 nm. In the latter case the nanocrystalline structure was expected since pulsed electrodeposition of pure nickel was already shown by other laboratories to form a nanocrystalline structure [10, 11].

As also can be seen, in the DC sample alumina particles with diameters of 100 nm and larger can be found whereas in the case of the PDC specimen no alumina particles larger than 80 nm are present. Obviously PDC plating promotes the incorporation of the smaller particles more than the incorporation of large particles. According to Faraday's law an average film growth of less than 7 nm per pulse under the employed experimental PDC condition can be expected. This means that within one pulse the largest particles might not be embedded sufficiently by growing nickel nuclei to be trapped in (on) the surface and might escape from the surface before the next pulse occurs. During the next pulse the same nuclei do not necessarily continue growing and new nuclei grow at other sites. This is consistent with the experimental findings.

Furthermore, as can be seen in Fig. 2 the Al$_2$O$_3$ particles in the composite plated under DC conditions are more agglomerated than in the case of the PDC sample. Particles in a electrolytic bath of high ion concentration, which is the case here, are expected to agglomerate due to van-der-Waals forces [12]. Therefore, when such agglomerates are adsorbed on the growing film under PDC conditions, small particles facing the growing film in such agglomerates will be sufficiently embedded to be trapped and the rest of the agglomerate left behind might be able to leave the film within the pulse pause similar as single large particles can do as discussed above. However in this case the attraction between particles of such agglomerates must be weakened e.g. due to physicochemical interaction at the end of the pulse and/or the beginning of the pulse pause when the Ni^{2+} and H$^+$ concentrations are the lowest in the electrical double

108

layer of the cathode. Under PDC plating the Ni^{2+} and H^+ concentrations in the double layer of the cathode will change according to the employed duty cycle. Therefore the electrical double layer around each nanoparticle and particle agglomerate close to the cathode will also be altered. In such an environment of temporary low ion concentration

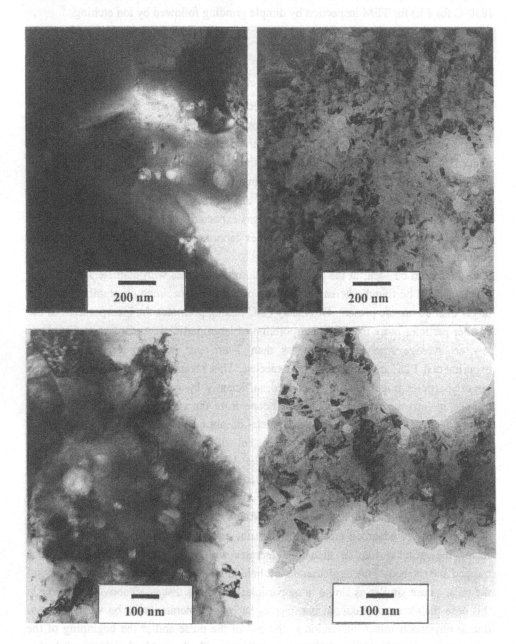

Figure 2. TEM pictures of Ni/Al₂O₃ composite produced at (left) DC and (right) PDC (Δt_{on} = 200 ms) conditions. The lower parts present the edge of the same specimens at a larger magnification.

close to the cathode dissociation of weakly bonded agglomerates might be possible directly on the deposited film or within the electrical double layer of the cathode. Under DC condition such low ion concentrations are not achieved and nanoparticles tend to agglomerate more. This scenario could explain the experimental findings. However, electrophoresis of colloid particles or particle agglomerates will contribute to the co-deposition behavior and electrophoresis will also be different under PDC.

Figure 3 shows the PDC plated Ni/Al_2O_3 composite after heat treatment at 850°C. As expected grain growth of the nickel matrix occurs during the heat treatment. However, the average matrix grain size does not exceed 1 µm. Pure electroplated nickel would show an average grain size of several µm after the same treatment. A lot of the alumina nanoparticles are distributed in the nickel grains. It is also remarkable that most of the alumina particles are distributed as single particles in the nickel matrix whereas under DC plating conditions the PVS particles are much more agglomerated in the metal as can be seen in the left part of Fig. 2. Earlier investigations on DC plated Ni/Al_2O_3 composites showed that large alumina particles or particle agglomerates will pin grain boundaries more efficiently than small particles [7, 8]. Since the PDC plated

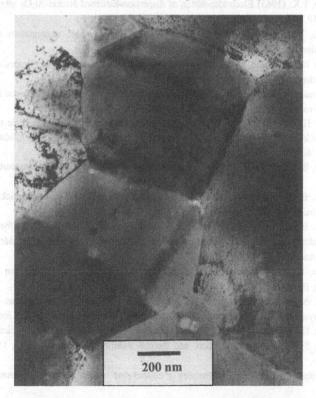

Figure 3. TEM picture of the Ni/Al_2O_3 composite plated under PDC (Δt_{on} = 200 ms) conditions after a heat treatment at 850°C for 1h.

110

composite of Fig. 3 contains none of the largest particles of Fig. 1 matrix grain size stabilization is less pronounced than in the case of samples plated at DC conditions in the same electrolytical bath [8]. The mechanical strength of the composite in Fig. 3 will be determined by the matrix grain size itself and also by dispersion strengthening due to the alumina nanoparticles in the matrix grains which will act as obstacles for dislocation glide.

In conclusion, it was shown that under PDC electroplating conditions nanocrystalline Ni/Al₂O₃ composites can be produced. In contrast to DC plating the incorporated alumina nanoparticles are less agglomerated in the deposited film. Furthermore, under the PDC plating conditions a particle size selection of the employed nanopowder occurs. This means here that PDC-plating promotes the incorporation of smaller particles more than the incorporation of large particles.

References

1. Sautter, F.K. (1963) Electrodeposition of dispersion-hardened Nickel-Al₂O₃ alloys, *J. Electrochem. Soc.* **110**, 557.
2. Nwoko, V.O. and Shreir, L.L. (1973) Electron micrographic examination of electrodeposited dispersion-hardened nickel, *J. Appl. Electrochem.* **3**, 137.
3. Oberle, R.R., Scanlon, M.R., Cammarata, R.C. and Searson, P.C. (1995) Processing and hardness of electrodeposited Ni/Al₂O₃ nanocomposites, *Appl. Phys. Lett.* **66**, 19.
4. Steinhäuser, S. and Wielage, B. (1997) Electrochemical plating of composite coatings, in: M.L. Scott (ed.), *Proceedings of ICCM-11* Vol. III, Gold Cost. Australia, pp. 495-504.
5. Ferkel, H., Müller, B. and Riehemann, W. (1997) Electrodeposition of particle strengthened nickel-films, *Eleventh Intern. Conf. on the Strength of Mater. ICSMA-11*, Prag 1997; *Mater. Sci. Eng.* **A234-236**, 474.
6. Müller, B. and Ferkel, H. (1999) Properties of nanocrystalline Ni/Al₂O₃ composites, *Z. Metallkd.* **11**, 868.
7. Müller, B. and Ferkel, H. (1998) Al₂O₃-nanoparticle distribution in plated nickel composite films, *Nanostructured Mater.* **10**, 1285.
8. Müller, B. and Ferkel, H. (1999) Studies on nanocrystalline Ni/Al₂O₃ films formed by electrolytic DC plating, in: *Proc. of Int. Symp. on Mechanically Alloyed and Nanostruct. Materials (ISMANAM-99)*, Dresden, in press.
9. Fischer, H. (1954) *Elektrolytische Abscheidung und Elektrokristallisation von Metallen*, Springer-Verlag, Berlin.
10. Wang, N., Wang, Z., Aust, K.T., and Erb, U. (1997) Room temperature creep behavior of nanocrystalline nickel produced by an electrodeposition technique, *Mater. Sci. Eng.* **A237**, 150.
11. Natter, H., Schmelzer, M., and Hempelmann, R. (1998) Nanocrystalline nickel and nickel-copper alloys: Synthesis, characterization, and thermal stability, *J. Mater. Res.* **13**, 1186; and references therein.
12. e.g.: Hiemenz, P.C. (1977) *Principles of colloid and surface chemistry*, Marcel Dekker Inc., New York.

FORMATION OF FINE-DISPERSION STRUCTURES BY ELECTRIC-SPARK, LASER AND MAGNETRON COATING WITH AlN-TiB$_2$/ZrB$_2$ COMPOSITE MATERIALS

A.D. PANASYUK, I.A. PODCHERNYAEVA, R.A. ANDRIEVSKY,
M.A. TEPLENKO, G.V. KALINNIKOV, L.P. ISAEVA
Institute for Problems of Materials Science of NAS, Kiev, Ukraine

Abstract

Structure, composition, tribological and corrosion properties of electric-spark, laser and magnetron coatings using advanced high-temperature AlN-TiB$_2$/ZrB$_2$ ceramics were studied. The similarity between the structures of electric-spark and laser coatings, showing the formation of globules after sintering of the finely dispersed composite material on a treated surface, was ascertained. The Al$_2$SiO$_5$, ZrSiO$_4$ and β-Al$_2$TiO$_5$ compounds, appearing in the coating directly during its formation, play a positive role in terms of tribological behavior for solid lubricants and of corrosion resistance as well. Under optimum conditions of electric-spark alloying of titanium VT6 and low-carbon 45 steel both friction coefficient and wear were decreased to 0.13–0.15 and 6–13 μm/km, respectively. The working temperature of the coated VT6 alloy was increased by 130 ^{0}C. The AlN-TiB$_2$ magnetron coating on monocrystalline silicon is characterised by a continuous structure of the ultra-dispersion even under corrosion at a temperature as high as 1400 ^{0}C.

1. Introduction

High thermal conductivity as well as resistance to corrosion of AlN in combination with hardness and electrical conductivity of TiB$_2$ and ZrB$_2$ enable one to consider AlN-TiB$_2$/ZrB$_2$ composites as prospective wear- and corrosion-resistant ceramics and as coatings on high-performance alloys. The AlN-TiB$_2$ ceramics was proposed by French researchers [1]. We have developed the new AlN-ZrB$_2$ composite material with a small content of ZrSi$_2$ additive. The peculiarity of these systems is the potential formation of such high-temperature compounds as β-Al$_2$TiO$_5$ tialite, Al$_2$SiO$_5$ mullite, ZrSiO$_4$ zircon and other ones directly during coating formation due to phase conversion under oxygen environment. The compounds mentioned above may be advantageously considered in corrosion as well as in tribological applications, and as solid lubricants if they wet a substrate material. As wetting is very important for the properties of coatings, we paid special attention to wetting effects in this study. It must be noted that under conditions of concentrated energy fluxes a dispersion of fine structures are formed. Electric-spark alloying (ESA), laser alloying and magnetron spraying are widely used for obtaining coatings. Among these methods ESA is characterized by both simplicity

111

M.-I. Baraton and I. Uvarova (eds.),
Functional Gradient Materials and Surface Layers Prepared by Fine Particles Technology, 111–118.

112

of the process and variety of physical-chemical phenomena taking place on the working surfaces. Therefore, it may serve as a model process to study the formation of wear- and corrosion-resistant coatings. Traditionally as an alloying electrode for ESA the hard alloys of "refractory compound–metallic binder" systems were used. In such systems a metallic binder ensures adhesion between coating and substrate, on the one hand, and material structure formation during the microprocesses of liquid-phase sintering – on the other hand [2]. At the same time a mechanical mixing between substrate material and electro-erosion products of the alloying electrode takes place in the micropool on the cathode surface. This suggests that such a metallic binder may play a role in the metallic alloying of the substrate. Differences in structure and properties of the coatings obtained by different methods from the same material can be essentially ascribed to differences in physical-chemical and thermal-mechanical effects on a material during its deposition. In this regard it is interesting: 1 – to carry out a comparative investigation of electric-spark, laser and magnetron coatings using the advanced $AlN\text{-}TiB_2$ composite material; 2 – to study the process of ESA of both titanium VT6 alloy and steels using the new $AlN\text{-}ZrB_2$-based composite material; 3 – to study composition, structure, tribological and corrosion properties of the obtained coatings.

2. Methods and materials

The $AlN\text{-}TiB_2$ (1:1) and $AlN\text{-}ZrB_2$ (with a small content of $ZrSi_2$ additive) electrodes/ targets were manufactured by hot pressing at 1680-1880 °C under a 150 MPa pressure. The $AlN\text{-}TiB_2$ power as a mechanical mixture of components was produced in a planetary mill. The size of power particles was 1-3 μm. As substrates titanium VT6 (Al-6%, V-4%) alloy, soft 45 steel, stamp 40X steel (C–0.36-0.45, Mn–0.5-0.8, Cr–0.8-1.1 mass. %) and heat-resistant Ni-alloy were used. The high-frequency ESA was carried out in air using an Elitron-21 installation in the following conditions: the short-circuit current was 0.9 A, the frequency of pulses was 1200 Hz, the energy of pulses was 0.08 J. The coating by laser evaporation of powder was carried out in Ar using a CO_2-laser with a power density of 25 kW/cm^2 and a gaussian power distribution in the heating spot. The laser beam scanning velocity was 0.5 m/min. Magnetron coatings were obtained using the method of radiofrequency-magnetron sputtering in an environment of ordinary purity Ar. The working surfaces were investigated by applying metallographic, XRD, EPMA, and SEM methods.

The tribological characteristics (friction coefficient f and linear wear I) of coatings were studied by using the "shaft–inset" scheme in the regime of dry friction in a pair with strengthened steel (HRA 58-62). The resistance to high-temperature oxidation was studied using thermogravimetry (TG) and differential thermal (DTA) analyses up to 1500 °C in a non-isothermal regime.

3. Results and discussion

The electrode materials are characterized by a heterogeneous structure with highly dispersed grains having a size of 1-3 μm, low porosity (3-5%) and homogeneous distribution of the nitride and boride components. The elemental distribution on the surface of electrodes shows that there is no interaction between the main phases. AlN is alloyed with a small amount of O and Fe. The latter is grinded in an amount smaller

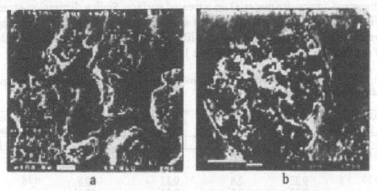

Figure 1. Microstructure of AlN-TiB₂ ESA-coating surface: a- general structure, b- a globule

than 2% in a planetary mill during the mixtures preparation. The materials developed have rather remarkable mechanical characteristics: the average microhardness is about 20 GPa, the bending strength is about 500 MPa.

During ESA a composite coating is formed resulting from interaction between electro-erosion products of the electrode and the substrate material in a cathode micropool. The microstructure of AlN-TiB₂ ESA-coating on heat-resistant Ni-alloy (Fig. 1, a) is an alloyed crystallization zone of substrate material with a typical island-like structure. Globules (Fig. 1, b) with a size of tens of microns and with a fairly spherical shape are formed on this surface as a sintered layer of composite material based on products of electrode electro-erosion. The grain size of this composite material is less than 1 μm and the layer thickness is about10 μm.

The coating phase composition is always different from that of the alloying electrode due to oxidation and phase conversion in air under conditions of electric-spark discharge. In accordance with XRD, the main phases of ESA-coating on Ni-alloy are TiB₂ and Al₂O₃. There are also Ni₃Al and Ni-Cr alloys. However, the elemental distribution on the coating surface at the "crystallization zone – globule" boundary (Fig. 2) shows β-tialite besides the phases above. The crystallization zone is Ni-Cr alloy alloyed with TiB₂ and Al₂O₃. The particles of crystallization zone have essentially a smaller size compared with the particles of a sintered layer. One can see that Al₂O₃ appears mainly in the sintered layer together with Ni-Cr inclusions while TiB₂ is distributed more homogeneously all over the surface. The presence of NiCr inclusions in a sintered layer shows that the metallic binder in the coating material leads to a substrate metallic alloying due to mass transfer from the micropool.

In the case of AlN-TiB₂ coating by laser evaporation a layer of 1 mm thickness is formed. It is reinforced with solid-phase inclusions of globules having a fairly spherical shape and a size of tens of microns. The globule is a sintered composite material with a grain size of 1-3 μm. The elemental distribution on a globule surface (Fig. 3) is practically the same as that of the electrode material. Therefore the average microhardness of the globule material is almost the same as that of the electrode material manufactured by hot pressing, that is about 20 GPa. Thus, there is a similarity between the structures of electric-spark and laser coatings showing the formation of globules of finely dispersed composite material on a treated surface. Such a dispersion-strengthened island-like structures has to improve the surface characteristics. The tribological parameters for AlN-TiB$_2$ ESA-coatings on different high-performance alloys under conditions of dry friction are presented in Table 1.

TABLE 1. Friction coefficient (f) and linear wear (I, μm/km) of AlN-TiB$_2$ electrode and ESA-coatings under different sliding velocities (v, m/s)

v, m/s	AlN-TiB$_2$ (1:1) electrode		For WC+6%Co alloy substrate		For high-speed P18 steel substrate		For low-carbon 45 steel substrate	
	f	I	f	I	f	I	f	I
5	0.40	2	0.39 0.39*	2.6 6.5*	0.31 0.30*	7.2 9.8*	0.19 0.35*	7.4 800*
10	0.30	6	0.37 0.37*	3.8 9.8*	0.27 0.24*	10.6 12.4*	0.16	11.6
15	0.24	7	–	–	0.22 0.24*	13.2 18.9*	0.15	13.1

*) for non-coated substrate

The best result was obtained for low-carbon 45 steel. Wear is almost the same compared with the non-coated high-speed steel and friction is essentially lower. This is very important for the problem of replacement of expensive alloyed steels by cheaper ones.

Magnetron coating on monocrystalline silicon substrate using the AlN-TiB$_2$ target is characterized by an ultra-dispersed structure which is probably amorphous. The elemental distribution on a coating surface confirms the presence of phases based on TiO$_2$ and Al$_2$O$_3$. After magnetron coating and high-temperature oxidation up to 1500^{0}C for 100 min., the ultra-dispersed structure of the oxidized coating is preserved. In accordance with EPMA, Al$_2$O$_3$ and TiO$_2$ as well as β-tialite are formed.

TG and DTA oxidation curves for AlN-TiB$_2$ magnetron coating show three oxidation stages. At the first stage for temperatures less than 1200 ^{0}C TiO$_2$ and Al$_2$O$_3$ are formed. At the second stage for temperatures higher than 1200 ^{0}C, the formation of Al$_4$B$_2$O$_9$ and Al$_{18}$B$_4$O$_{33}$ aluminium borates is likely to take place [3], and the peak of rate of these reactions is 1240 ^{0}C. The temperature of the last (third) peak on the DTA oxidation curve is 1320 ^{0}C. The formation of an oxide film is completed at 1400 ^{0}C. The total mass gain during 100 min of non-isothermal specimen heating to 1500 ^{0}C is only 0.66 mg/cm^2, that is comparatively small.

Under ESA with AlN-ZrB$_2$-based electrode material of both titanium alloy and steels the same dispersion of fine island-like structures is also formed on the surface. The parts of the sintered layer based on products of the electrode electro-erosion are on the surface of the alloyed crystallization zone (Fig. 4). The grain size of this sintered layer is 0.4-0.8 μm. The phase composition of different parts of the ESA-coatings is presented in Table 2.

The main phases of the sintered layer for the coating on VT6 alloy were ZrB$_2$ and Al$_2$O$_3$. There are also phases corresponding to additives such as Al$_2$SiO$_5$ mullite, ZrSiO$_4$ zircon, TiB$_2$, Ti$_3$Al intermetallic and traces of Fe$_2$TiO$_5$ brukite. Ti-containing phases in the sintered layer appeared due to Ti mass transfer from the substrate. These phases are mainly near the boundary between sintered layer and substrate. In the crystallization zone are the titanium alloy alloyed with ZrO_2, Al_2SiO_5 and $ZrSiO_4$.

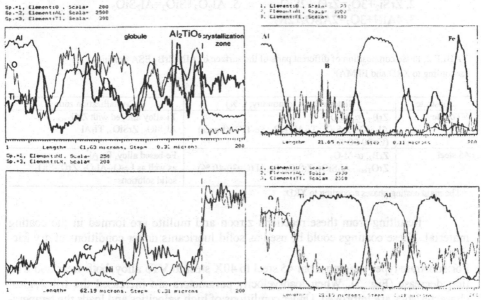

Figure 2. Elemental distribution on the AlN-TiB₂ coating surface for the "crystallization zone-globule" boundary

Figure 3. Elemental distribution on a globule surface in a cross section of AlN-TiB₂ laser coating

In accordance with XRD measurements, the main phase of electric-spark coating on 45 steel is Fe_3Al. This is correlated with the fact that the coating is poorly continuous (only 30-40% ; see Table 2).

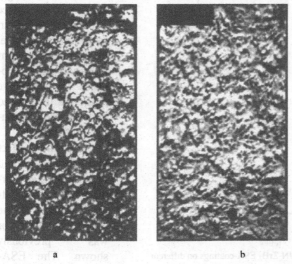

Figure 4. Microstructure of AlN-ZrB₂ ESA-coating surface on VT6 alloy: a- sintered layer, b- crystallization zone

The sintered layer mainly consists of the same phases as those for coating on titanium alloy. For both substrates the sintering additive in the coating material formation leads to Ti_3Al and Fe_3Al intermetallics for titanium alloy and low-carbon steel, respectively. For titanium and iron substrates both size and amount of particles in the crystallization zone are essentially smaller compared with the sintered layer zones. Thus, the following main oxidation and solid-phase reactions take place under non-equilibrium conditions during ESA in air :

1. $2ZrB_2 + 5O_2 = 2ZrO_2 + 2B_2O_3$ 4. $ZrO_2 + SiO_2 = ZrSiO_4$

2. $ZrSi_2 + 3O_2 = ZrO_2 + 2SiO_2$ 5. $Al_2O_3 + SiO_2 = Al_2SiO_5$

3. $4AlN + 3O_2 = 2Al_2O_3 + 2N_2$

TABLE 2. Phase composition of different parts of the surface of AlN-ZrB$_2$ ESA-coatings (according to XRD and EPMA)

Substrate	Sintered layer (continuity, C %)	Crystallization zone
VT6 alloy	$ZrB_2{}^*$, α-$Al_2O_3{}^*$ Al_2SiO_5, $ZrSiO_4$, Ti_3Al, TiB_2, Fe_2TiO_5 (traces) (C=80-90 %)	Ti-alloy alloyed with ZrO_2 cub., Al_2SiO_5, $ZrSiO_4$, Ti_3Al
45 steel	ZrB_2, α-Al_2O_3 ZrO_2 cub., Al_2SiO_5, $ZrSiO_4$, Fe_3Al (C=30-40 %)	Fe-based alloy, $Fe_3Al{}^*$ as well as Fe_2O_3-Al_2O_3, ZrO_2-SiO_2 solid solutions

*) The main coating phases according to XRD

Resulting from these reactions zircon and mullite are formed in the coating material. These coatings could be used as solid lubricants under conditions of dry friction.

For substrate types ranging from 45 steel to 40X steel to VT6 alloy, both wear and friction decrease (Fig. 5). When load-velocity parameters are increased the tribological characteristics are improved. Under conditions of high velocities and loads the temperatures increase in the contact zone during friction [4]. This leads to thin oxide film formation. These oxide films are probably more strongly bonded to a surface of ceramic layer than to a metallic surface. So, the improvement of tribological properties in the above range is correlated with a more continuous coating.

The effect of Al_2SiO_5 and $ZrSiO_4$, formed during ESA in the coating material, on the tribological behavior was previously shown. The ESA-coating obtained under high-frequency electric-

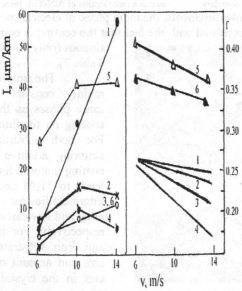

Figure 5. Tribological parameters of AlN-ZrB$_2$ ESA-coatings on different substrates: 1-45 steel (P=2MPa), 2-40X steel (P=2MPa), 3-VT6 alloy (P=2MPa), 4-VT6 alloy (P=2,56MPa), 5-initial VT6 alloy (P=2MPa), 6-AlN-ZrB$_2$ —based electrode material (P=2MPa)

spark pulses has essentially worse characteristics: wear is increased from 13 μm/km (at 1200 Hz) to 110 μm/km (at 1600 Hz) and the friction coefficient is increased from 0.2 (at 1200 Hz) to 0.29 (at 1600 Hz) under 14 m/s sliding velocity. However, under optimum ESA regimes, when high-temperature Al_2SiO_5 and $ZrSiO_4$ compounds are formed in the coating, the coating tribological characteristics under conditions of dry friction are good enough: for coated VT6 alloy subjected to a load of 2.56 MPa and sliding velocity of 14 m/s, the friction coefficient was 0.13 and wear was 6 μm/km.

ESA also increases the resistance to high-temperature oxidation. The working temperature corresponding to 2.0 mg/cm^2 mass gain is 130^{0}C higher for VT6 coated alloy compared with initial one due to mullite and zircon formation at the surface layer which prevents oxygen diffusion into the base.

4. Conclusions

1. The similarity between the structures of electric-spark and laser coatings, showing the formation of globules in the sintered finely dispersed composite material on a treated surface, was ascertained. The Al_2SiO_5, $ZrSiO_4$ and $β-Al_2TiO_5$ phases, appearing in the coating directly during its formation, play a positive role in the tribological behavior of the coating which finds applications as a solid lubricant and in protection against corrosion.

2. The ESA of titanium VT6 alloy as well as low-carbon 45 steel decreases the friction coefficient to 0.13 and 0.15, and wear to 6 and 13.1 μm/km, respectively. The working temperature of coated VT6 alloy was increased by 130 °C.

3. Metallic alloy substrates can play the role of metallic binder in the formation of a sintered layer coating during ESA where the electrode material is used without binder. The material wetting is very important to ensure the high level of service characterizing ESA coatings.

4. The magnetron coatings of $AlN-TiB_2$ target on monocrystalline silicon substrates are characterized by a continuous ultra-dispersed structure as well as a high resistance to high-temperature corrosion: the total mass gain for 100 min of non-isothermal heating up to 1500 ^{0}C of the coated specimen was only 0.66 mg/cm^2.

5. The developed $AlN-TiB_2/ZrB_2$ composite materials can be recommended for improvement of tribological and corrosion properties of performance alloys through protective coatings deposition.

118

5. References

1. Schneider S.V., Desmaison-Brut M., Gogotsi Y.G. and Desmaison J. (1996) Oxidation behavior of a hot isostatically pressed TiB_2-AlN composite, *Key Engineering Materials* **113**, 49-58.
2. Klaffke D., Hartelt M., Washe R. (2000) Non-oxide ceramics as tribo-materials-analysis of wear mechanisms under oscillation sliding condition in various atmospheres, *Tribology 2000 Plus, 12th Int. Colloquium, January 11-13, 2000* **III**, 1755-1763.
3. Ravikiran A. (1994) Effect of interface layers formed during dry sliding of zirconia toughened aluminia against steel, *Wear* **171**, 129-134.
4. Kloss H., Woydt M., Skopp A. (2000) Wear Prediction in Ultra-High Speed and High Temperature Applications, *12th Int. Colloquium "Tribology 2000 – Plus". January 11-13, 2000* **III**, 1639-1646. Technische Akademie Esslingen.

OPTICAL GRADIENT GLASSES OBTAINED USING THE METHOD OF HIGH-TEMPERATURE ISOTHERMAL CENTRIFUGATION

A.O. MYLYANYCH, M.A. SHEREDKO,
Yu.Ya. VAN-CHYN-SYAN and T.A. GANYSHNYUK

Department of Physical Chemistry,
State University "Lviv Polytechnic"
S.Bandery St. 12, Lviv-13, 79013 Ukraine

1. Introduction

Important elements in optical systems are often the gradient optics. Such elements include optical lenses with a gradient of the refractive index both at the surface and in the volume, reflectors made of a glass with a tunable refractive index,[1] geodesic lenses, etc. The common property shared by these products and the technological process to produce them is the presence of a gradient in the refractive index. The quality of such products depends on the accuracy of this gradient. Most frequently, these optical elements are made of glass, and the method of iono-exchange diffusion[2] is applied. However, this complicated, time consuming and expensive method very often demands additional operations to increase the mechanical properties of the product surface.[3]

It has been proved that the oxide glasses are microscopically inhomogeneous.[4] This means that their microstructure presents clusters or groups near atom positions. Such inhomogeneous micro-domains are formed already in the melts, and due to their

119

M.-I. Baraton and I. Uvarova (eds.),
Functional Gradient Materials and Surface Layers Prepared by Fine Particles Technology, 119–125.

quick cooling they are preserved by the change from liquid to glass. The structure of such domains most frequently depends on the chemical compositions and the temperature of the melts. Assuming that such micro-inhomogeneities are characterized by features (including density) which are different from those of the matrix in which they are distributed, one can predict the possibility of their dispersion (or concentration) in melts under the action of gravity or a centrifugal field. This should promote changes in the chemical compositions in cross sections of the samples after cooling, and thus yield gradient glasses. To this end the high-temperature isothermal centrifugation (HTIC) method was developed.[5] The effectiveness of such a method will be demonstrated in the case of the glass based on the PbO-SiO_2 system, which is at the root of the production of many types of optical glass.[6]

2. Experimental

2.1. SAMPLE PREPARATION

The glasses under study were synthesized in a platinum crucible from chemically pure PbO and SiO_2 oxides by fusion in an electrical muffle furnace at 1100 ^{0}C. The obtained glass samples were abruptly cooled by immersion of the crucible into water. The pieces of glass were thoroughly mixed and melted again.

The uniformity of the chemical composition of the synthesized initial glass samples was controlled in different parts of the crucible by hydrostatic measurement of the density[7] of six pieces of glass each having a mass greater than 5 g. The error was ± 0.003 g cm^{-3}. The glass was considered as appropriate for further investigation if the scattering of the measurements did not exceed 0.008 g cm^{-3}. Since glass density is a typical volume-additive quantity that depends considerably on glass chemical composition and structure, and in contrast to chemical analysis its measurement is quick and accurate[1] (0.05% in this paper), this parameter has been chosen as the principal criterion for the control of possible change in the chemical composition of the glass samples. In addition, all initial samples have been analyzed in terms of PbO and SiO_2

content by a chemical method[7]. The error was ± 0.3 weight-% as can be seen from the density measurements and chemical analysis data in Table 1.

TABLE 1. Chemical composition and physical properties of glass

Glass No	Samples	Chemical composition (mol.-%)		Density, d ($g\ cm^{-3}$)	Refractive Index, n	Microhardness, P (MPa)
		PbO	SiO$_2$			
1	Initial glass	59.6	40.4	6.717	2.021	1640$\pm$50
	Top sections of centrifuged glass	57.9	42.1	6.571	1.937	1860$\pm$50
	Bottom sections of centrifuged glass	61.5	38.5	6.810	2.046	1560$\pm$50
2	Initial glass	50.3	49.7	6.004	1.911	2080$\pm$50
	Top sections of centrifuged glass	49.5	50.5	5.959	1.859	2210$\pm$50
	Bottom sections of centrifuged glass	51.6	48.4	6.069	1.928	1990$\pm$50
3	Initial glass	39.0	61.0	5.059	1.797	2270$\pm$50
	Top sections of centrifuged glass	38.7	61.3	5.028	1.789	2330$\pm$50
	Bottom sections of centrifuged glass	39.4	60.6	5.065	1.802	2240$\pm$50

2.2. APPARATUS AND PROCEDURE

HTIC has been carried out in a setup consisting of a centrifuge, a control unit and a power supply. The diagram of the centrifuge with the heating element in the rotor as well as the methodology of HTIC are shown in detail in ref.[8] Melts were centrifuged in platinum crucibles with a cover. During HTIC the crucible with the studied melt was in a horizontal position. HTIC of all melts of glasses was carried out at a temperature of 1000 ± 20 ^{0}C for 1 hour, and at a centrifuge rotor speed of 1050 rpm. Melts were cooled in the field of centrifugal forces at a speed of about 2 ^{0}C s^{-1} to 230 ^{0}C.

As a result of centrifugation, the glass samples were obtained in the form of cylinders with a height of 30 mm and a diameter of 18 mm. The samples were cut, and their top (the part closest to the rotation axis) and bottom (the part furthest from the axis) sections, each with a height of 4 mm, were studied. Thus, all further results are expected to be average values in the obtained sections. Some physical properties (density, refractive index, mechanical properties, light absorption) and chemical composition of the centrifuged samples have already been given in Table 1.

The refractive index (n) has been measured by ellipsometry (ellipsometer LEF-3M-1, λ = 632.8 nm)

The change in mechanical properties of the glass samples after HTIC was determined by measuring the microhardness (apparatus PMT-3, loading on Wicker's diamond pyramid is equal to 20 g)[7].

The light absorption spectra were recorded by means of an automated device based on an MDR-23 spectrophotometer. As optical glasses are commonly used in the visible spectrum (1000-400 nm), the spectra were recorded in the 750-450 nm wavelength range.

3. Results and Discussions

As can be seen in Table 1, the chemical composition and physical properties of the glass samples change after HTIC. This can be explained as follows. The abundant inhomogeneities in lead oxide enrich the bottom sections of the melts during HTIC and thus increase their density; at the same time the silica-oxygen formations enrich the top sections. It is known[9] that after adding PbO to a glass its refractive index increases; whereas an increased content of SiO_2 leads to a decrease in n. It was found in Ref. 10 that an increase of PbO concentration in lead-silicate glasses causes a monotonous decrease of their microhardness.

Light absorption spectra were recorded for the samples of glass No. 1, the chemical composition and physical properties of which underwent the most substantial changes. As it can be seen on Figure 1, curves 2 and 3 which correspond to the initial glass and bottom section after HTIC respectively, are similar and greatly differ from curve 1 which characterizes the top section. The obtained spectra demonstrate the ability of top sections to maximize absorption and dispersion of light in the visible range of the spectrum whereas bottom sections of the centrifuged glass decrease in light absorption. Besides, we can see in the spectrum of the top sections a clear minimum approximately at 611.54 nm, which can be the evidence of the presence in their structure of a comparatively large amount of ordered silica-oxygen groups.

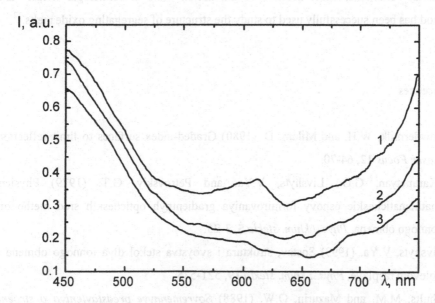

Figure 1. Light absorption spectra of glass No 1: 1, top sections after HTIC (thickness - 3.1 mm); 2, initial sample (thickness - 3.1 mm); 3, bottom sections after HTIC (thickness - 3.0 mm).

It is also determined that the differences in the structure of various parts of the centrifuged glass lead to various processes of phase-formation during their crystallization. An example of such a change is thoroughly described in ref. 8.

It is important that the sections of the glasses thus obtained are free from gaseous and defect inclusions.

124

4. Conclusions

The conducted experiments have shown that the samples of glass obtained after HTIC exhibit changes in their chemical composition and physical properties in their cross sections. Hence, HTIC can be used to obtain materials with novel properties, for example, optical gradient glasses.[5] Therefore, the use of such glasses in optical technology has significant advantages. Besides, this research is also interesting from the point of view of the comparison of processes of glass structure formation under terrestrial conditions, under absence of gravitation[11] and in centrifugal fields.[8] This method has been successfully used to study the structure of segregating oxide melts.[12]

References

1. Lowdermilk, W.H. and Milam, D. (1980) Graded-index surfaces to limit reflection, *Laser Focus* **12**, 64-70.

2. Karapetyan, G.O., Livshyts, V.Ya. and Petrovskiy, G.T. (1979) Physico-mathematicheskie osnovy formirovaniya gradientnyh opticheskih sred methodom ionnogo obmena, *Phys. i khim. stekla* **5**, 3-25.

3. Livshyts, V.Ya. (1993) Sostav, struktura i svoystva stekol dlya ionnogo obmena v solevom rasplave, *Phys. i khim. stekla* **19**, 521-535.

4. Shults, M.M. and Mazurin, O.W. (1988) *Sovremennyye predstavleniya o stroenii stekol i ih svoystvakh*, Nauka, Leningrad.

5. Sheredko, M.A., Bek, Yu.M., Dygdalovitch, A.M. and Mylyanych, A.O. (1996) Method of obtaining of gradient glasses. *Patent N 9549A (C04B19/04) Ukraine*

6. Demkina, L.I. (1976) *Physical-chemical bases of manufacture of an optical glass.* Chemistry, Leningrad.

7. Pavlushkin, I.I., Sentyurin, G.G. and Khodakovskaya, R.Ya. (1970) *Praktikum po technologiy stekla i sitallov*, Stroyisdat, Moscow.

8. Mylyanych, A.O., Sheredko, M.A. and Melnyk, S.K. (1999) Study of glass structures and crystalline phases in the PbO-Al$_2$O$_3$-SiO$_2$ system, *J. Anal. At. Spectrom.* **14**, 513-521.

9. Demkina, L.I. (1958) *Issledovaniye zavisimosti svoystv stekol ot ih sostava*, Oborohgis, Moscow.

10. Ashizuka, M. and Bradt, R.C. (1982) Fracture toughness in the PbO-GeO$_2$ and PbO-SiO$_2$ binary glass systems, *J. Mater. Sci. Lett.* **1**, 314-315.

11. Petrovskiy, G.T. and Voronkov, G.L. (1984) *Opticheskaya tehnologiya v kosmose*, Mashynostroenie, Leningrad.

12. Ganyshnyuk, T.A., Sheredko, M.A. and Mylyanych, A.O. (1999) Doslidzhennya likvuyuchyh rozplaviv systm MeO-B$_2$O$_3$ metodom centryfuguvannya, *Visnyk DU "Lvivska politekhnika"* **361**, 29-30.

GRADED STRUCTURES DEVELOPED BY LASER PROCESSING OF POWDER MATERIALS

N.TOLOCHKO
YU.SHIENOK
Institute of Technical Acoustics, NASB, Vitebsk, Belarus
T.LAOUI
L.FROYEN
University of Leuven, Heverlee, Belgium
M.IGNATIEV
Université de Perpignan, France
V.TITOV
Institute of Metallurgy, RAS, Moscow, Russia

Abstract. The principles of fabrication of functionally graded materials (FGMs) by laser processing of powders and powder metallurgical (PM) materials are discussed. Some results of FGMs fabrication are presented. The potential applications of FGMs are considered.

1. Introduction

Today powder metallurgy is the most widely-distributed method for manufacturing of the FGMs [1-3]. One of promising ways to improve this method is the use of novel technical approaches based on the laser processing of powders and PM materials [4-9]. The main aim of the paper is to study the principle possibilities to produce the FGMs of powders and PM materials using various types of laser processing technologies, in particular, such as laser sintering of powders as well as laser surface heat treatment of PM materials.

As it is known, there are laser-assisted rapid prototyping technologies utilising powders as the feed materials. They are selective laser sintering (SLS) and some modified variants of laminated object manufacturing (LOM) [5,6] which are especially promising to produce the FGMs. In principle, various ways may be proposed in order to fabricate the FGMs by these technologies. In particular, in case of SLS it is possible to develop techniques related to powder deposition and laser sintering capable of producing components with a gradient in particle size and/or in porosity directly by SLS process. Besides, it is also possible to use the specific powder depositing systems allowing to form powder layers with different composition. In its turn, in case of LOM it is possible to develop techniques related to formation of sheet powdered materials possessing different structure and composition. Due to laser cutting and following

127

M.-I. Baraton and I. Uvarova (eds.),
Functional Gradient Materials and Surface Layers Prepared by Fine Particles Technology, 127–134.
© 2001 *Kluwer Academic Publishers. Printed in the Netherlands.*

packing of such materials the components with a gradient in particle size or porosity, as well as composition, can be fabricated directly by LOM process. Some of these technical approaches are considered in the paper.

2. FGMs` fabrication by laser sintering technique

2.1 GRADED STRUCTURE OF SINGLE POWDER LAYERS SINTERED BY LASER.

As a rule, only part of radiation is absorbed by the outer surface of particles of a loose powder. The other part penetrates through the interstitial spaces (pores) into a loose powder depth interacting with the underlying particles. The thickness of the zone subjected to direct laser heating is rather small. Usually it is equal to 3-4 times particle diameter. The heat extends from this zone into the rest of powder by heat transfer mechanisms. Due to inhomogeneous temperature field developed in the heat affected zone the graded porous structure of the sintered zone is formed as it is shown in Fig. 1 (here the beam of CW- Nd:YAG laser was motionless above the target surface).

150 μm

Figure 1. Single layer of Ni-alloy powder with graded porous structure formed during laser sintering. laser power $P = 18$ W; laser spot size $D = 7$ mm; irradiation time $t = 10$s; atmospheric air.

2.2 MULTILAYER PARTS WITH GRADED STRUCTURE FORMED BY SLS PROCESSING.

Ni-alloy (chemical composition (wt.%): 10-20 Cr; 2.8-3.4 B; 4-4.5 Si; 0.6-1 C; balance Ni) powder and Fe-Cu powder mixture were used in the study. The loose powder was deposited layer-by-layer onto the ceramic substrate with the use of powder feeding system based on the slot feeder [7] and subjected to sintering by CW- Nd:YAG laser ($\lambda = 1.06$ μm). The beam was fixed at the target surface in case of Ni-alloy powder and scanned across the target surface in straight lines in case of Fe-Cu powder mixture. The particles` size, d, was changed for different layers (of 300-400 μm thickness) of Ni-alloy powder. In its turn, the portions of components, γ_{Fe} and γ_{Cu}, were changed for different layers of Fe-Cu powder mixture. It was allowed to form the gradient of porous structure in the first case and the gradient of composition as well as porous structure in the second one (Fig.2). It should be noted that in case of Fe-Cu powder mixture the values of laser power, P, were changed for different layers in accordance with variation of their absorptivity.

 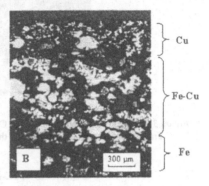

Figure 2. Graded structure of samples formed by SLS (cross section).
(A) Ni-alloy powder; $P = 18$ W, $D = 7$ mm; $t = 10$ s; atmospheric air.
(B) Fe-Cu powder mixture, $P = 35$-50 W, $D = 7$ mm, beam scanning speed $V = 1$ mm/s, $d = 150$-200 μm for Fe and 63-100 μm for Cu; vacuum.

2.3 GRADED STRUCTURE OF INTERPARTICLES` CONTACTS FORMED DURING LASER SINTERING.

In case of two-component systems when the components possess high ability to mutual solubility in both liquid and solid states the formation of rather strong interparticles` contacts with a gradient in composition can take place. Laser sintering of Ni-alloy-Cu systems is a typical sample of such case.

Ni-alloy-Cu powder mixtures and Cu-coated Ni-alloy powders were subjected to sintering under influence of CW- Nd:YAG laser. Particle sizes were 100-150 μm for Ni-alloy and 63-100 μm for Cu. The beam was moved across the target surface in straight lines. Both components were almost of the same melting point. In case of Ni-alloy-Cu powder mixtures the surface melting of Ni-alloy particles took place due to higher absorptance. Cu particles dissolved into molten layer aroused on the surface of Ni-alloy particles. Fig.3 shows a good example of such interaction between Ni-alloy and Cu particles (here Ni-alloy particle is spherical and Cu particle is irregular in shape). As a result, a smooth boundary between Fe and Cu phases was observed. In case of Cu-coated Ni-alloy powders the Cu coatings melted due to direct laser heating. When Cu coatings melted completely the partial dissolution of Ni-alloy particles into the melt took place. As a consequence, the boundary between Ni-alloy particles and Cu coatings was smooth too. Thus, in both cases the necks formed between the Ni-alloy particles consisted of both Ni-alloy and Cu components. The distribution of different components in the zone of interparticles` contact formed during SLS is shown in Fig.4.

130

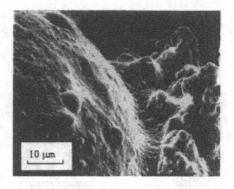

Figure 3 Interaction between Ni -alloy and Cu particles during SLS of Ni-alloy-Cu powder mixture. Fragment of Ni –alloy particle is on the left and fragment of Cu particle is on the right.

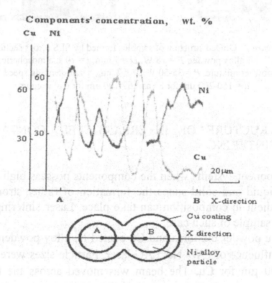

Figure 4. Components distribution in the interparticles' contact zone formed during SLS of Cu-coated Ni-alloy powders.

2.4 POROUS SURFACE LAYER GRADED STRUCTURE FORMED BY LASER SINTERING OF POWDER ON A SUBSTRATE.

The layer of Ni powder (with particle size 0,5 μm) was deposited (with the use of a slot feeder) onto the surface of Ni substrate manufactured by traditional powder metallurgy and subjected to sintering by fixed beam of CW-Nd:YAG laser. Layer thickness was about 1-2 μm. The joining of powder layer with the PM substrate was provided during sintering. As a result, the graded porous structure was formed on the surface as it is shown in Fig.5 (here the porosity is about 60% for substrate and 30% for sintered layer).

Figure 5. Graded structure on the surface of PM substrate (fracture of PM substrate with sintered powder layer on the surface; top view). $P = 40$ W; $D = 10$ mm; $t = 60$ s; vacuum.

3. FGMs' fabrication by laser surface heat treatment of PM materials

PM samples of iron (with porosity about 20% and particle sizes 100-200 μm) were used in the study. Laser processing was accomplished with CW-CO_2 and pulse-periodic Nd: YAG lasers. The beam scanned across the target surface in straight lines. Three different regimes of the processing were realised: (1) CW laser processing in argon and air mediums, (2) CW laser processing under the action of air flow and (3) pulse-periodic laser processing in argon and air mediums. In all the cases the processing caused the modification of superficial layer of the material. This layer possessed a gradient of both porous structure (Fig.6) and mechanical properties (Fig.7). As a rule, the density and the microhardness were maximum on the surface and decreased with the increase of distance from it. Some peculiarities of specified types of laser processing were investigated. In the first case the radiative heating was comparatively low. The surface did not melt. The essential modification of surface structure was not observed. In the second case radiative heating was comparatively low too. However, the air flow effect caused the intensive exothermic reactions. As a result, the modified surface layer consisting mainly of iron oxides high was formed. The reactions led to the additional heating and, as a consequence, more significant modification of surface structure (curve 3 in Fig.7). At last, in the third case radiative heating was comparatively high. The deep melting of the material could take place.

4. FGMs' fabrication by modified LOM technique

The sheets of feedstock based upon a polymer binder were used in the study. The feedstock contained the powders of quartz, gypsum and fusible glass (softening point

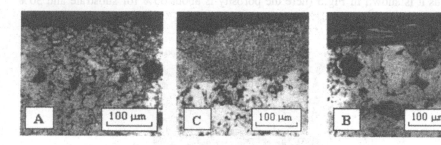

Figure 6. Surface structure of processed samples (cross section).
(A) CW-CO_2 laser, P = 100 W, D = 0.8 mm, V = 2 mm/s, air medium.
(B) CW-CO_2 laser, P = 100 W, D = 0.8 mm, , V = 10 mm/s, air flow.
(C) pulse-periodic Nd:YAG laser, pulse duration τ = 4 ms, pulse repetition rate
f = 15 Hz, energy content per pulse q = 4 J, D = 1.5 mm, V = 2 mm/s, air medium.

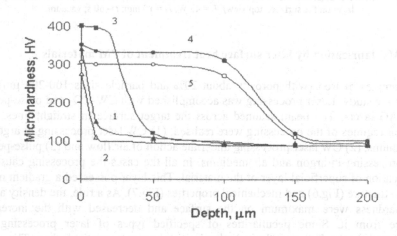

Figure 7. Changes of microhardness in depth of processed samples.
CW-CO_2 laser, P = 100 W (1,2,3); pulse-periodic Nd:YAG laser, τ = 4 ms, f = 15 Hz, q = 4 J (4,5);
V = 2 (1,2) and 10 mm/s (3,4,5); D = 0.8 (1,2), 10 (3) and 2 mm (4,5); argon (1,4), atmospheric air
(2,5) and air flow (3).

about 950 °C). Particle sizes were 5-30 μm for both quartz and glass and 10-50 μm for gypsum. The thickness of sheets produced by casting of slip was about 150-200 μm. The sheets were cut by CW- CO_2 laser (λ= 10.6 μm). The laser-cut sheets in shape of square were stacked into rectangular blocks subjected to heat treatment in the furnace under an applied low pressure. During treatment the polymer burnt out, and quartz and gypsum particles were bound by the softened glass. A variation of the portions of powder components in different sheets led to the gradient of composition and porous structure of sintered samples (Fig.8). It should be noted that the samples had a gradient of shrinkage, too.

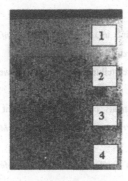

Layers' No	1	2	3	4
γ (quartz)	50	50	50	50
γ (gypsum)	20	15	10	5
γ (glass)	30	35	40	45

Figure 8. Graded structure (cross section) and powder components' portions, γ (mass.%), for different layers of sample formed by modified LOM technique.

5. FGMs' applications

The potential applications of FGMs fabricated by laser processing of powders or PM materials are:
- biomedical implants, in particular, tooth's root implants (cp-Ti, Ti6Al4V, Ti-6Al-7Nb, 316L stainless steel) having a porous surface to provide sufficient bone in-growth and dense core to provide the strong mechanical joint between tooth's root and crown implants;
- oil-impregnated bearings (Fe/316L stainless steel-Cu/bronze) possessing friction surface with higher volume portion of copper (or bronze) and lower pores' size at the same time (here graded porous structure ensures the incessant ingress of oil from the main part of bearing to the friction surface);
- filters for both medical (cp-Ti, Ti6Al4V, 316L stainless steel) and non-medical (Cu, brass, bronze) applications having high throughput and filtration fineness of filter-elements due to structure graded in pores' size in thickness of filtering barrier;
- metal-ceramic coatings (WC/TiC-Co/Ni, ZrO_2-NiCrAlY/CoCrAlY) in which the gradient in composition in thickness prevent their destruction under conditions of severe thermal stresses; etc.

6. Conclusion

Laser processing of powders and PM materials is the promising way to produce the FGMs. In particular, using some of the laser-assisted rapid prototyping technologies it is possible to fabricate the parts of complex geometry of FGMs directly during processing. Besides, it is possible to create the FGMs due to laser surface modification of conventional materials manufactured by powder metallurgy.

134

References

1. Koizumi, M. (1992) Recent progress of functional gradient materials in Japan, *16ᵗʰ Annu. Conf. Compos. and Adv. Ceram. Mater., Cocoa Beach, Fla/ Ceram. Eng. And Sci. Proc.*, **13**, 7/8, 333-347.

2. Delfosse, D. and Ilschner, B. (1992) Pulvermetallurgische Herstellung von Gradientenwerkstoffen, *Materialwiss. und Werkstofftechn.*, **23**, 7, 235-240.

3. Gasik, M.M., Lilius, K.R. and Ostrik, P.N. (1997) Developments of functional gradient materials fabricated by powder metallurgy methods, *Abstr. Int. Conf. "Novel Proc. and Mater. in Powder Met"*, Kiev, 345.

4. Tolochko, N.K. (1999) About some principles of graded materials' formation, *Poroshkoaya Metallurgiya (Powder metallurgy), Kiev*, 11/12, p.1-9.

5. Tolochko, N.K., Khlopkov, Yu.V., Mozzharov, S.E. and Mikhailov, V.B. (1999) About possibilities to fabricate the parts of graded materials using rapid prototyping technologies, *"Nowe kierunki technologii i badan materialowych", Redakcja naukowa J.Ranachoowski, J.Raabe, W.Petrovski, Warszava, Atos*, 349-352.

6. Tolochko, N.K., Mozzharov, S.E. and Mikhailov, V.B. (1999) Rapid prototyping technologies: current state and development·prospects, *Int. Conf. "Novel computer technologies in industry, power and education". Abstr., Alushta*, 45.

7. Tolochko, N.K., Yadroitsev, I.A., Mozzharov, S.E. and Mikhailov, V.B. (1998) Selective laser sintering: some questions of physics and technology, *Proc. PM 98 World Congress, Granada, Spain*, **5**, 407-412

8. Tolochko, N.K., Khlopkov, Yu.V. Mozzharov, S.E., Laoui, T., Titov, V.I. and Ignatiev, M.B. (1999) Measurement of Powder's Absorptance with Nd:YAG and CO_2 lasers, *Science of Sintering*, **31**, 3, 187-194.

9. Tolochko, N.K., Mozzharov, S.E., Yadroitsev, I.A., Titov, V.I. and Ignatiev, M.B. . (1999) Structure of Sintered Materials Fabricated by Laser Beam, *Science of Sintering*, **31**, 2, 91-96.

MICROWAVE PROCESSING OF NANOSTRUCTURED AND FUNCTIONAL GRADIENT MATERIALS

Y.V. BYKOV, S.V. EGOROV, A.G. EREMEEV, K.I. RYBAKOV,
V.E. SEMENOV, A.A. SOROKIN

Institute of Applied Physics of the Russian Academy of Sciences
46 Ulyanov St., Nizhny Novgorod 603600 Russia

1. Introduction

Microwave processing is being extensively investigated as a promising method for development of advanced materials. The research is aimed on the purposeful utilization of unique features of the microwave heating in the processes that involve thermal treatment of materials. This approach is pursued to obtain materials possessing novel properties which cannot be achieved by conventional methods.

The main feature of the microwave heating is its volumetric nature. Physically, volumetric absorption of microwave radiation eliminates the need in energy transfer by thermal conduction which is relatively slow in dielectric materials. Technically, the absence of massive furnaces and heaters leads to significant reductions in process duration and energy consumption. Another specific feature is selectivity of microwave heating, i.e., the dependence of the deposited power distribution on the microwave absorption properties of materials. Microwave heating is easily controllable by changing the power of the microwave source. Finally, numerous experiments suggest that microwave treatment results in a lower temperature of process activation and shorter processing times.

In recent years, application of microwave energy to high-temperature processing of materials, such as sintering and joining of ceramics, deposition of functional coatings, annealing of semiconductor structures, *etc.*, has been extensively studied [1]. The interest drawn to these applications is not even due to potential energy savings (although these can be especially substantial for high-temperature processes) but mostly due to the prospects of creating materials with novel properties. For example, densification of ceramic materials during sintering is always accompanied by recrystallization grain growth, which leads to the coarsening of microstructure and deterioration of mechanical properties [2]. A reduction in the sintering time due to volumetric heating results in obtaining ceramic materials with fine and uniform microstructure and enhanced properties.

Most experimental teams use microwave furnaces operating at the standard microwave frequency, 2.45 GHz. Use of this frequency is adequate for thermal processing of materials possessing relatively high dielectric loss, such as polymers, food, pharmaceuticals etc. Yet, serious difficulties arise when standard microwaves are applied to processing of low-loss materials which include many important ceramics, such as alumina

135

M.-I. Baraton and I. Uvarova (eds.),
Functional Gradient Materials and Surface Layers Prepared by Fine Particles Technology, 135–142.
© 2001 *Kluwer Academic Publishers. Printed in the Netherlands.*

and silicon nitride. Poor absorption of microwave radiation in these materials leads to severe temperature nonuniformity and/or inability to heat the materials up from the room temperature. Since the dielectric losses grow with frequency, radiation of higher frequency, the so-called millimeter waves (24 GHz and higher), is preferred for high-temperature processing of many ceramic materials.

Another advantage of the millimeter-wave processing is higher uniformity of the electromagnetic energy distribution within the applicator [3]. Because of productivity requirements, for most industrial purposes it is necessary to utilize multimode cavities as applicators. The degree of uniformity of electromagnetic field in such cavities depends upon the ratio of cavity dimension to wavelength, L / λ. In the millimeter-wave range it is possible to use untuned multimode applicators with $L / \lambda \geq 100$. High uniformity of the electromagnetic energy distribution is achieved as a result of the superposition of hundreds of modes excited simultaneously. Therefore millimeter-wave processing is feasible for producing large-size and complex-shape ceramic parts.

The Institute of Applied Physics (IAP) has long been involved into development of high power sources of millimeter-wave radiation, gyrotrons. More recently, specialized gyrotron systems for the processing of materials have been developed [4, 5]. The systems include the following major components:
- gyrotron with variable output power, complete with a magnet system. Gyrotrons for frequencies in the range 24 – 83 GHz with output cw power 3 – 30 kW are available;
- three-mirror quasioptical transmission & focusing line;
- millimeter-wave furnace 50 cm in diameter and 60 cm in height;
- furnace vacuum system for operation in the pressure range $1 - 2 \cdot 10^5$ Pa;
- temperature measurement and process control system. The millimeter-wave power is controlled during the process in accordance with a preset temperature-time schedule by a computer-based feedback loop which monitors the signal from the temperature sensor.

Today IAP has a unique set of facilities for the feasibility studies in the potential application areas of the millimeter-wave power. The diversity of the gyrotron system parameters is of paramount importance whenever the research pursues either of the two goals: to clarify the specific effect inherent to the millimeter-wave processing, and to determine optimal conditions for yielding high-quality final products.

2. Millimeter-wave sintering of nanostructured ceramic materials

High thermal treatment rates and short sintering times make millimeter-wave heating a promising method for obtaining nanostructured ceramic materials. The prospects of nanoceramics are associated with their unique physical and mechanical properties [6]. According to recent experimental data, nanoceramics exhibit increased strength, microhardness, fracture toughness, ductility, wear and thermal oxidation resistance, as compared to fine ceramics with micron-size grains.

The experiments performed at IAP suggest that millimeter-wave sintering of nanophase powders results in obtaining ceramics with improved microstructure and enhanced mechanical properties compared to the materials sintered conventionally from the same powders [7, 8]. In particular, the microstructure of the millimeter-wave sintered ceramic materials exhibits a grain size on the order of 100 nm.

The improved microstructure correlates well with enhanced microhardness of the millimeter-wave sintered materials. As seen in Fig. 1, in the millimeter-wave sintered nanoceramic titania samples a smaller grain size generally corresponds to higher values of microhardness. The data shown in Fig. 2 suggest that there is correlation between microhardness and the absorbed millimeter-wave power, which is approximately proportional to the sum of the rates of temperature change at the heating and cooling stages (there was no hold at the maximum temperature in these sintering experiments). Different values of the absorbed millimeter-wave power were obtained by varying the conditions of thermal insulation at the surface of the samples and using different heating rates. The maximum temperature was 1100 °C in all experiments.

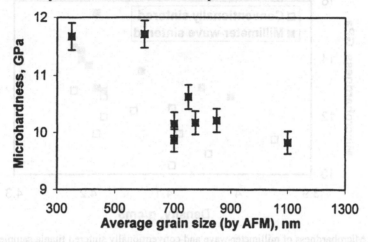

Figure 1. Microhardness vs. average grain size for titania samples sintered by millimeter waves under different heating rates and thermal insulation conditions. Note that AFM measurements give only approximate results for the grain size.

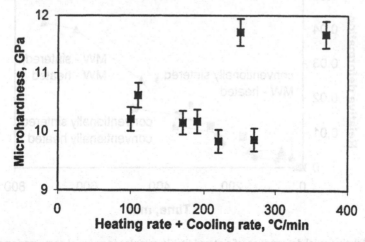

Figure 2. Microhardness vs. sum of the rates of temperature change at the heating and cooling stages for titania samples sintered by millimeter waves under different heating rates and thermal insulation conditions.

138

The mechanical properties of the millimeter-wave sintered materials were compared with those of the materials sintered conventionally. As seen in Fig. 3, the microhardness of the millimeter-wave sintered nanoceramic samples of titania systematically exceeds the microhardness of the conventionally sintered material. At higher densities it is even larger than the microhardness of the single crystalline rutile, which is about 13 GPa [9]. Figure 4 shows the data from the uniaxial compression strength test performed on the millimeter-wave and conventionally sintered nanoceramic titania samples under the millimeter-wave and conventional heating. It can be seen that the millimeter-wave sintered sample has withstood larger deformation before failure, i.e., its ductility is better.

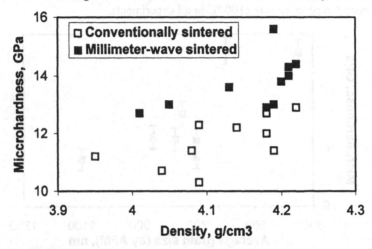

Figure 3. Microhardness of millimeter-wave and conventionally sintered titania samples vs. their final density after sintering.

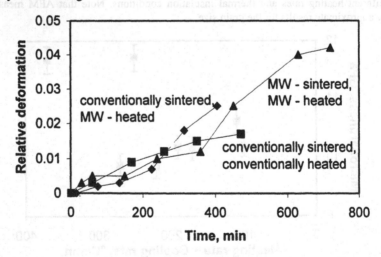

Figure 4. Relative axial deformation of sintered titania samples in uniaxial compression strength tests at 850 °C.

3. Millimeter-wave processing of functional gradient materials

Besides volumetric heating, millimeter-wave power beams can be used for surface treatment of materials. If for a given material and frequency the microwave penetration depth is small enough, then the deposition of the microwave energy takes place only in the near-surface layer of the material. This regime can be viewed as one more method for surface treatment of materials by intense energy flows, which can be alternative or additional to such widely used in practice techniques as electron and ion beams, plasma processing, and laser irradiation. As a rule, relatively high intensity of radiation is required to make high-temperature surface processing possible. As intensity is determined by the power of the radiation source divided by the area on which this radiation falls, in order to maximize the intensity one should increase the power and focus the radiation into a spot of minimum possible size. The latter is limited from below by the radiation wavelength. Therefore, microwave surface processing is most feasible in the millimeter-wave range. An intensity as high as $2 \cdot 10^5$ W/cm^2 is achievable when the output power of the gyrotron sources is focused into a spot of size on the order of a wavelength.

It appears that the millimeter-wave beam processing has much in common with the methods of infrared laser treatment. At the same time, many features inherent in gyrotron wavebeam systems differ them advantageously from the laser-based systems:
- higher output power per one device;
- higher efficiency of the radiation source (about 0.4 for regular gyrotrons and above 0.6 for the gyrotrons with depressed collector);
- much more robust system for the transport of radiation;
- negligibly low loss of radiation at transportation.

One of the promising application areas for the microwave surface processing techniques is development of functional gradient materials (FGM). These materials which have spatially inhomogeneous microstructure and gradient functional properties are expected to exhibit improved performance in such fields as energy production, ceramic engines, gas turbines, nuclear fusion, etc. [10]. The principal idea of the research in the field of FGM is to combine dissimilar materials in a way that takes advantage of each. In particular, metal-ceramic FGM are especially adequate to high-temperature applications. One of the approaches to FGM fabrication is based upon the sintering of metal-dielectric powder mixtures with a gradually varying composition. Unfortunately, due to the mismatch in thermal expansion the residual thermal stresses in the metal-ceramic FGM are often prohibitively large. Microwave absorption in the compositionally graded materials results in a graded deposition of the microwave energy. Thus, compositional selectivity of microwave energy absorption makes it in principle possible to purposely create such profiles of the microwave energy deposition and, as a result, such temperature distributions that prevent excessive thermal stresses in the FGM. Currently, the research into the use of microwave heating for the synthesis of FGM is in the early stage. However, the fruitfulness of the approach has been demonstrated for many metal-ceramics compositions of practical importance, such as Al_2O_3 – steel, Al_2O_3 – Mo, ZrO_2 – $Ni_{80}Cr_{20}$ [11].

To access feasibility of processing thermal barrier gradient metal-ceramic coatings with millimeter waves, a computer simulation was undertaken at IAP. The structure of the simulation framework includes a database of material properties, a solver for the electromagnetic problem, a solver for the thermal conduction equation, an automatic power control routine, and a sintering simulator. One-dimensional geometry of the model space is used. A plane electromagnetic wave falls on the ceramic side of the (initially porous) composite with gradually varied metal-to-ceramic concentration ratio. The effective complex dielectric permittivity and thermal conductivity of the composite are calculated at each point on the basis of the 3-component (ceramic, metal, and vacuum) effective medium approximation [12]. The distribution of the millimeter-wave power over the thickness is calculated using the impedance matching method. Since the loss factor of the metal component is very high, there is a percolation transition in a narrow region at some distance from the coating surface. A prevailing portion of the electromagnetic energy is absorbed within this region, which requires careful handling by the impedance matching routine and imposes a limitation on the mesh spacing.

The distribution of temperature in the composite material is obtained by solving the heat conduction equation. Its boundary conditions assume presence of thermal insulation with generally different heat loss at the front and rear surfaces of the material. The intensity of the incident millimeter-wave beam is automatically adjusted so that the temperature at a selected mesh point evolves in accordance with a prescribed schedule.

The obtained evolution of temperature is used to compute densification due to sintering. For this purpose, the master sintering curve approach [13] is utilized. Master sintering curves are the experimentally obtained dependencies of relative density versus logarithm of the temperature evolution function

$$\Theta(t, T(t)) \equiv \int_0^t \frac{1}{T} \exp\left(-\frac{Q}{RT}\right) dt'.$$

The curves for a composite of arbitrary metal-to-ceramic concentration ratio needed to calculate densification at each point within the composite are computed as

$$\log \Theta_{eff}(t, T(t)) = \Phi_{eff}(\rho); \quad \Phi_{eff}(\rho) = \frac{c_m}{c_m + c_c} \Phi_m(\rho) + \frac{c_c}{c_m + c_c} \Phi_c(\rho),$$

where $\Phi_m(\rho)$ and $\Phi_c(\rho)$ are the experimentally obtained master sintering curves for pure metal and pure ceramic powders, respectively, c_m and c_c are the relative concentrations of metal and ceramic at a given point within the composite, and the function Θ_{eff} is computed using the effective value of the activation energy

$$Q_{eff} = \frac{c_m}{c_m + c_c} Q_m + \frac{c_c}{c_m + c_c} Q_c,$$

where Q_m and Q_c are apparent activation energies for the sintering of metal and ceramic, respectively. The length of each mesh interval is scaled as an inverse cubic root of the computed local relative density of the composite at each time step.

A sample simulation result, intentionally not optimized, is shown in Fig. 5. It can be seen that maximum energy is released in the dielectric loss percolation region. As densification proceeds, this region moves towards the front surface ($z = 0$). The rest of the material is heated mostly due to thermal conduction, therefore it is important to thermally insulate surfaces. By varying thermal insulation properties it is also possible to compensate, to a certain extent, different sinterability of the ceramic and metal powders.

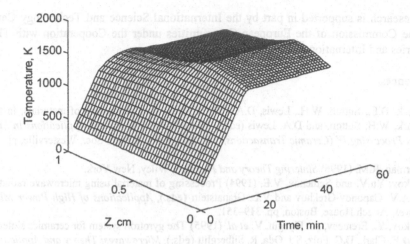

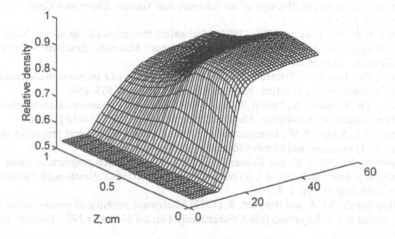

Figure 5. Simulated distributions of temperature (above) and density (below) during millimeter-wave beam sintering of an Al_2O_3 – Ni gradient composite. The wavebeam falls on the $z = 0$ surface, i.e., on the ceramic side of the composite. Initial relative density 0.53, millimeter-wave frequency 90 GHz, heating rate 1.5 °C/s, maximum temperature 1500 °C (temperature controlled at 0.4 of the coating thickness), hold time at maximum temperature 30 min. Temperature insulation intentionally made weaker than optimal. Master sintering curves for Al_2O_3 and Ni from Ref. 13 are used.

142

Acknowledgment

This research is supported in part by the International Science and Technology Center and the Commission of the European Communities under the Cooperation with Third Countries and International Organizations programme.

References

1. Clark, D.E., Sutton, W.H., Lewis, D.A. (1997) Microwave processing of materials, in D.E. Clark, W.H. Sutton, and D.A. Lewis (eds.), *Microwaves: Theory and Applications in Material Processing IV* (*Ceramic Transactions*, Vol. **80**), Amer. Ceram. Soc., Westerville, pp. 61–96.
2. German, R.M. (1996) *Sintering Theory and Practice*, Wiley, New York.
3. Bykov, Yu.V. and Semenov, V.E. (1994) Processing of material using microwave radiation, in A.V. Gaponov-Grekhov and V.L. Granatstein (eds.), *Applications of High Power Microwaves*, Artech House, Boston, pp. 319–351.
4. Bykov, Y., Eremeev, A., Flyagin, V. *et al.* (1995) The gyrotron system for ceramics sintering, in D.E. Clark, D.C. Folz, S.J. Oda, R. Silberglitt (eds.), *Microwaves: Theory and Applications in Material Processing III* (*Ceramic Transactions*, Vol. **59**), Amer. Ceram. Soc., Westerville, pp. 133 – 140.
5. Bykov, Y., IAP team, GYCOM team (2000) CW gyrotrons and gyrotron systems for applications in technology, in *Abstracts of the International Vacuum Electronics Conference 2000*, Monterey, p. 21.3.
6. Siegel, R.W. and Fougere, G.E. (1994) Mechanical Properties of Nanophase Materials, in G.C. Hadjipanayis and R.W. Siegel (eds.), *Nanophase Materials: Synthesis – Properties – Applications*, Kluwer, Dordrecht.
7. Bykov, Yu., Gusev, S., Eremeev, A. *et al.* (1995) Sintering of nanophase oxide ceramics by using millimeter-wave radiation, *NanoStructured Materials*, **6**, 855–858.
8. Bykov, Yu., Eremeev, A., Egorov, S. *et al.* (1999) Sintering of nanostructural titanium oxide using millimeter-wave radiation, *NanoStructured Materials*, **12**, 115–118 (1999).
9. Mayo, M.J., Siegel, R.W., Narayanasamy, A. et al. (1990) Mechanical properties of nanophase TiO2 as determined by nanoindentation, *J. Mat. Res.*, **5**, 1073.
10. Miyamoto, Y., Niino, M. and Koizumi, M. (1997) FGM research programs in Japan – from structural to functional uses, in I. Shiota and Y. Miyamoto (eds.), *Functionally Graded Materials 1996*, Elsevier, pp. 1–8.
11. Willert-Porada, M. A. and Borchert, R. (1997) Microwave sintering of metal-ceramic FGM", in I. Shiota and Y. Miyamoto (eds.), *Functionally Graded Materials 1996*, Elsevier, pp. 349–354.
12. Bergman, D.J., Stroud, D. (1992) Physical Properties of Macroscopically inhomogeneous media, in H. Ehrenreich and D. Turnbull (eds.), *Solid State Physics: Advances in Research and Applications*, Vol. **46**, Academic Press, New York, pp. 147–269.
13. Su, H. and Johnson, D.L. (1996) Master sintering curve: a practical approach to sintering, *Journal of the American Ceramic Society*, **79**, 3211–3217.

APPLICATION OF ADVANCED CCD CAMERA SYSTEM FOR DIAGNOSTICS OF SPRAYING PARTICLES IN-FLIGHT

M. IGNATIEV
Université de Perpignan
52, avenue de Villeneuve, 66860 Perpignan, France
I. SMUROV
Ecole Nationale d'Ingénieurs de Saint-Etienne
58, rue Jean Parot, 42023 Saint -Etienne Cedex 2, France
V. SENCHENKO
Pyrolab Ltd., Izhorskaya str., 13/19, 127412 Moscow, Russia

Abstract. Industrial type fully digital diagnostic system was developed in the frame of BRITE Project BE 97-5040 for monitoring of particles velocity and temperature in thermal spraying. The system is based on the new CCD image sensor providing higher sensitivity in near infrared spectral region and higher saturation signal level in comparison with conventional devises. The diagnostic system and special software for image treatment were tested under actual industrial conditions of plasma spraying (plasma gun power was in the range 20-50 kW) of Al_2O_3 and ZrO_2+wt.8%Y_2O_3 powders. Diagnostic system demonstrates high performance in plasma spraying monitoring: 20 μm particle size detection limit; 1500 K minimum detectable temperature; 700 m/s maximum particle velocity.
Key words: thermal spraying, optical diagnostics, CCD camera.

1. Introduction

Thermal Spraying (TS) is a rapidly growing part of materials engineering with its high-performance products applied in more than 35 modern industrial sectors. The spreading range of new spray materials and different substrates are opening promising potential for TS wide applications, if strict requirements for process stability, reproducibility and efficiency would be ensured.

Efficiency of optical methods for in-flight particle state monitoring in thermal spraying have been confirmed by different authors [1,2]. A few years ago, commercial diagnostic system based on high speed pyrometer appears on the market [3]. One may note, that high cost of this system limits its wide dissemination among the majority of the potential users (in particular SME).

One of the most promising ways of diagnostic systems development is the application of the advanced CCD cameras [4,5]. However, existing prototypes of CCD

143

M.-I. Baraton and I. Uvarova (eds.),
Functional Gradient Materials and Surface Layers Prepared by Fine Particles Technology, 143–150.
© 2001 *Kluwer Academic Publishers. Printed in the Netherlands.*

camera systems still remain expensive for wide industrial application (in particular systems using intensified image sensors).

2. Diagnostic system design

Highly stable and sensitive CCD image sensor with acceptable cost recently has appeared on the market. It enables to build an inexpensive and reliable diagnostic system for monitoring of particle in-flight state. The system is based on non-intensified Sony Ex-view HAD CCD image sensor. The main feature of this camera is high quantum efficiency in near infrared region (900-1000 nm). This region is practically free from plasma jet radiation that creates favourable conditions for measurement of particles own radiation. The number of effective pixels is 732x282, unit cell size (HxV) is 8.6x16.6 μm. Achromatic objective provides spray jet monitoring within 9x7 mm area. The basic diagnostic system is monochromatic one. Operating wavelength is 900 nm with bandwidth 15 nm.

The digital data flow is controlled by specially developed Plasma Spray Software (PSS) allowing control of CCD camera supporting electronics, automatic camera calibration, process monitoring, data recording and image analysis.

The software for automatic image analysis includes two main modules: (1) extraction of representative particles tracks and binary image analysis (background correction, exclusion of fringe tracks, definition of tracks length and width, construction of particles velocity histogram, etc); (2) analysis of representative tracks intensity and definition of particles brightness or true temperature.

A blackbody radiation source (temperature range is 1000-2500 K, accuracy 0.5 K) was used to calibrate the developed diagnostic system. Rotating disk with small hole (40 μm) was used to imitate particle passing through a focal plane of the diagnostic system. Calibration results are recorded as a file and are used for brightness temperature calculation in a measurement cycle.

The specifications of a diagnostic system are shown in the Table 1.

Table 1 Specifications of CCD camera system

1	Temperature range, K	1500 - 3800
2	Number of brightness temperature channels	1-2(optionally)
3	Wavelengths λ, nm	905, 850 (optionally)
4	Instrumental error, %	± 1
5	Effective distance to the target L, mm (This parameter could be adjusted in accordance with customer requirements)	147 ± 2
7	Sampling rate, μs	1
8	Type of computer interface	PCI or PCMCI card
9	Start-up time, min	15
10	Weight, kg	3
11	Power consumption, W	15
12	Optical head dimensions, mm	220×170×150

3. Experimental conditions

The developed diagnostic system has been tested with three different types of industrial plasma spraying installations applied for deposition of Al_2O_3 and $ZrO_2+8\%Y_2O_3$ powders. The CCD camera was sighted on the centre of a plasma jet at different distance from the plasma gun nozzle (Figure 1). Spraying parameters were typical for industrial conditions. The main aims of the instrument test were: (1) to check up instrument performance; (2) to study the influence of process parameters variation on particle velocity and temperature. General view of CCD camera system adaptation with industrial plasma spraying equipment is shown in Figure 2.

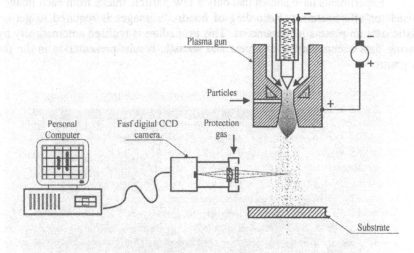

Figure 1. Scheme of CCD camera system trials.

Figure 2. General view of CCD camera system adaptation with industrial plasma spraying equipment.

4. Results of diagnostic system test in plasma spraying

Figure 3 shows live image of spray jet recorded by diagnostic system. CCD camera detects radiation of flying particles while an electronic shutter is opened. Thus, on the graphics window, particles are visible as flashing tracks on a dark field. Depending on how far the particle is located with respect to the focal plane of camera objective, there is some degree of blurring of the track image. This type of tracks are excluded from further image analysis. An automatic selection of tracks suitable for proper definition of particle velocity and temperature is realized by Plasma Spray Software after image binarization.

Experiments have shown that only a few particle tracks from each image could be used for aftertreatment. Recording of hundreds images is required to get reliable statistic data on plasma jet parameters. This procedure is realised automatically by PSS allowing data accumulation, treatment and statistic results presentation in the form of histograms.

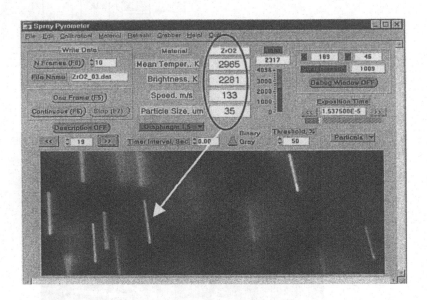

Figure 3. Image of spray jet (ZrO_2+wt.8%Y_2O_3). Exposition time is 15 μs.

Short exposition time of CCD camera provides possibility for real-time monitoring of particles collision with substrate. The instrument should be switched to "Substrate" operation mode allowing temperature measurements in any point of the image chosen by operator. Figure 4 shows that average substrate temperature is relatively low and could not be detected by CCD camera system. Nevertheless, measurement of temperature of individual lamella forming as a result of the impact can be carry out. This information is very important for selection of an optimum temperature conditions allowing to avoid cracks generation.

Only a few particles impinges on the substrate during selected exposition time (5 µs). Some particles having high velocity are crushed after collision with substrate and smaller droplets scatter in different directions. This phenomenon could cause some defects in coating structure and its monitoring could be important for process optimisation.

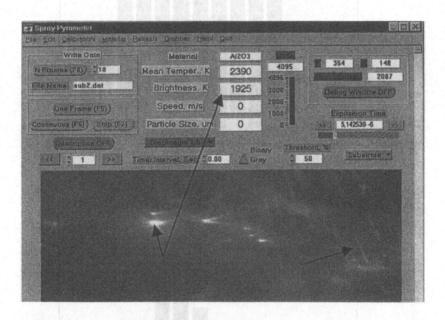

Figure 4. Image of Al_2O_3 particle collision with substrate. Exposition time is 5 µs.

Results of statistic analysis of the recorded images are presented in Figure 5 and 6 in the form of histograms. The majority of Al_2O_3 particles (Powder dispersion +50-100 µm) has velocity in the range 150-250 m/s for chosen spraying conditions (arc current 775 A, voltage 55 V; Ar consumption 70 l/min; H_2 consumption 15 l/min; Powder flow rate 12 g/min). Particles temperature histogram shows wide spread in values with relatively sharp maximum near 2000 K. Complete data array is used to define mean values of particle velocity and temperature.

Mean values of spray jet can be used for real time monitoring of process stability. Figure 7-A demonstrates that increase of plasma arc current (power) leads to more intensive particles heating and acceleration. Increase of hydrogen flow rate leads to increase of particle mean temperature and does not influence on particle velocity (Figure 7-B).

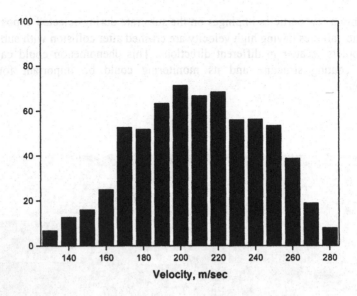

Figure 5. Particle velocity histogram. Spraying of Al₂O₃ (+50-100 μm).

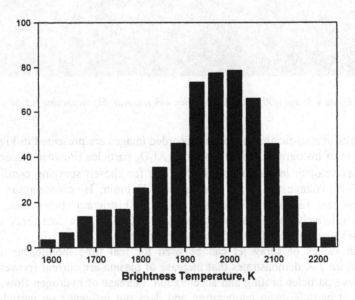

Figure 6. Particle brightness temperature histogram. Spraying of Al₂O₃ (+50-100 μm).

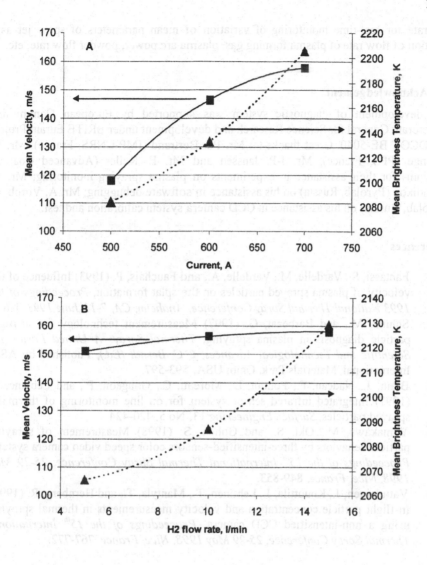

Figure 7. Influence of plasma arc current (A) and hydrogen flow rate (B) on mean values of particle velocity and brightness temperature (Al₂O₃, diameter 40 μm).

5. Conclusion

The developed diagnostic system based on inexpensive CCD camera demonstrates good performance in plasma spraying: particle size detection limit is 20 μm; temperature sensitivity is 1500 K; maximum particle velocity is 700 m/s. The particle flux is visualized in real time and can be used for monitoring of particle flux density, velocity, direction and homogeneity of particle distribution. The diagnostic tool is sufficiently accurate for real-time monitoring of variation of mean parameters of spray jet as a function of flow rate of plasma forming gas, plasma arc power, powder flow rate, etc.

6. Acknowledgement

The development of diagnostic system was supported by European Commission Directorate General for Science Research and development under BRITE Euram Project INODCOT BE-5040. Great thanks to: Mr. Ph. Bertrand (IMP-CNRS, France), Mr. G. Colonge (PPS, France), Mr. J-P. Janssen and Mr. E. Keller (Advanced Coating, Belgium) for their assistance in experiments on plasma spraying monitoring; Mr. V. Dozhdikov (Pyrolab, Russia) on his assistance in software debugging; Mr. A. Vorobyev (Pyrolab, Russia) on his assistance in CCD camera system calibration and test.

References

1. Fantassi, S., Vardelle, M., Vardelle, A., and Fauchais, P. (1993) Influence of the velocity of plasma sprayed particles on the splat formation, *Proceedings of the 1993 National Thermal Spray Conference, Anaheim, CA, 7-11 June 1993*, 1-6.
2. Schutz, M., and Barbezat, G. (1997) Measurement technology for in-flight particle diagnosis in plasma spraying, *Thermal Spray: A United Forum for Scientific and Technological Advances, C.C. Berndt (Ed.)*, Published by ASM International, Materials Park, Ohio: USA, 593-597.
3. Blain, J., Nadeau, F., Pouliot, L., Moreau, C., Gougeon, P., and Leblanc, L. (1997) Integrated infrared sensor system for on line monitoring of thermally sprayed particles, *Surface Engineering* 13, No 5, 420-424.
4. Yamakawa, M., Oki, S., and Gohda, S. (1998) Measurement of spraying particle behaviors by three-intensified-sensors color speed video camera system, *Proceedings of the 15th International Thermal Spray Conference, 25-29 May 1998, Nice, France*, 849-853.
5. Vattulainen, J., Knuuttila, J., Lehtinen, T., Mantyla, T., and Hernberg, R. (1998) In-flight particle concentration and velocity measurements in thermal spraying using a non-intensified CCD camera, *Proceedings of the 15th International Thermal Spray Conference, 25-29 May 1998, Nice, France*, 767-772.

LASER SINTERING OF MULTILAYER GRADIENT MATERIALS

A. V. RAGULYA and O.B. ZGALAT-LOZYNSKYY
Frantsevich Institute of Problems in Materials Science NAS of Ukraine
3, Krzhizhanovsky St., 03142 Kiev, Ukraine

1. Introduction

Selective laser sintering (SLS) is an unconventional sintering process [1], which consists in depositing a thin powder layer on a substrate and sintering it under incident laser beam scanned over the surface. The selective laser sintering is commonly used as a method of rapid prototyping of parts and articles i.e. for production of templates by their computer aided design and following layer-by-layer manufacturing [1,2]. The main requirement to the final product is the accordance between projected geometrical dimensions and ones of the real product [3,4]. Such kind of SLS application is widely used in industry as soon as the DTM Corp. designed and produced several types of laser plants, working at 50 metallurgical companies [5]. Materials commonly used for rapid prototyping are polycarbonates, wax, pure and glass-filled up nylon, and metals. Ceramic prototypes are produced by SLS of mixtures containing binder, the low-melting point component (polymer), and ceramic powder, the high-melting point component. Then the binder is removed by burning out in air.

Recently, an alternative laser sintering technique suitable for manufacturing of multicomponent ceramics was explored. Instead of polymer or polycarbonate binders, some low-melting point eutectic or pritectic compositions were used instead [6]. Such a process was based on the fundamental principle of laser sintering as a liquid-phase sintering process with high amount of liquid (up to 50 vol.%). Thus, the components whose melting temperatures are different by hundreds of degrees considered suitable for SLS.

The advantage of this process is that any part can be produced without a press tool. The shape can be quite complex, including internal cavities, and is imparted with the aid of a computerized system. Another attractive feature of laser sintering is the possibility of combining materials and composites with different types of chemical bonding. It appears possible to obtain a gradient of properties by producing concentration and density gradients, as well as gradients of grain and pore sizes [7,8]. The laser has a number of advantages as a heat source: purity, ability to generate high temperatures, and to provide localized heating; possibility to simplify control of a gas atmosphere (to protect the material from oxidation or to carry out a chemical reaction on the surface of the material). These advantages permit the use of laser heating in the sintering of thin ceramic layers [9,10].

Among disadvantages, low accuracy in controlling the sintering temperature, uselessness of unstable, volatile, sublimating substances should be mentioned as well as

M.-I. Baraton and I. Uvarova (eds.),
Functional Gradient Materials and Surface Layers Prepared by Fine Particles Technology, 151–159.
© 2001 *Kluwer Academic Publishers. Printed in the Netherlands.*

cracking on cooling due to high thermal stresses, incompatibility of substances, pure wetting and low rate of liquid spreading. This limits the choice of systems and compositions since for example, the difference between the liquidus and solidus temperatures of the given material must be substantially larger than the error in temperature control and mutual solubility of components becomes desirable factor for substantial improvement of wetting and liquid spread in the pore channels of sintered layers. The present paper is an overview of recent research in sintering of multilayer ceramic slabs using layer-by-layer methodology of manufacturing when applied to nanocrystalline powders and nanostructured bulk materials.

2. Fabrication of nanostructured ceramics by SLS

Selective laser sintering has not been used previously to consolidate nanocrystalline powders in order to obtain dense materials with ultrafine grain structures. It was assumed that liquid-phase sintering, which particular case is the laser sintering, becomes largely unsuitable for this because of the rapid mass transport through the liquid and the extensive grain growth. However, there are at least three ways of making a dense material containing nanosized grains. Firstly, one can use laser sintering of a composite made of nanodispersed powders, where the sintering occurs during the short time that the liquid exists (less than 1 sec), which is sufficient for the consolidation but very short for appreciable grain growth, apart from the case of liquid-like coalescence. Secondly, eutectics formed in various multicomponent systems crystallize to form a structure with ultrafine grains. Thirdly, the rapid cooling (quenching) of the melt can lead to a metastable or amorphous structure, which can be transformed to a nanocrystalline structure by annealing at moderate temperatures.

We have studied the laser sintering of multilayer composites consisting of a mixture of refractory compounds (TiN, TiB$_2$), a TN20 hard-alloy mixture, and a steel substrate. It has been shown that the method is effective in making dense sintered materials having fine-grained gradient structures. High relative density and small grain size favor improved hardness in the refractory layer by comparison with standard characteristics obtained for hot-pressed composites with the same composition due particularly to the high heating and cooling rates [7].

The basis is to use the temperature gradient produced by the laser and the differences between the melting points of the substances in the layers. The top layer should have a higher melting point than the lower layers (the eutectic in the TiN—TiB$_2$ system has a melting point of 2600°C [11]). Laser sintering is effective only in the presence of a liquid phase, therefore the top layer should be heated above the fusion point of pure nickel, that is above 1450°C. The hard-alloy mixture has a high thermal conductivity and a relatively low melting point, and if the temperature gradient is adequate, this should favor consolidation and adhesion of the second layer to the substrate. Then the second (lower) layer is also heated to temperatures exceeding the melting points of the readily fusible components. For example, the eutectic temperature in the TiC—Ni-Mo system is 1280 °C, while the melting point of eutectics in the TiN—TiB$_2$ system is higher than 2600 °C. The phase diagram of this system is shown in Fig. 1 [11] and shows that the components have restricted mutual solubility, quite close to the melting points and not so wide supersolidus field, which is not appropriate for SLS. We studied the sintering of multilayer powder composites by laser heating, the main purpose being to demonstrate the effectiveness of selective laser sintering for obtaining multilayer materials having gradient properties.

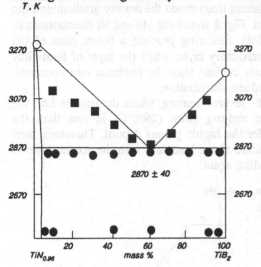

Figure 1. TiN—TiB₂ phase diagram [11].

We used nanocrystalline powders of the refractory compounds TiN, TiB₂, and TiC together with those of the metals Ni and Mo as raw-materials. The specific surface was measured by the BET method, while the mean particle size was calculated from the surface area assuming a spherical shape for the particles, and the oxygen content was determined by neutron activation. We also determined the pycnometric density of each powder and the poured density of the mixtures (mass %):70 TiN—30 TiB₂, 50 TiN—50 TiB₂, 79 TiC—6 Ni—15 Mo.

The layer thicknesses were close to 200-300 μm, and the porosities varied from 70-75% (poured) to 35-40% (for example, after pressing or rolling). The poured-layer thicknesses were monitored with a special device equipped with a micrometer head and mounted on the laser unit. A poured layer was carefully flattened. We used an LTN-103 permanent Nd-YAG laser. The protective medium in the working chamber was nitrogen or argon at 0.6 MPa. The input power was varied over the 30-250 W range, the defocus spot size was 1-3 mm, the scan speed varied from 1 to 20 mm/sec, and track overlap 0-50% of the beam diameter. The composition of the upper refractory layer was also varied over the range 25-50 mol. % TiN. To prevent the refractory composition from cracking on intensive heating, preliminary mild heating was used for the entire composite material using the laser in moderate regime of generation.

The X-ray microprobe phase analysis was performed with a Camebax instrument, which employed section scanning to determine the penetration depth of the Mo—Ni liquid into the refractory layer and also the likely composition change in the refractive compounds with depth. The microhardness depth distribution was examined with a PMT-3 instrument employed with a standard load of 50 gf.

The comparison of some experiments with the simulation data has been recently developed [6] for the "mild" case (moderate laser power) and for the "severe" case (elevated power). Full-scale experiments were performed with similar parameters for the laser operation and scanning device in order to define the best conditions for making a multilayer composite.

In the "mild" state, a molten Mo—Ni eutectic appeared at the boundary between the first and second layers and rose through capillaries into the top layer, and caused it to consolidate, thus connecting the layers together. On subsequent heating,

154

the upper layer melted, which substantially accelerated the consolidation. Simultaneously, the second layer was depleted in molybdenum and nickel in the region adjoining the refractory layer and consequently the consolidation rate was markedly reduced. The steady state in temperature front migration along the specimen together with the corresponding consolidation gradient determined the density gradient and the distribution of the Mo—Ni bonding agent. Fig. 2 shows the Mo and Ni distributions in a cross section from the microprobe data. Sintering produce a broad zone of the Mo—Ni melt penetration into the top refractory layer, while the layer of hard-alloy composite was porous, as it was indirectly evident from the hardness measurements. Fig. 3 shows the corresponding microhardness distribution.

The best results were obtained with "severe" heating, where the time to heat the surface of the refractory layer to the melting point (2600°C) is less than the characteristic thermal-conduction time for the highly porous deposit. Therefore, melt appears at the surface when the temperature at the boundary with the second layer is below the melting point of the metal bonding agent.

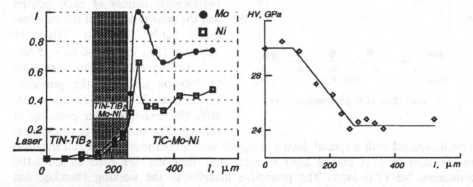

Figure 2 Concentration patterns for Mo and Ni indicated by microprobe ("mild" laser heating); 0 on the X axis corresponds to the surface treated by the laser.

Figure 3 Microhardness gradient produced by "mild" laser sintering.

In the subsequent stages, a liquid metal phase is also formed at the boundary between the layers, but the capillary transfer into the top layer is restricted because of the rapid consolidation in that layer. Rapid consolidation occurs in the nitride-diboride mixture because of liquid-phase sintering of TiN (solid) in a large volume of the TiN—TiB$_2$ molten eutectic. The "severe" mode produces complete consolidation in the two upper layers during a single laser pass. Also, the heat is sufficient to raise the temperatures in the two layers close to the substrate and provide the conditions for bonding these two layers. The nanocrystalline titanium nitride passes partially into the eutectic liquid, but it also remains partially solid when TiN is in excess, which provides an ultrafine-grained microstructure of crystallization. In any mixture where there are approximately equal contents of titanium nitride and diboride, a complete melting is observed, and then the TiN—TiB2 eutectic crystallizes at fairly high cooling rates, which may favor the formation of a fine-grained or nanocrystalline structure. Experiments in that sintering mode gave a dense layer of refractory composite 200 μm thick, whose hardness attained 33-34 GPa. The molybdenum and nickel distributions together with the hardness gradient related to the metal concentration gradient are shown in Figs. 4 and 5. The hardness level attained is obviously the same in the similar composition but made by hot pressing [12].

Figures 6 and 7 show the microstructures for composites having component ratios of 70:30 and 50:50 after "severe" sintering; Fig. 6 shows a fragment in the region of the boundary between the refractory layer (dark region) and the hard alloy (light region).

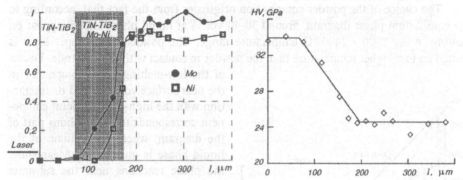

Figure 4. Distributions of Mo and Ni indicated by microprobe ("severe" heating);

Figure 5. Microhardness gradient produced by "severe" laser sintering.

The refractory layer consists of sintered groups of titanium nitride particles (gray grains), between which there is crystalline eutectic (dark gray inclusions). The framework structure of the TiN linked by the eutectic has the hardness level quoted above. The platy structure for a 50:50 eutectic is shown in Fig. 7.

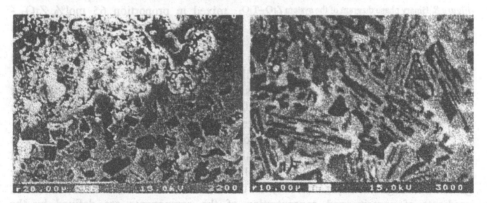

Figure 6. Part of the boundary between the refractory layer (dark gray) and the hard-alloy layer (pale gray).

Figure 7. TiN—TiB$_2$ eutectic structure.

3. SLS of oxide materials

The model ceramic system ZrO_2-TiO_2 was selected for laser sintering experiments due to large difference in fusion temperatures and mutual solubility of components, eutectic

156

reaction and, therefore, good wetting. Phase diagram is presented in Fig. 8. The difference between the fusion temperatures of the components is quite large, around 870 °C (T_{ZrO2} =2700 °C and T_{TiO2} = 1830 °C). Wide supersolidus field of cubic zirconia solid solutions (the field of melt and solid zirconia coexistence), low-melting temperature eutectic (T_e= 1720 °C), and peritectic (T_p = 1830 °C) are required preconditions for laser sintering [13].

The choice of the powder composition originates from the fact that, according to the equilibrium phase diagram, around 30-40 vol.% of liquid phase should appear on heating in the 2000 - 2400 °C temperature range. The powder of an upper layer is heated up to a higher temperature than the powder in contact with the substrate. Fusion of the low-melting point component in the near surface volumes and its interaction with the high-melting point component corresponds to supersolidus part of the diagram, where the volume of the liquid phase is quite large. Meanwhile, the phase reactions near the substrate proceed at lower temperatures and with a smaller amount of liquid having a composition different from that of the upper surface. Thus, the temperature gradient through the layer leads to a gradient of the chemical composition. Titania and zirconia powders were mixed in proportion 65 mol% ZrO_2 - 35 mol% TiO_2 in ethanol and then dried at 250 C. The powder layer was deposited over a substrate or one layer was deposited over another layer using a special powder-feeder. The green layer was carefully flattened.

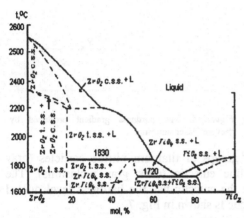

Figure 8. Binary phase diagram of the system ZrO_2-TiO_2 [13].

The laser beam regimes were explored as follows: power around 140 W, defocused spot size at 3 mm, scan speed varying in the 1-20 mm/sec range, every track overlapping each other by 0— 50% of the beam diameter, poured height ranging from 0,9 to 1.1 mm.

As a result of the laser beam action on a powder layer, the near surface volumes heat up most intensively as soon as they absorb the laser radiation. Thermal conductivity is responsible for the thermal wave propagation and for the temperature gradient through the layer normally to the substrate. Gradients of viscous fluxes, gradients of porosity and concentration of the components are defined by the temperature gradient. On the first stage of heating titania melts and reacts with solid zirconia by the eutectic and peritectic reactions.

During sintering, the surface of the ZrO_2-TiO_2 powder layer is heated over liquidus temperature because of the low thermal conductivity of the porous oxide ceramics. Below this temperature the melt pool contains the rest of solid zirconia and quickly eliminating pores. Experimental evidence of the described sintering mechanism was found from microstructural observations. Quenching of the surface melt pool results in crystallization of primary zirconia dendrites. Meanwhile the matrix microstructure is appropriate to the internal volumes of every layer performing zirconia crystals surrounded by the solidified eutectics (Fig. 9).

Image analysis of the dendrite structure showed that the mean grain size is 5.5 μm, general dendrites length is 20 μm, volume fracture of the primary zirconia dendrites is about 33% and volume fraction of the secondary solidified eutectics is about 67%, respectively (Fig. 9b). In those volumes of the layer, where the temperature did not rise liquidus temperature (~2300 °C), the volume fraction of ZrO_2 grains was 63% and solidified eutectic mixture 37 % (Fig. 9d).

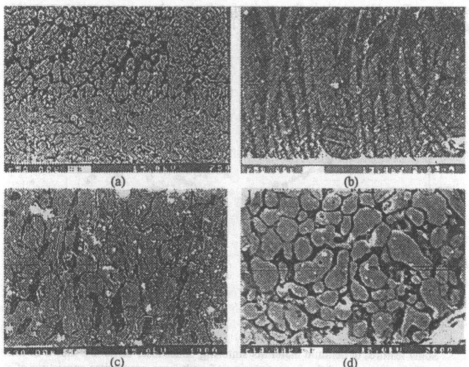

Figure 9. SEM images of microstructure in ZrO_2 -TiO_2 samples sintered under laser beam in air: general view of interlayer boundary (a), near surface dendrite crystallization (b), zirconia crystals in peritectics (c,d; c is closure to the surface).

Note, the zirconia grain size was found in the 10-13 μm range, that is one order of magnitude larger than the initial particle size, but twice less than the agglomerate size. Probably, the particles in the agglomerates coalesce forming coarse grained structure. One can see small intragrain pores (Fig. 9d), which are the consequence of the powder agglomeration. The mechanism of ultrarapid intraagglomerate sintering was not studied yet, but as suggested, rapid shrinkage becomes possible due to fast impregnation of agglomerate by melted titania, rearrangement of particles within the agglomerate and following reaction dissolution of titania in zirconia. Obtained ceramics are near fully dense.

The distribution of Zr and Ti along the cross section of the specimen changes quite sharply over a distance of 280 μm from the top surface: content of zirconium decreases

158

whereas titanium increases with the distance from the surface (Fig. 10). Microhardness across one layer varies according to the change in phase composition: the highest hardness is measured at the layer center. The microhardness evolution within one layer follows a sinusoidal variation across the section as a whole (Fig. 11). The harder volumes correspond to the zirconia based solid solutions, and softer fields contain mainly titania. Theoretical hardness of zirconia and titania are ~ 12 GPa, and ~ 10 GPa, respectively, both correspond to minimum and maximum on the curve. Within the periodical change of composition and hardness from layer to layer, there is a general gradient of hardness directed from the bulk to the surface of the multilayered specimen (Fig. 11), which is the result of a further phase homogenization in the previously sintered layer while a novel layer is under preparation.

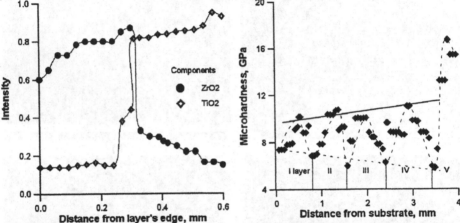

Figure 10. Distribution of zirconium and titanium

Figure 11. Spatial distribution of hardness along the cross section of the specimens

4. Conclusion

Some features of ceramic materials sintering under laser beam are shown in this paper. We have examined the laser sintering of multilayer composites consisting of a mixture of refractory compounds (TiN, TiB$_2$), a TN20 hard-alloy mixture, and a steel substrate, which has shown that the method is effective in making densified sintered materials having fine-grained gradient structures. High density and small grain size favor improved hardness in the refractory layer by comparison with standard characteristics obtained for hot-pressed composites with the same composition, due to the high heating and cooling rates.

The multilayer ceramics was obtained by laser sintering of powder mixtures. The hardness has been revealed to vary across one layer from 7 to 16 GPa and across the entire specimen from 9 to 12 GPa. This results from inhomogeneous phase redistribution during ultrarapid liquid phase sintering. Such a behavior is a consequence of heat transfer in the powder multilayer body undergoing laser radiation.

5. Acknowledgement

Athors thank NATO Science for Peace Program (Grant No 973529) for support of the present research.

6. References

1. Bourell, D. R., Marcus, H. L, Barlow, J. W., and Beaman, J. J. (1992) Selective laser sintering of metals and composites, *Int. J. Powder Metall.*, **28**, No. 4, 369-383.
2. Weiss, W. L., and Bourell, D. L. (1993) Selective laser sintering of intermetallics, *Metall. Trans. A*, **24A**, No. 3, 757-759.
3. Tolochko, N. K., Mikhailov, V. B., Mozzharov, S. E., et al. (1995) Laser selective layer sintering of powders: problems and prospects, *Poroshk. Metall.*, Nos. 3-4, 32-37.
4. Tolochko, N. K., Mikhailov, V. B., Mozzharov, S. E., et al., (1997) Kinetics of interparticle contact formation in the laser sintering of one-component metallic powders, *Ibid.*, Nos. 1-2, 54-61.
5. Dollmeiter, K., Petsold, F.(1997) *Int. J. Powder Metallurgy*, 3/1, 35-43.
6. Ragulya, A.V. (1998) Selective laser sintering. Part 1. Principles and continuum model, *Poroshk. Metall.*, Nos. 7-8, 16-26.
7. Ragulya, A. V., Stetzenko, V. P., Vereshak, V. M., Klimenko, V. P., Skorokhod, V. V. (1998) *Powder Metallurgy & Metal Ceramics*, Nos. 11/12, 10-17.
8. Chavez P. (2000) Optomec demonstrates deposition of functionally graded titanium alloys. *Metal Powder Report*, **55**, Nos. 7/8, 7
9. Raether, F., Baber, J. (2000) Laser sintering of thin ceramic coatings on glassy substrates. *Proc. Sintering '99 Intern. Conf. , University Park, Nov. 1-3, 1999* in press.
10. Baber, J., Raether, F., (1998) Laser densification of sol-gel derived ceramic thin films. in *Proc. ECLAT'98 (European Conference on Laser Treatment of Materials)* B.L. Mordike (ed.) pp. 105-110.
11. Chupov, V.D., Unrod, V.I. and Ordan'yan, S.S. (1981) Interactions in the TiN—TiB$_2$ system, *Poroshk. Metall.*, No. 1, 62-66.
12. Andrievski, R. A., Kalinnikov, G.V., Potafeev, A.F.and Urbanovich, V.S. (1995) Synthesis, structure and properties of nanocrystalline nitrides and borides, *Nanostructured Materials*, No. 6, 353-356.
13. Shevchenko, A.V., Lopato, L.M., Mayster, I.M., Gorbunov, O.S., (1980), The system ZrO$_2$ -TiO$_2$. *J. Unorg. Chem.*, **25**, 2496 - 2499.

RATE-CONTROLLED SINTERING OF NANOSTRUCTURED TITANIUM NITRIDE POWDERS

O. B. ZGALAT-LOZYNSKYY[*], A. V. RAGULYA[*],
M. HERRMANN[**]
[*] Institute of Materials Science Problems NAS of Ukraine,
3, Krzhyzhanivs'kyy St., 03142 Kyiv, Ukraine
[**] Fraunhofer Institut Keramische Technologien und Sinterwerkstoffe
Dresden, Germany

1. Abstract

Sintering behavior of nanocrystalline (NC) TiN powders (45 nm, 15 nm prepared by PVD and ~80 nm prepared by plasmachemical synthesis) are studied under both linear heating rate and shrinkage rate-controlled regimes in vacuum. The non-linear temperature-time path of rate-controlled sintering (RCS) results in more uniform pore and grain structures than the linear heating schedule. The final grains size distribution (50 – 75 nm) and residual porosity around 1.9 % are the best evidence of the RCS advantages over linear heating rate regime (600 - 1100 nm grain size and ~6 %, respectively). The near fully dense and fine-grained rate-controlled sintered specimens present the highest hardness of 25 GPa. The dependence of hardness versus heating rate, grain size, porosity and sintering regime is discussed in the present work.

2. Introduction

Nanostructured materials based on high-temperature melting compounds have been in the focus of research activity for many years. In particular, nanosized titanium nitride powders of different origins have been used in pressureless sintering [1,2,3] and high pressure sintering [4]. Properties of TiN such as strength, hardness, fracture toughness, are very sensitive with respect to reduction of the grain size. Nanograined ceramics are expected to lead to the improvement of these properties compared to conventional coarse-grained ceramics.

·Sintering seems to be suitable to consolidate the nanosized powders into dense polycrystalline ceramics at temperatures much lower than those conventionally used for micron-sized powders due to extraordinary high driving forces in the nanograined-nanoporous compacts [2-4]. However, the achievement of a highly dense material is incompatible with retaining its nanograined structure because of strong competition between grain growth and densification. Conventional sintering regime, such as ramp-and-hold, leads to intensive grain growth, which in turn prevents full densification at the final sintering stage and therefore, cannot be successfully applied [2,3]. Meanwhile, several authors showed the advantages of rate-controlled sintering to achieve the dense material while retaining the nanograined structure [5-8]. Rate-controlled sintering (RCS)

161

M.-I. Baraton and I. Uvarova (eds.),
Functional Gradient Materials and Surface Layers Prepared by Fine Particles Technology, 161–167.

is known as a non-linear, non-isothermal sintering regime. Moreover, there are no clear physical reasons to sinter under isothermal or constant heating rate. There is no clear explanation either on the fact that the non-isothermal sintering regimes and especially RCS promote reducing the microstructure. The concept of rate controlled sintering was first introduced by Palmour III and Johnson, who postulated that a finer-grained microstructure could be developed in a nearly fully dense sample if, instead of controlling temperature or heating rate, one controlled the densification rate instead [6-8]. Thus, during RCS, temperature becomes a function of density and densification rate, in direct role, reverse from conventional sintering. The feedback between density and temperature allows the computer control of the optimized non-linear heating rate regime.

For this reason, the aim of this work was to investigate the densification kinetics and microstructural evolution resulting from sintering of the nanocrystalline TiN powders under rate-controlled and conventional linear heating rate conditions.

3. Experimental

Titanium nitride powders with grain size 45 nm (TiN25) and 15 nm (TiN33) produced by H C Starck GmbH (PVD) and ~80 nm (TiN14) prepared by plasmachemical synthesis were used in the present study. Powder with grain size 45 nm was preliminary coated by surfactant (alkylsuccinimid) to preserve it from oxidation during storage. Crystallite size and lattice parameters have been estimated by XRD, specific surface areas of all the powders measured by means of nitrogen absorption (BET). Oxygen and nitrogen contents were determined by chemical analysis.

As-prepared TiN14 powder obtained by plasmachemical synthesis contains around 25 vol% fraction of sintered agglomerates and a small fraction of coarse metallic titanium particles The agglomerates and titanium particles were separated by sedimentation from the main fraction. Suspension of the powder in heavy hydrocarbons (i.e. i-decan) was stirred and ultrasonicated and then left for sedimentation for 24 h. A fraction of the finest particles was extracted from a centrifuged suspension of the supernatant. Before compaction, the powder was purified by annealing at 700 °C for 1 hour in a stream of extra-dry hydrogen. The residual oxygen content decreased down to 0.5% after annealing, and the specific surface area slightly increased up to 14,8 m^2/g. The nitride stoichiometry remained unchanged.

Titanium nitride powders TiN33 and TiN25 with or without binder has been pressed in cylinder tablets having 6 mm in diameter and 3 mm in thickness. Powders TiN25 and green pellets TiN25 were treated in the stream of dry nitrogen or hydrogen to remove the CUM, adsorbed oxygen and to activate the surface. TiN33 green pellets were treated in dry hydrogen only. After treatment, all specimens were sintered in 2 steps under linear heating rate regimes without isothermal hold controlled by a high-temperature vacuum dilatometer (~10^{-5} torr) equipped with CAD: from 20 to 1000 ^{0}C with a heating rate of 2 ^{0}C/sec; and from 1000 to 1300 ^{0}C (1600-1700 °C for TiN25 and TiN14) the heating rate was set in the range 0.05-2 ^{0}C/sec to collect the kinetic response and to optimize the process. Selected specimens were chemically analyzed to specify the oxygen content and the stoichiometry of TiN. XRD was used to determine the lattice parameters of TiN. Microstructures of samples selected among those having the highest density and hardness were studied on polished surfaces by Field Emission Scanning

Microscope (FESEM). Grain sizes have been estimated using x-ray technique of line broadening (TiN33) and image analysis of scanning electron micrographs (SIAMS 340).

Characteristics of TiN nanopowders are shown in Table 1. The finer is the powder, the higher is the amount of oxygen both chemically bonded and adsorbed despite of using a protective polymer coating, special techniques of surface purification in different gas media allow one to reduce the percentage of oxygen adsorbed on the powder surface.

The linear densification curves are presented in Fig. 1a. The one-stage densification is an important consequence of the near-uniform structure evolution from green body to final dense ceramics. A high fractional density is achieved when the green specimen is heated up to 1600 °C linearly. The final density is dependent on the heating rate highly and reaches the maximum value of 98% at 1.4 °C/s. The data obtained on TiN sinterability have been converted into 'kinetic field of response' to calculate RCS schedules and to optimize the sintering process (Fig 1b).

TABLE 1. Properties of TiN nanopowders

Powder	crystallite size, [nm]	specific surface area, [m^2/g]	lattice parameter, [Å]	[O], wt.%	[N], wt.%
TiN33	15	33	4.2334± 0.001	3.8	20.3
TiN25	45	25.7	4.2439 ± 0.001	1.81	19.2
TiN14	82	14	4.2400± 0.001	2.5	-

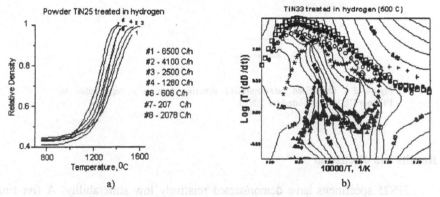

Figure 1 a - Linear schedules of sintering; b – 'Kinetic fields of response' for TiN powders

The optimization of the temperature-time path for the rate-controlled sintering was carried out based on the model of the kinetic field of responses as it was described elsewhere by Palmour III at al. [6-8] and shown in Fig. 2a. Optimal path is shown in Fig 2b. It results in a non-linear temperature-time path and in an uniform microstructure development. There is no noticeable pore coalescence at low temperatures and the average pore size monotonously decreases at high temperatures (Fig.2b). It should be noted that the calculations of the RSC-schedule were performed only for those series of sintering runs, which correspond to high hardness achieved. New series of samples have been sintered according to this RCS-schedule to maximise their density and hardness.

164

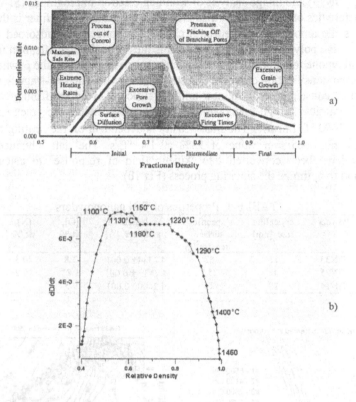

Figure 2 - Optimized rate-controlled sintering path: a – represented by Palmour III; b – calculated for TiN25

4. Results and discussion

TiN25 specimens have demonstrated relatively low sinterability. A five-time increase of the grain size has been established after the RCS. The sintering of TiN14 under rate controlled conditions lead to the increase of the fractional density up to 98.2-99.3%. The grain size remains around 150-200 nm at the maximal density (Fig.3).

The best results have been achieved for TiN33 specimens only. They exhibited a perfect sinterability and the final temperature of sintering was around 1200 °C. The grain size of the material was preserved within the range below 100 nm in near fully dense ceramics (Fig.4). By XRD, not only the TiN phase was detected but also lines of TiON were observed (Table 2).

An oxynitride phase in the TiN33 specimens was formed during the sintering process due to oxygen dissolution in the lattice of TiN. This phase obviously aided the densification process. The measured lattice parameters of the TiON phase are identical to those found in the literature.

TABLE 2. Properties of TiN sintered bodies.

	TiN25 treated in H₂	TiN25 treated in N₂	TiN33 treated in H₂	TiN14 treated in H₂
Relative density	0,96	0,953	0,981	0,993
Lattice parameter, nm	4.2411	4.2421	4.2362	-
Oxygen, wt.%	2.28	2.95	5.34	-
Phase analysis	TiN only	TiN only	TiN, TiON phase	TiN only

This result along with the high hardness of the samples is the proof of the presence of the oxynitride in the samples after the sintering of nanodispersed powders [9]. We certainly need to perform a more severe preliminary purification of the powders to totally eliminate oxygen and to work under perfectly clean vacuum conditions.

Figure 3 - TiN33 specimens sintered under RCS regime. Grain size (detected by XRD) ~ 50nm; porosity of 1.9% and pore size of 35 nm % (both estimated by SIAMS);

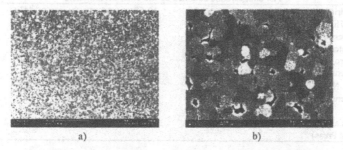

Figure 4 - TiN25 specimens sintered under RCS regime: grain size of 250 nm; porosity of 4 %; pore size of 136 nm (all estimated by SIAMS)

Hardness higher than 20 GPa has already been observed for hot pressed nano-grained titanium nitride specimens (bulk ceramic) (24 GPa) although this value of 20GPa remains lower than the hardness of TiN thin films (28-35 GPa) [1,4]. Presently the high hardness of 25 GPa has been achieved for sintered bulk ceramics at fractional density of 98.2%. The better results obtained under RCS process, compared to those obtained under linear heating or hot pressing, might be explained by both higher density and finer grain size. According to Hall-Petch type empirical equation $H=H_0+kL^{-0.5}$, the hardness should increase in pore-free titanium nitride from 20 GPa at grain size $L=10^3$ nm (experimental value) up to 33 GPa at $L=70$ nm (expected value from the best fit of the experimental data). Grain and density growth results in simultaneous decrease of hardness.

To establish a relationship between microstructure and some mechanical properties we have measured the hardness of samples as a function of grain size and porosity. Hardness is a very structure sensitive characteristic and can be theoretically estimated by the modified Hall-Petch equation (1):

$$H = H_0 + 10^{-4}/L^{1/2} + 10 \, exp(- \, 40 \cdot \theta) \qquad (1)$$

This approximating equation is only valid at porosity less than 5%. There is a good correlation between practical and theoretical hardness data for samples with grain sizes above 1 μm. In the nanoscale range, however the equation (1) gives substantial deviations, because theoretically the hardness is inversely proportional to the grain sizes. (Tab. 3)

TABLE 3. Mechanical properties of TiN nanocrystalline ceramic

Material (Technique)	Conditions of Sintering	Fractional Density (%)	Grain Size (nm)	HV measured (GPa)	HV calculated (GPa)	HK 0.1 (GPa)
TiN bulk ceramic obtained by RCS						
TiN 14	vacuum T = 1600°C	99.3	150-200	23±1	26	-
TiN 25	vacuum T = 1460°C	96	250	18 ±1	21	14±0,5
TiN 33	vacuum T = 1200°C	98.1	50	25 ±1.5	24.5	18±0,8
TiN bulk ceramic obtained by the other investigators						
Andrievski (High Pressure Sintering)	4 – 7.7 GPa, 1300 °C	95	30-60	26 ±1		[4]
Castro, Ying preserureless	0.1 MPa N₂, 1200 °C	99	140	23.2±1.9		[1]
R. Nass preserureless	1400 °C colloidal processing	98	400	21.5		[1]
T. Yamada (Hot Press)	40 MPa, 1600 °C	95	1,450	18		[1]

The plasticity of the nanocrystalline materials has been estimated using equation (2) [10] and the results of calculations are presented in the table 4.

$$\delta_H = 1 - 14.3 \, (1 - v - 2v^2) \, HV/E, \qquad (2)$$

where δ_H is a non-dimensional parameter (portion of plastic deformation into the common elasto-plastic deformation under an indentor); v is the Poisson's ratio and E the Young's modulus [10].

Parameter δ_H varied from 0 at the perfectly elastic indentation to 1 at the absolute plastic deformation. According to calculated volumes, we conclude that TiN25 deformed under indentor more plastically than TiN33. Thus, the finer is the grain size and the harder is the sintered material, the lower is the parameter δ_H and therefore the material is less plastic.

TABLE 4. Plasticity estimation
of TiN nanocrystalline ceramic

Material	HV, GPa	E, GPa	v	δ_H
TiN25	18	440	0.25	0.63
TiN33	25	440	0.25	0.49

5. Conclusions

Rate controlled sintering of nanocrystalline powders allows authors to reach the near fully dense titanium nitride ceramics with grain size within nanometer scale.
Sinterability of TiN nanopowders increases with reduction of their size. Dense TiN specimen obtained under RCS (final temperature 1200^0C) shows grain size of 50 nm and top hardness properties (25 ± 1.5 GPa).
Formation of TiON phases during sintering aids densification process.
Residual porosity of 2-4 % in TiN samples might be explained by dissociation of titanium nitride on heating: $2TiN \rightarrow 2Ti + N_2$
According to an estimation of plastic characteristics, TiN33 is more brittle material than TiN25.

Acknowledgements

Authors O.Zgalat-Lozynskyy and A.Ragulya thank NATO Science for Peace Programme for support via Project # 973529 and IKTS (Dresden, Germany) for HRSEM images and powder supply.

6. References

1   D.T.Castro, J.Y.Ying, (1998) Fine, Ultrafine and Nano Powders'98 Conference
2   T.Rabe, R. Wasche (1995) Nanostructured Materials 6, 357-360
3   T. Rabe and R. Wasche, ibid., 357.
4   R.A. Andrievsky, G.V.Kalinnikov, A.F.Potafeev and V.S.Urbanovich (1995). NanoStructured Materials, 6, 1-4, 353
5   Ragulya A.V.and Skorokhod V.V. (1995) ibid. 835-844
6   Palmour III H. and Hare T.M. (1985) Proc. of the Sixth World Round Table conference on sintering, Hercig-Novi, New-York, Plenum Press, pp. 16-23
7   Palmour III H. (1988) Powder Metal Report, 9, 572-580.
8   Palmour III H. (1989) Proc. of the Seventh World Round Table conference on sintering, Hercig-Novi, New-York, pp. 337-342
9   Alyamovsky, Zaynulin Yu. G., Shveykin G.P. (1981) [in Russian], Nauka, Moscow, pp. 58-62
10   Milman Yu. V. (1999) [in Russian]: Powder Metallurgy, 7-8, 85-91

PROPERTIES OF NANOCRYSTALLINE TITANIUM NITRIDE-BASED MATERIALS PREPARED BY HIGH-PRESSURE SINTERING

V.S. URBANOVICH
Institute of Solid State and Semiconductor Physics
National Academy of Sciences of Belarus
17, P. Brovka Str., Minsk, 220 072, Belarus

1. Introduction

Recently, the interest in nanocrystalline materials based on high-melting point compounds has increased. It is hoped that new superhard materials can be obtained. The main problem in the consolidation of particulate nanostructured materials is to combine full density with conservation of the nanostructure, that is keeping the grain size of the initial ultrafine powder (UFP). Conventional temperature regimes in sintering and hot pressing are not suitable due to intense recrystallization. Nowadays, high-energy methods of consolidation of nano-particulate materials are the most promising ones [1, 2]. A review of methods for the preparation of nanocrystalline materials by using high pressures can be found in [3]. High-pressure sintering is one of the promising methods. Many superhard polycrystalline materials based on diamond and cubic boron nitride have been produced by this method [4, 5]. High-pressure sintering is especially effective for fabricating ceramics based on high-melting point compounds because dense materials can be produced without any additive while at the same time the duration of sintering can be reduced. For example, dense AlN ceramics with high thermal conductivity and highly dense α-Si_3N_4 ceramics with high level of corrosion resistance have been obtained by this method [6, 7].

Earlier we have shown that high-pressure sintering allows to decrease the sintering temperature of ultra fine TiN powder in the 600-800 °C range in comparison with the pressureless sintering, and to prepare dense ceramics with a nanocrystalline structure [8]. The high hardness of this ceramics is caused by the small grain size. Dense TiN/TiB_2 composites with high hardness have been obtained from powders prepared by plasma-chemical synthesis [9].

In this work we present the properties of nanoceramics obtained by high-pressure sintering of ultrafine TiN and TiN/TiB_2 powders prepared by mechanical alloying. Preliminary results of our investigations have been reported earlier [10, 11].

2. Experimental procedure

Ultrafine TiN and TiN/TiB_2 composite powders with a mean particle diameter of 40 nm and 270 nm respectively, prepared by mechanical alloying were used as starting materials. For comparison we also used ultrafine TiN powder with larger particle size (~ 80

169

M.-I. Baraton and I. Uvarova (eds.),
Functional Gradient Materials and Surface Layers Prepared by Fine Particles Technology, 169–176.
© 2001 *Kluwer Academic Publishers. Printed in the Netherlands.*

nm) prepared by nitriding titanium hydride in dc-plasma. Table 1 shows their properties. The particle size of all powders has been determined from surface specific area measurements.

TABLE 1. Characteristics of initial TiN and TiN/TiB$_2$ powders.

Powder	Composition, wt. %							a (TiN)	S	d_s	$\rho_{theor.}$
	Ti	B	N	O	Fe	Ni	Cr	nm	m^2/g	nm	g/cm^3
TiN (40)			18.3	0.5	3.8	0.5	0.96	0.4241	26.8	42	5.48
TiN (80)	78.3		18.7	2.5	≤ 0.5 (met. impurities)			0.4240	14	80	5.32
TiN/Ti B$_2$*	82.54	9.63	11.0	-	0.09	0.01	0.02	0.4240	4.7	270	4.88

* TiB$_2$ (in Ar) 300 h + Ti (in N) → (45.49 wt.% TiN - 43.46 wt. % TiB$_2$-11.05 wt. % Ti)

Figure 1 shows the modified high pressure and high temperature anvil-type apparatus used for sintering. Such type of apparatus is simple to manufacture and widely used for both industrial synthesis of superhard materials and research. Details of the experimental procedure have been described earlier [8, 10]. The sample structure has been studied by XRD, FE-SEM (Field Ion-SEM, Hitachi, S-900) and TEM (Hitachi H800 associated with Analytical System AN10000, FDC) methods.

Figure 1. High pressure and high temperature apparatus for sintering of nanoceramics.

3. Results and discussion

Figure 2 shows the evolution of properties of TiN samples versus sintering temperature at pressures of 3 and 4 GPa. On can see that density and hardness of samples prepared at higher pressure are much higher. For both pressures the dependence of microhardness on sintering temperature is identical, with a maximum appearing at 1100-1200 °C. However, at a pressure of 4 GPa the microhardness clearly reaches a maximum. The larger

171

change of microhardness versus sintering temperature at a pressure of 4 GPa is correlated with the higher sample density. It can be observed that, for an increase of sintering
temperature up to 1400 °C, the density of TiN samples increases, reaches saturation, and
then sharply decreases at a temperature of 1600 °C. A characteristic feature is that the
reduction of microhardness versus sintering temperature begins at a lower temperature
(~1300 °C), before the maximal value of density (at 1400-1500 °C) is reached. The
highest values of microhardness agree satisfactorily with results related to samples prepared from plasmachemical powder [12].

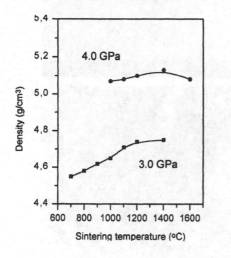 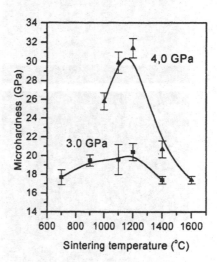

Figure 2. Density and microhardness H_v (Load 0.5 N) of sintered TiN (40) samples
vs. sintering temperature and pressure.

Table 2 shows additional results for TiN samples subjected to high-pressure sintering
at different temperatures. Parallel changes in microhardness (Fig. 2) and one half of the
broadening of the (422) peak versus temperature demonstrate that the increase in microhardness is accompanied by a deformation of the lattice, which is caused by an increase
of the level of internal stresses in the samples. Then, a reduction in microhardness is
related to the beginning of the recrystallization processes and removal of internal
stresses. The lattice parameter of TiN increases at higher sintering temperature. This is
likely to be related to both impurity (e.g. oxygen atoms) and change in the nitrogen content.

TABLE 2. Properties of TiN (40) samples, after sintering at pressure of 4 GPa
(ρ - density, W - one half of broadening (422) peak; α - lattice parameter; L - crystallite size).

Temperature (°C)	ρ (g/cm³)	$W \cdot 10^3$ (rad)	α (nm)	L (nm)
1000	5.07	10.5	0.4230_6	~30
1100	5.08	11.6	0.4230_3	~30
1200	5.10	10.5	0.4234_1	~60
1400	5.13	8.1	0.4249_6	~200
1600	5.08	7.6	0.4248_5	~300

172

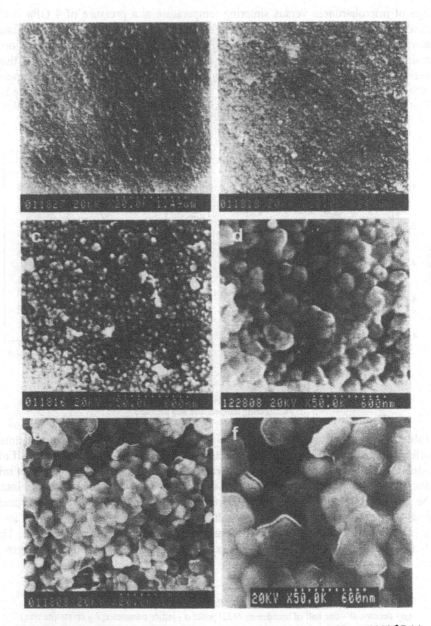

Figure 3. Fracture surfaces of TiN (40) specimens sintered under 4 GPa at 1000 °C (a),
1100 °C (b), 1200 °C (c), 1400 °C (d), and 1600 °C (e, f).

Figure 3 demonstrates FE-SEM micrographs of fracture surfaces of TiN (40) samples
prepared at different temperatures. One can see the recrystallization. The crystallite size
begins to grow from a temperature of 1200 °C. Fracture surface of TiN samples is intergranular.

Figure 4 shows microhardness versus sintering temperature of TiN samples from different particle size powders and of TiN/TiB$_2$ composites, sintered at a pressure of 4 GPa. One can see that a decrease of the particle size of TiN powder leads to an increase of the maximum value of microhardness. However, the temperature of recrystallization is lowered. The microhardness of the TiN-phase in the TiN/TiB$_2$ composite increases with sintering temperature up to 1200 °C, and then decreases like for samples of pure TiN (40). Thus the maximum value of microhardness is higher than that for TiN.

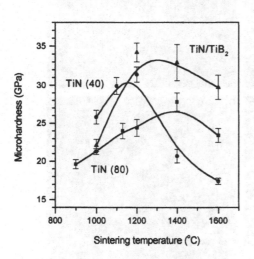

Figure 4. Microhardness of TiN samples and TiN-phase of TiN/TiB$_2$ composite vs. sintering temperature and particle size of initial powder.

Microhardness decrease after reaching a maximum is not so significant and is an indication of retarded recrystallization processes in the composite in comparison with pure titanium nitride (Table 3, Fig. 5). This points to a higher recrystallization temperature of TiN/TiB$_2$ composites. Fracture of the surface of TiN/TiB$_2$ composite samples is intergranular like in the case of TiN samples.

The density of TiN/TiB$_2$ composites increases monotonously with temperature (Table 3) in contrast to microhardness. The TiN lattice parameter values in Table 2 and Table 3 at T=1400-1600 °C are close but there is a slight difference at lower temperatures. The highest values of microhardness have been obtained at 1200-1400 °C and these results are superior than those for TiN/TiB$_2$ composites obtained from plasmachemical and carbothermal powders [9].

TABLE 3. Properties of TiN/TiB$_2$ composites.

Temperature (°C)	Density (g/cm^3)	Relative density (%)	TiN lattice parameter (nm)	Crystallite size (nm)
1000	4.43	90.8	0.4247$_1$	150-300
1200	4.72	96.7	0.4244$_5$	150-300
1400	4.82	98.8	0.4241$_7$	200-400
1600	4.85	99.4	0.4244$_5$	200-550

Table 4 shows the properties of TiN-based nanoceramics obtained by various methods. The highest microhardness is reached for high-pressure sintering at temperatures up to 1400 °C. Our results for powders prepared by mechanical alloying are superior in comparison with those of other authors. Microhardness of TiN/TiB$_2$ ceramic composites is on the same level as that of thin-layer coatings, prepared by selective laser sintering from melted state [13].

174

Figure 5. TEM micrographs of fracture surfaces of TiN/TiB$_2$ samples sintered under 4 GPa
at 1000 °C (a), 1200 °C (b), 1400 °C (c) and 1600 °C (d).

TABLE 4. Properties of nanocrystalline titanium nitride-based ceramics.

Material (particle size, nm)	Preparation method	Relative density	Grain size, nm	H_v(GPa) at load (N)
TiN (80) [14]	MIP (*P*=10 GPa, *T*=600 °C)	0.92	-	20.0
TiN (80) [15]	MIP (*P*=10 GPa, *T*=600 °C)→SCSW	0.91	30	25.0
TiN (15) [16]	RCS (*T*=1200 °C)	~1.0	50	25
TiN (8-25) [17]	SGPF (*P*=5 MPa T=1300 °C)	~1.0	15-100	-
TiN (50) [18]	HPS (*P*=5 GPa, *T*=1500 °C)	0.95	1500	18
TiN (70-80) [12]	HPS (*P*=4 GPa, *T*=1400 °C)	~1.0	50-80	26 (0.5)
TiN (16-18) [12]	HPS (*P*=4 GPa, *T*=1400 °C)	~1.0	30-40	29 (0.5)
TiN (40)	HPS (*P*=4 GPa, *T*=1100-1200 °C)	~1.0	30-60	30-31.5 (0.5)
TiN/TiB$_2$ (270)	HPS (*P*=4 GPa, *T*=1200-1400 °C)	~1.0	250-300	33-34 (TiN, 0.5)
TiN(80)/TiB$_2$ (350) [13]	SLS (*T*=2600 °C)	~1.0	-	33-34 (0.5)

MIP - magnetic impulse pressing, SCSW - shock compaction by spherical waves, RCS - rate-controlled sinter-
ing, SGPF - sintering in a gas pressure furnace, HPS - high-pressure sintering, SLS - selective laser sintering.

4. Conclusions

The high-pressure sintering of TiN and TiN/TiB$_2$ composite ultrafine powders prepared by mechanical alloying was performed at pressures up to 4 GPa and temperatures in the 1000-1600 °C range. Effect of such sintering on density, structure, microhardness and lattice parameters was studied. It is found that the initial nanocrystalline structure is preserved at temperatures up to 1100-1200 °C under a pressure of 4 GPa.

The highest hardness for TiN (31.5 GPa) has been obtained without significant grain growth. This value is higher than that for samples from coarser titanium nitride powder prepared by plasmachemical synthesis. However, microhardness change with sintering temperature indicates a more intense recrystallization.

The recrystallization process is retarded in TiN/TiB$_2$ composite in comparison with pure titanium nitride. The samples have a nearly full density and a higher hardness (34 GPa). These results are superior than those for TiN/TiB$_2$ composites obtained from plasmachemical and carbothermal powders.

5. Acknowledgements

The author would like to express his deep appreciation to Prof. R.A. Andrievski for support of this work and the useful discussion of results and to Dr. K.I. Yanushkevich for the XRD analysis. The author also would like to express his gratitude to Prof. Y. Ogino and Dr. T. Yamasaki from Himeji Institute of Technology (Japan) for the preparation of TiN and TiN/TiB$_2$ ultrafine powders by mechanical alloying and for the SEM study of TiN samples. This research is sponsored by NATO's Scientific Affairs Division in the framework of the Science for Peace Programme (Project No 973529) and Belarussian Programme "New Materials and Engineering of Surface".

References

1. Andrievski, R.A. (1994) Fabrication and Properties of Nanocrystalline High-Melting Compounds, *Successes of Chemistry* **63**, 431-448 (in Russian).
2. Andrievski, R.A. (1998) The-state-of-the-art of nanostructured high melting point compound-based materials, in G.M. Chow and N.I. Noskova (eds.), *Nanostructured Materials. Science and Technology*, Kluwer Academic Publishers, Dordrecht, pp. 263-282.
3. Urbanovich, V.S. (1998) Consolidation of nanocrystalline materials at high pressures, in G.M. Chow and N.I. Noskova (eds.), *Nanostructured Materials. Science and Technology*, Kluwer Academic Publishers, Dordrecht, pp. 405-424.
4. Frantsevich, I.N., Gnesin, G.G., Kurdumov, A.V. et al. (1980) *Superhard Materials*, Naukova Dumka, Kiev (in Russian).
5. Novikov N.V. et al. (1986) *Synthetic Superhard Materials,* Naukova Dumka, Kiev (in Russian).
6. Urbanovich, V.S. (1996) Sintering at high pressures and properties of aluminum nitride ceramics, in W.A. Trzeciakowski (ed.), *High Pressure Science and Technology*, World Scientific Publishing Co. Pte. Ltd, Singapore, pp.112-114.

176

7. Urbanovich, V.S., Gogotsi, Y.G., Nickel, K.G. et al. (1999) Properties of highly dense α-Si$_3$N$_4$ ceramics, sintered at high static pressure, in *British ceramic proceeding No 60. Book 718, vol. 2*, IOM Communications Ltd, Cambridge, pp. 9-10.

8. Andrievski, R.A., Urbanovich, V.S., Kobelev, N.P., and Kuchinski, V.M. (1995) Structure, Density and Properties Evolution of Titanium Nitride Ultrafine Powders under High Pressures and High Temperatures, in A. Bellosi (ed.), *Fourth Euro Ceramics, Basic Sciences - Trends in Emerging Materials and Applications*, Gruppo Edit. Faenza, Printed in Italy **4**, pp. 307-312.

9. Andrievski, R.A., Kalinnikov, G.V., and Urbanovich, V.S. (1997) Consolidation and Evolution of Physical and Mechanical Properties of Nanocomposite Materials Based on High-Melting Compounds, in S. Komarneni, J.C. Parker and H.J. Wollenberger (eds.), *Nanophase and Nanocomposite Materials II*, **457**, MRS, Pittsburgh, pp. 413-418.

10. Andrievski, R. A., Ogino, Y., Urbanovich, V. S. and Yamasaki, T. (1998) Consolidation of TiN/TiB$_2$ Ultrafine Powder Composites under High Pressure and Properties of Obtained Ceramics, in P. Vincenzini (ed.), *"Ceramics: Getting into the 2000's-Part C", Advanced in Science and Technology, 15. Proc. of the World Ceramics Congress, part of the Ninth CIMTEC - World Ceramics Congress & Forum on New Materials*, held in Florence, Italy, on June 14-19, 1998, Techna Srl, Faenza, pp. 435-440.

11. Andrievski, R. A., Urbanovich, V. S., Ogino, Y. and Yamasaki, T. (1999) Consolidation processes in nanostructured high melting point compound-based materials, in *British ceramic proceeding No 60. Book 718, vol.1*, IOM Communications Ltd, Cambridge, pp. 389-390.

12. Andrievski, R. A., Urbanovich, V. S., Kobelev, N.P., Torbov, V.I. (1997) *Reports of Russian Academy of Sciences* **356**, 39-41 (in Russian).

13. Ragulya, A.V., Stetsenko, V.P., Vereschak V.M. et al. (1998) Selective laser sintering, *Powder Metallurgy* **11/12**, 9-15 (in Russian).

14. Ivanov, V.V., Paranin, S.N., Vikhrev, A.N. (1997) Densification of nanoscaled powders of hard materials by magnetic-impulse method, in G.G. Taluts and N.I. Noskova (eds.), *The Structure, Phase Transformations and Properties of Nanocrystalline Alloys*, UD RAS, Ekaterinburg, pp. 46-56 (in Russian).

15. Noskova N.I., Korznikov, A.V., Idrisova, S.R. (1999) Structure, hardness and destruction peculiarities of nanostructured materials, in G.G. Taluts and N.I. Noskova (eds.), *The Structure and Properties of Nanocrystalline Materials*, UD RAS, Ekaterinburg, pp. 138-146 (in Russian).

16. Zgalat-Lozynskyy, O.B., Ragulya, A.V., Herrmann, M. (2000) Rate-controlled sintering of nanostructured titanium nitride powders, in M.I. Baraton and I.V. Uvarova (eds.). *NATO ASI Conference "Functional Gradient Materials and Surface Layer Prepared by Fine Particles Technology"*, June 18-28, 2000. Kiev, Ukraine. Poster P-23. Kluwer Academic Publishers, Dordrecht.

17. Rabe, T., Wäsche, R. (1995) Sintering behaviour of nanocrystalline titanium nitride powders. *Nanostructured Materials* **6**, 357-360.

18. Yamada, T., Shimada M., and Koizumi M. (1980) *Amer. Ceram. Soc. Bull.* **59** (6), 611-616.

COMPACTION AND ELASTIC RELAXATION OF FULLERITE POWDER

L. L. KOLOMIETS, S. M. SOLONIN, and V. V. SKOROKHOD
Institute for Problems of Materials Science, Ukraine National Academy of Sciences, 3,Krzhizhanovsky str., 03142 Kiev, Ukraine.

1. Introduction

Fullerene is a new allotropic form of carbon that exists as closed spherical molecules C_n $(n > 4)$. C_{60} and the other higher fullerenes develop surprising physical and chemical properties. The condensed crystal phase from fullerene molecules connected by Van der Waals forces is called fullerite.

The molecules of C_{60} are highly symmetrical and they have the shape of truncated icosahedra. The molecule diameter equal 0.71 nm [I]. C_{60} crystallizes in a FCC-lattice (a = 1.417 nm). The distance between the closest neighbors is 1.0 nm. A crystal may be compressed at low pressure along any axis unlike from graphite, which can be compressed along the c axis only.

The crystal structure of C_{70}, obtained by evaporation from solution carbon in toluene at normal temperature, is HCP (parameters $a = 1.01 \pm 0.05$ nm, c = 1.70 ± 0.08 nm [2-4]). The other form C70 can be prepared by sublimation on a silicon base (004) substrate by heating to 500°C, has a FCC-structure (1.489 nm [3]). The authors of these works found the presence of packing defects and amorphous component in the fullerite structure.

At normal pressure and temperature, fullerites are crystals in which molecules are linked by weak Van der Waals forces. The density of these fullerites is about 1.6 g/cm³ and the bulk modulus of all-round compression is B=18 MPa [5]. Other phases with high density and high mechanical properties have been obtained after treatment at high pressure and temperature [6-10]. In [6], C_{60} was studied at pressures up to 35 GPa by Raman spectroscopy. It was observed that at a pressure of 23 Gpa, even traces of C_{60} were absent from the spectrogram and a sharp peak appeared in the region of diamond stability. However, it was impossible to identify this phase only on the basis of the optical spectra. The phase transformations that occur in fullerite C_{60} up to 22-25 GPa are reversible. But at higher pressures, there is the irreversible transformation into an amorphous phase amount of which increases with pressure increasing. The authors [8] consider that fullerite compressed under hydrostatic conditions at high temperature undergoes transformation into sp^2-graphite, and under non-hydrostatic conditions it transforms into amorphous sp^3-phase and polycrystalline FCC-diamond.

Therefore, at pressures above 20 GPa and at temperature T = 1000°C, molecules of C_{60} break down and an irreversible transformation into amorphous graphite and diamond occurs, whereas in the lower pressures region there are structure transformations connected with orientation ordering and polymerization. The

M.-I. Baraton and I. Uvarova (eds.),
Functional Gradient Materials and Surface Layers Prepared by Fine Particles Technology, 177–182.
© 2001 *Kluwer Academic Publishers. Printed in the Netherlands.*

interpretation of the polymerization mechanism and study of the densest modifications of fullerite was carried out in [9, 10].

2. Results and discussions

It is interesting to evaluate the compressibility of the fullerite powders in a closed die, and to study plastic deformation and elastic relaxation of stresses after unloading.

In our work, we have studied the structure changes that occur in loading fullerite powder under free settlement conditions at room temperature. Shear modulus is evaluated, the relaxation of elastic distortions is demonstrated, and the nature of fullerite deformation under pressure is discussed.

The starting fullerene material [4] was a mixture of C_{60} and C_{70} (about 5% C_{70}). The x-ray diffraction pattern of the original fullerite powder, obtained in Co-radiation, was interpreted as a FCC-structure with spacing $a=1.416$ nm and absence of reflections for (200) and (400) that are resolved under x-ray diffraction conditions (Fig.la). The absence of these reflections is connected with the orientation disorder of the rotation type in a FCC-structure [2].

The original fullerite powder was compacted at pressures of 0.3, 1, and 7 GPa. The powder is readily amenable to briquetting.in a die. The specimen compacted at 1 GPa broke into fragments after two days. It indicates on a high level of stored elastic energy. The deformation at high pressure was carried out in two stages: at first, a specimen was briquetted in a die at 0.3 GPa and then this compact was deformed at 7 GPa in conditions of free compression. After unloading of the pressure, an explosive breaking of the specimen as a result of a partial elastic relaxation was observed.

X-ray diagrams of the specimens compacted at different pressures are presented in Fig 1b, c, d. When pressure is increasing, reflections (200) and (400) appear and their intensity increases as the result of the ordering in the structure of FCC-fullerite. This type of transformation has been observed when temperature is decreasing [2] and also at high pressure and elevated temperature [9]. The most appropriate model for describing of ordering in this system is "freezing" of fullerite molecules in the FCC-lattice [2].

It was also established that after action of pressure 7 GPa, the lattice parameter decreases (a = 1.396 nm) [9].

It is interesting to note that the width of x-ray peaks increases when compaction pressure is increasing apart from (200) and (400) peaks whose intensity increases and half-width decreases (Table 1).

TABLE 1- Change of X-ray half-width peaks (β, min)
for fullerite specimens versus compaction pressure

| hkl | θ | Pressure, GPa | | | |
		0	0.3	1	7
200	7°17′	-	-	15	12
220	10°19′	10	20	30	38
311	12°02′	10	20	30	45
400	14°36′	-	-	25	18
422	18°00′	20	45	55	-

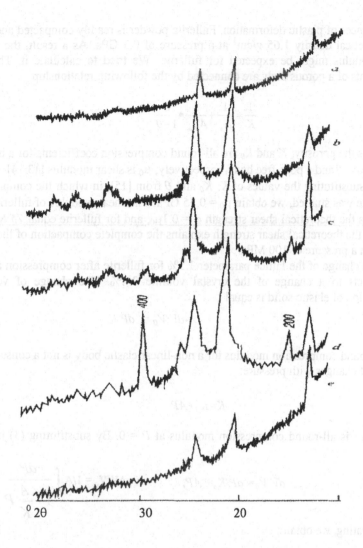

Figure 1. X-ray diffraction pattern (CoKₐ) of original fullerite powder (a), pressed at 0.3 (b), 1 (c), 7Gpa (d), and of the powder obtained after breaking of briquette, pressed at 7 GPa (e).

The approximate analysis of the width ratio indicates that it is close to the ratio of the tangents. Therefore it is possible to assume that the peak broadening is caused by microdistortions. Microdistortions can be evaluated by the equation

$$\Delta a/a = \beta/4tg\theta$$

and equal $\Delta a/a =(5\text{-}7)10^{-3}$. This is a relatively high value comparing with ductile metals. For example, microdistortions for nickel pressed at 0.8 GPa equal $1.6 \cdot 10^{-3}$ [11]. Microdistortions for MgO after explosive compaction equal $\Delta a/a =(5\text{-}7)10^{-3}$ [12]. Therefore, it is necessary to study whether these changes are reversible or irreversible consequences of plastic deformation. Fullerite powder is readily compacted and reaches the theoretical density 1.65 g/cm^3 at a pressure of 0.3 GPa. As a result, the very low shear modulus might be expected for fullerite. We tried to calculate it. The elastic coefficients of a porous body are connected by the following relationship:

$$\frac{1}{K} = \frac{1}{K_o} + \frac{3}{4\mu_o} \cdot \frac{\theta}{1-\theta}$$

(1)

where θ is the porosity; K and K_0 are all-round compression coefficients for a body with the porosity θ and a pore-free body, respectively; μ_0 is shear modulus [13,14].

By substituting the values of K, K_0 and θ from [15], in which the compressibility of fullerite was studied, we obtain $\mu_0 = 0.75$ GPa. The shear modulus of fullerite is very low. Thus the theoretical shear strength $\sigma = 0.1\mu_0$, and for fullerite equal 75 MPa. This value for the theoretical shear strength explains the complete compaction of the fullerite powder at a pressure of 100 MPa.

The change of the lattice parameter 1.5% for fullerite after compression at 7 GPa, corresponds to a change of the crystal volume 2-4%. The change of volume at compression of elastic solid is equal

$$-dV/V_0 = dP/K.$$

(2)

The all-round compression modulus for a non-linear elastic body is not a constant value because it changes with pressure:

$$K = K_0 + AP$$

(3)

where K_0 is all-round compression modulus at $P = 0$. By substituting (3) in (2), we obtain:

$$dV/V_a = dP/K_o + AP, \qquad -\Delta V/V_o = 1/A \int \frac{dP}{1 + \frac{A}{K}P}$$

By integrating, we obtain:

$$-\Delta V/V_0 = 1/A \ln(1 + AP/K)$$

(4)

The dependence of all-round compression modulus for fullerite at high pressures (up to 20 GPa) was determined in [16]:

$$K = (18.1 + 5.7P) \text{ GPa.}$$

Then, at a pressure of 7 GPa, the change in volume calculated by Eq. (4) should be

$-\Delta V/V_0$ = 0.205 which markedly exceeds the experimental value ($-\Delta V/V_0 = 0.042$). This is explained by the fact that elastic stresses stored within a specimen during compaction relax during unloading, which we have also observed visually as the explosive breaking of a briquette. Microdistortions and a change in the lattice parameter measured by x-ray diffraction are the residual values retained by macrostresses during compaction. The value of these macrostresses may be determined by equation (2). If $-\Delta V/V_0 = 0.042$, then $P = 850$ MPa. At breaking of a compact body into powder, a relaxation of macrostresses is taken place. As can be seen from x-ray diffraction patterns, disintegration of compact leads to producing of original fullerite, i.e. a total relaxation of elastic stresses occurs within the lattice. It is possible to confirm that deformation of polycrystalline fullerite in our test there is no plastic deformation, i.e. the deformation is elastic and reversible. This observation is not very evident since there is experimental evidence of plastic deformation in a freshly-prepared single crystal of fullerite [17].

3. Conclusion

The density of 100% is achieved by compaction of fullerite powder at very low pressure (of the order of 100 MPa). This may be explained by the low theoretical shear strength of fullerite, equal to about 75 MPa. In a fullerite compact significant residual distortions have been found and the lattice parameter is reduced. This is caused by residual macrostresses evaluated as 850 MPa. Disintegration of compacts into powder leads to the total relaxation of elastic macrostresses and consequently to the disappearing of the corresponding lattice microdistortions. It demonstrates the elastic and reversible nature of fullerite deformation at pressure.

182

4. References

1. Andreoni W., Gygi F., and Pamnello M. (1992). Structural and electronic properties of C_{70}, *Chem. Phys. Lett.*, **189**, No. 3, 241-244.
2. Heiney P. A., Fischer J. E., McGhie A. R. (1991) Orientational ordering transition in solid C_{60}, *Phys. Rev. Lett.*, **66**, No. 22, 2911-2914.
3. Valsakumar M. C., Subramanian N., Yjusuf M., et al., (1993). *Pramana-Journal of Physics*, **40**, No. 2, L137-L144
4. Dravid V. P., Lin X.,. Zhang H, (1992). Transition electron microscopy of C_{70} single crystals at room temperature, *J. Mater. Res.*, **7**, No. 9, 2440-2446 .
5. Zhang Y., Gao H., and Weaver M. J., (1992). Scanning tunnelling microscopy of C_{60} and C_{70} on ordered Au (111) and Au(110): molecular structure and electron transmission, *J. Phys. Chem.*, **96**, No. 2, 510-513 .
6. Moshary F., Chen N. H., Silvera I. F, et al. (1992). Gap reduction and the collapse of solid C_{60} to a new phase of carbon under pressure, *Phys. Rev. Lett.*, **69**, No. 3, 466-469.
7. Snoke D. W., Raptis Y. S., and Syassen K., (1992). Vibrational modes, optical excitations, and phase transition of solid C_{60} at high pressures, *Phys. Rev. B*, **45**, No. 24, 14419-14422.
8. Hodeau J. L., Tonnerre J. M., and Bouchet-Fabre B., et al., (1994-II).High-pressure transformation of C_{60} to diamond and sp -phases at high temperature, *Phys. Rev. B*, **50**, No. 14, 10311-10314 .
9. Brazhkin V. V, Lyapin A. G., and Popov S. V., (1996) Mechanism of three-dimensional polymerization of fullerite C_{60} at high pressure, *Letters to ZhETF*, **64**, No. 11,755-759.
10. Brazhkin V. V, Lyapin A. G., Lyapin S. G., (1997). New crystal and amorphous modifications of carbon obtained from fullerite at high pressure, *Konferentsi Simpoziumy*, **167**, No. 9, 1019-1022.
11. Martynova I. F., Skorokhod V. V., and. Solonin S. M., (1974). Plastic deformation during compaction of ductile metal powders, *Poroshk. Metall.*, No. 3, 40-46.
12. Skorokhod V. V., Savvakin G. I, Solonin S. M., and. Kolomiets L. L, (1974). Study of the substructure and certain production properties of refractory material powders treated by explosion, *Poroshk. Metall.*, No. 8, 80-84.
13. Mckenzie J. K., (1950). The elastic constants of solid containing spherical holes, *Proc. Phys. Soc. (B)*, **63**, No. 1, 2-11.
14. Skorokhod V. V., (1961). Calculation of elastic isotropic moduli for fine hard mixtures, *Poroshk. Metall.*, No. 1, 50-55.
15. Weisheng Li, Sherwood R. D., Cox D. M., and Maceij Radosz, (1994). PVT measurements on fullerite from 30 to 330°C and at pressure 200 MPa, *J. Chem. Eng. Data*, No. 39, 467-469.
16. Duclos S. J., Brister K., Haddon R. C., et al. (1991). *Nature*, **351**. 380-382 .
17. Orlov V. I., Nikitenko V. I., Nikolay R. K., et al., (1990). Experimental study of dislocations in crystals of fullerene C_{60} and mechanisms of their plastic deformation, *Pis'ma Zh. Eksp. Teor. Fiz; **59**, No. 10, 667- 670 .

NANOSTRUCTURED MATERIALS FOR MICROSTRUCTURAL CONTROL OF THERMAL PROPERTIES

J.A. Eastman, G. Soyez[1], G.-R. Bai, and L.J. Thompson

Materials Science Division

Argonne National Laboratory

Argonne, IL 60439 USA

1. Introduction

Compared to many other properties of nanostructured materials such as their mechanical behavior, thermal transport properties have received considerably less attention. Recently, however, this area has begun to attract attention. It is well known that conventional polycrystalline materials typically exhibit lower thermal conductivity than low-defect single crystals of the same material. Several investigators have realized recently that this could result in significantly reduced thermal conductivities in nanostructured materials that could lead to important improvements in behavior for applications such as thermal barrier coatings (TBC's). Recent studies of grain-size-dependent thermal conductivity in yttria-stabilized zirconia (YSZ), the most commonly utilized thermal barrier material, have begun to investigate the feasibility of using nanostructured components in future-generation TBC's [2-4].

While a grain-size-dependent reduction in thermal conductivity may be desirable for thermal barrier applications, there are other applications involving nanometer-scale microstructures where retention of high heat transfer rates is desirable. For example, the continual trend towards miniaturization in the micro-electronics industry is resulting in

183

M.-I. Baraton and I. Uvarova (eds.),

Functional Gradient Materials and Surface Layers Prepared by Fine Particles Technology, 183–197.

increased localization of heat loads with concomitant increasing thermal management challenges. Cooling systems based on fluids pumped through heat exchangers become unfeasible when component sizes are reduced to the nanometer scale. The development of new solid-state high thermal conductivity "heat pipes" synthesized, for example, from thin film diamond nanostructures would provide important advantages if the nanostructure processing can be optimized to maintain sufficient heat transport rates.

The role of microstructure in controlling heat transfer properties in nanostructured materials will be discussed in this paper, with emphasis on discussing the possible effects of grain size in controlling conductive heat transfer via changes in sound velocity, specific heat, or phonon mean-free-path.

2. Heat Transport Fundamentals

There are two fundamental mechanisms of heat transfer in materials [5]: 1) radiative heat transfer, which is transfer of heat by photon emission, and 2) conductive heat transfer, which is transfer of heat by molecular or atomic motion. Convection is sometimes considered a third heat transfer mechanism, but more appropriately should be considered to be simply correlated conductive heat transfer [5].

Radiative heat transfer is particularly important at high temperatures, such as may be encountered in TBC applications. A detailed study of the effects of grain size, if any, on radiative heat transfer at high temperatures has not been carried out to-date, however, and thus this mechanism will not be discussed further in this brief review. At room temperature or below, where limited studies of grain size effects on thermal conductivity have been performed (e.g., [4, 6]), contributions from radiative heat transfer are expected to be negligible.

There are three mechanisms of conductive heat transfer. Uncorrelated molecular collisions transfer heat via a ballistic process. Ballistic heat transfer is dominant in non-crystalline materials, including liquids, gases, and amorphous solids. Heat transfer by correlated atomic motion (phonons) occurs in all crystalline solids. In electrical conductors, significant heat transfer also occurs through the motion of conduction electrons. The focus here is on heat transfer in materials where electron motion contributions to the thermal transport properties are negligible, however, such as for materials including yttria-stabilized zirconia (YSZ) and diamond.

3. Expected Effects of Grain Size on Thermal Conductivity

For situations where the number of phonon scattering events is large and random, as is the case in nanostructured materials with randomly oriented and spaced grain boundaries, a particle transport approach is appropriate for understanding heat transfer via phonons (as opposed to an approach taking into account the wavelike nature of the phonons, which may be more appropriate in other cases). A kinetic theory approach based on particle-based energy transport can be used to derive a simple relationship between thermal conductivity, k, specific heat, C, sound velocity, v, and phonon mean-free-path, l [7]:

$$k = \frac{Cvl}{3}$$
(1)

By examining the potential grain size effect on each of the quantities in the numerator of the right-hand-side of Eq. 1 it should be possible to predict a relationship between grain size and thermal conductivity for nanostructured materials.

Reports in the literature to-date indicate little or no effect of grain size on specific heat in nanocrystalline materials. Lu et al. [8] observed that for measurements from ~300-500K, nanocrystalline Ni-P alloys exhibited approximately 10% larger specific heat values than either amorphous or coarse-grained polycrystalline alloys of the same composition. In the case of diamond, Moelle et al. [9] found that for temperatures of approximately 50-300°C, no measurable differences were observed between the specific heats of single crystal, coarse-grained polycrystalline, and nanocrystalline diamond samples. Based on these reports, it is expected that changes in specific heat will have little contribution to any grain-size-dependent changes in thermal conductivity.

To the best of our knowledge, at the present time there are no reported direct measurements of sound velocity as a function of grain size in nanocrystalline materials. Since v is related to the bulk modulus, B, and the density, ρ, according to

$$v = \left(\frac{B}{\rho}\right)^{1/2} \tag{2}$$

grain-size-dependent changes in either B or ρ could affect v. Reduced elastic moduli compared to bulk values have been reported for both nanocrystalline ceramics [10] and metals [11]. However, this reduction has generally been interpreted as resulting from porosity due to sample processing rather than being an intrinsic grain-size effect. While it is possible that a real reduction in moduli could occur in extremely fine-grained nanocrystalline materials, this would be expected to occur simultaneously with a lower expected density in grain boundary regions; thus, the net effect of grain size on v is still expected to be negligible.

In contrast to the expectations of minimal effects of grain size on specific heat and sound velocity, grain size is expected to have a significant effect on the phonon mean-free-path, l. Debye [5] derived the relationship

$$l = \frac{20 d T_m}{\gamma^2 T} \tag{3}$$

where T_m is the absolute melting temperature, d is the lattice constant, and γ is the Gruneisen constant, which is ~2 for most materials [5] and is given by [12]

$$\gamma = \frac{3\alpha B}{C} \tag{4}$$

where α is the thermal expansion coefficient. Particularly at low temperatures, the single crystal l value can be very large compared to nanocrystalline grain sizes. For example, for single crystal diamond at 10K, $l \approx 1$ mm [6]. For samples with very low defect concentrations, this can lead to a sample-size dependence of k, such as has been observed at low temperatures in LiF [13].

Due to scattering of phonons by grain boundaries, reducing the grain size below the calculated single-crystal l value should result in a reduced phonon mean-free-path and consequently a reduction in k. Based on this type of argument, theoretical predictions have been made of grain-size-dependent reductions in k in nanocrystalline materials such as YSZ [2]. A lower limit can be predicted for k by substituting the interatomic spacing for l in Eq. 1 [14]. A wide range of amorphous and highly defective crystalline materials have been found to exhibit thermal conductivities well-represented by this minimum thermal conductivity criterion [14-15].

4. Experimental Studies of the Effect of Grain Size on Thermal Conductivity

Relatively few experimental studies of the grain-size-dependence of thermal conductivity have been performed to-date. Results for both nanocrystalline YSZ [3,4] and diamond [6, 16-18] will be briefly reviewed in this section. Since YSZ is highly defective and exhibits low thermal conductivity even in single crystal form due to the large oxygen vacancy concentration required to maintain charge neutrality, the effect of grain size on k is expected to be much smaller than for a material like diamond which can exist as highly perfect single crystals with extremely large thermal conductivity.

4.1 THERMAL CONDUCTIVITY IN NANOCRYSTALLINE YSZ

Raghavan et al. were first to experimentally investigate the thermal conductivity of nanocrystalline YSZ [3]. In that study, bulk samples were prepared from sintered nanocrystalline powders with grain sizes as small as 30 nm. The interpretation of their results was complicated by the significant grain-size-dependent porosity found in all samples with grain sizes less than 70 nm. In particular, they observed no effect of grain size on thermal conductivity for dense samples, but a significant dependence of the thermal conductivity on varying amounts of porosity for smaller-grained samples.

A more recent study of nanocrystalline YSZ by the present authors and co-workers [4] observed a grain-size-dependent reduction in the room-temperature thermal conductivity of nanocrystalline yttria-stabilized zirconia for the first time. YSZ films with thicknesses of 0.5 - to - 1.2 μm and yttria compositions of 8 to 15 mol.% were grown on polycrystalline α-Al$_2$O$_3$ substrates by metal-organic chemical vapor deposition with controlled grain sizes from 10 to 100 nm. The measured room temperature thermal conductivity values for YSZ thin films as a function of grain size are shown in Figure 1. For all grain sizes larger than 30 nm, the thermal conductivity is approximately constant and ~20% smaller than the literature value of 2.3 W/m-K for

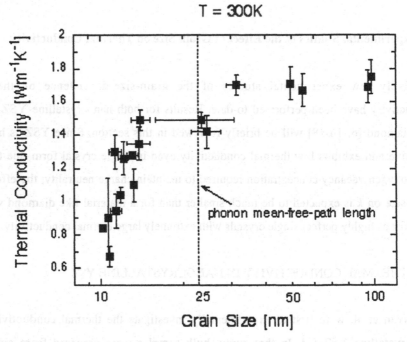

Figure 1. Measured room temperature thermal conductivity values for YSZ thin films as a function of grain size (after [4]). Note that for grain sizes larger than the single crystal phonon mean-free-path, the thermal conductivity is approximately independent of grain size, while for smaller grain sizes a rapid decrease is observed.

high-density, polycrystalline YSZ with a yttria-content of 10 mol.% at 20°C [19]. For smaller grain sizes, the thermal conductivity was observed to decrease rapidly with decreasing grain size, reaching a value of less than one-third the bulk value for the smallest grain-sized samples.

While the measured thermal conductivity values from Ref. [4] are in good agreement with theoretical predictions of Klemens and Gell [2] for grain sizes larger than 30 nm, the observed rapid decrease in thermal conductivity for smaller grain sizes was not predicted. At the smallest grain sizes, the observed thermal conductivity values are approximately half the predicted values for the minimum conductivity amorphous limit of YSZ calculated using an Einstein model [15].

The results or Ref. [4] are in good agreement with Ref. [3] in that both studies found no effect of grain size on thermal conductivity for nanocrystalline YSZ samples with grain sizes larger than ~30 nm. The strong grain-size-dependent effect seen by Soyez et al. [4] was observed only for samples with significantly smaller grain sizes than measured by Raghavan et al. [3]. From observations that indicated a lack of any systematic change in porosity with changes in grain size, it was concluded in Ref. [4] that the variations in thermal conductivity observed in that study were more likely attributable to changes in grain size than changes in porosity.

A possible explanation for the observed grain size dependence of the thermal conductivity of YSZ in Ref. [4] can be provided by considering the relationship between the grain size of the samples and the phonon mean-free-path. Using Equations 3 and 4, l for single crystal YSZ is calculated to equal 25 nm at 300K. It is reasonable to expect that, consistent with the present observations, reductions in thermal conductivity due to phonon scattering by grain boundaries will only become significant for grain sizes smaller than l. Since l is temperature dependent, this explanation predicts that the critical grain size below which the thermal conductivity drops rapidly compared to the bulk conductivity will also depend on temperature.

190

4.2 GRAIN-SIZE DEPENDENCE OF THERMAL CONDUCTIVITY IN DIAMOND

At room temperature, the thermal conductivity of high quality single crystals of diamond exceeds 2200 W/m-K [16-18] and can be even larger in isotopically-purified single crystals [20]. Since this is approximately three orders-of-magnitude larger than the thermal conductivity of YSZ, there is a potential for a much larger grain-size effect. Two recent studies have begun to address this issue.

Graebner and co-workers were the first to investigate grain-size dependent thermal conductivity in diamond [16-18]. In those studies, polycrystalline diamond was produced by chemical vapor deposition and the thermal conductivity was measured at room temperature. Their samples were found to exhibit a columnar morphology that resulted in an increasing in-plane grain size with increasing film thickness (distance from the silicon substrate). By examining the thermal conductivity of samples of several thicknesses, the authors were able to infer an effect of grain-size. They concluded that samples with grain sizes larger than approximately 20 μm exhibited essentially single-crystal-like conductivity, while below that grain size the conductivity decreased monotonically to a value of approximately 25% of the single-crystal value for a grain size of about 2 μm. It is interesting and not understood why samples with grain sizes as large as 2-20 μm would show deviations from single crystal thermal conductivity. Using Eq. 1 and literature values for k, C, and v, l for single crystal diamond would be predicted to be approximately 100 nm at room temperature. The possible role of other microstructural influences were not considered in Ref. [16] (e.g., variations in strain or porosity with changing film thickness).

Efimov and Mezhov-Deglin [6] performed a more recent study of thermal conductivity in diamond films prepared by microwave plasma enhanced chemical vapor deposition. At room temperature, their data indicated no difference in the thermal conductivities of single crystal diamond and a 1-10 μm grain-sized polycrystalline film, but a significantly reduced room temperature thermal conductivity in a 100 nm grain-sized sample. They also obtained data at temperatures as low as 8 K and observed significant reductions in thermal conductivity in both grain-sized polycrystalline samples at low temperatures compared to single crystal diamond. These results seem in line with expectations based on Eq. 1; however, additional factors such as the presence of an amorphous carbon component in their samples may have influenced the results. Further studies with phase-pure samples are needed to definitively determine the grain-size dependence of thermal conductivity in diamond.

5. Processing of Thin Film Nanostructures by Metal-Organic Chemical Vapor Deposition (MOCVD)

Since nanostructured coatings and films are desirable for potential thermal management applications, processing techniques that provide optimal control of microstructure are needed. Nanocrystalline thin films and coatings have been previously produced by several techniques including thermal spray [21,22], electrodeposition [23], sputter deposition [24], and MOCVD [25,26]. Of these techniques, only thermal spray and MOCVD appear feasible for applications where cost-effectiveness, scalability, and field servicing are major concerns. Previous studies of MOCVD processing of nanocrystalline coatings of TiC, TiN, and SiC [25], and AlN/TiN [26] have established that several strategies can be employed in obtaining and stabilizing nanocrystalline microstructures via MOCVD processing. For example, Liu and co-workers [26] successfully used high speed deposition techniques to suppress relaxation processes believed to result in grain growth. These same workers [26] also established that co-depositing insoluble components to form a composite is an effective means of stabilizing a nanocrystalline microstructure.

Results from a recent study of the processing of nanocrystalline YSZ are described here. Nanocrystalline YSZ films were grown by MOCVD using a low-pressure, horizontal, cold-walled deposition system. Yttrium b-diketonate ($Y(thd)_3$) and zirconium t-butoxide ($Zr(OC(CH_3)_4)$) [27] were chosen as precursor materials. High-purity nitrogen was used as the precursor carrier gas. The precursors were mixed with high-purity oxygen and nitrogen in the quartz deposition chamber.

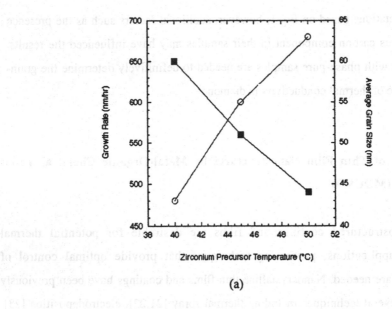

(a)

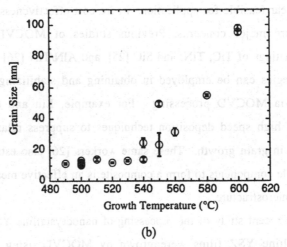

(b)

Figure 2. Grain size can be controlled by varying MOCVD processing parameters such as (a) the precursor bubbler temperature(s), or (b) the substrate temperature. In (a), the open symbols denote growth rate, while the closed symbols indicate the average grain size.

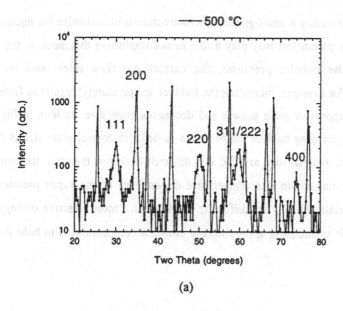

(a)

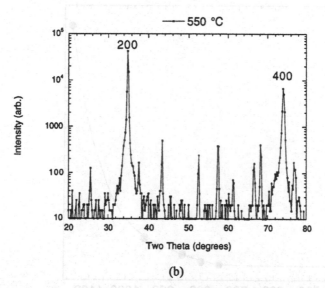

(b)

Figure 3. For nanocrystalline YSZ films grown on polycrystalline alumina substrates x-ray diffraction results indicate that a random texture was observed for all substrate temperatures < 550°C (a), while a strong (100) texture develops for substrate temperatures ≥ 550 °C. The sharp unlabelled peaks in these figures are from the alumina substrate.

194

A goal in producing a small-grained microstructure is to maximize the nucleation rate. Several process parameters may play a role in accomplishing this, such as the substrate temperature, the bubbler pressures, the carrier gas flow rates, and the bubbler temperatures. For example, increasing the bubbler temperature(s) results in faster growth rates, which suppresses grain growth and decreases grain size, as seen in Fig. 2(a). A difficulty with changing bubbler temperatures in order to control grain size is that, for a multicomponent material such as YSZ, it is difficult to change the grain size without also changing the composition (the temperature dependence of the vapor pressures of the precursor materials typically are different). As a result, a more effective strategy for vary grain size while maintaining approximately constant composition is to hold the bubbler

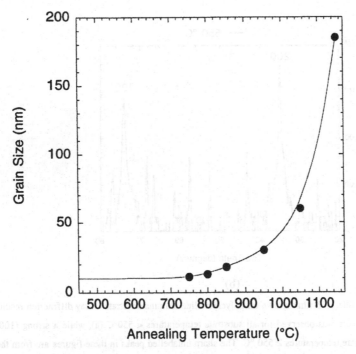

Figure 4.   The effect of annealing on grain growth is shown. YSZ samples grown on polycrystalline alumina substrates were annealed at each temperature for two hours in Ar-0.2% O_2. The heating and cooling rates were 200°C/hr.

temperatures constant while varying the substrate temperature. In this case, while making only small additional changes in the pressure in the Zr precursor bubbler, samples with controlled grain sizes and approximately constant composition can be prepared, as shown in Fig. 2(b). With increasing substrate temperature, the grain size increases smoothly and reproducibly from 10 nm for growth at 500°C to about 100 nm at 600°C, as seen in Fig. 2(b). YSZ samples produced with alumina substrate temperatures < 550°C were found to be non-textured, while a (100) texture was observed to develop at higher growth temperatures as seen in Fig. 3

The resistance of the nanocrystalline microstructure to grain growth at high temperatures is also an important issue when considering potential TBC applications of nanocrystalline YSZ. The onset of rapid grain growth in the YSZ films of the present study was observed to occur at temperatures above approximately 900°C, as seen in Fig. 4. For usage in potential applications above this temperature it will be necessary to develop strategies to suppress grain growth. Co-depositing an insoluble second phase (e.g., alumina) may be one technique for improving the grain size stability.

6. Summary and Conclusions

A review of the possible effects of grain size in phonon-based conductors such as YSZ or diamond indicates that a reduction in phonon mean-free-path below the bulk value is likely to have a significant effect on thermal transport. While the grain size dependences of specific heat and sound velocity have not yet been thoroughly characterized, the expected grain size effects on these quantities is small.

A new type of functional gradient may be desirable for optimizing thermal properties. For example, a grain-size graded YSZ coating could consist of a dense nanometer-grain-sized layer to provide phonon scattering and improved mechanical properties, plus a coarser-grained porous layer that would scatter photons.

Studies of thermal properties in nanostructures comprised of normally high thermal conductivity materials such as diamond have only begun recently and require further investigation. Process optimization may lead to optimal microstructures that would provide highly localized heat transfer in future microelectronic applications.

This work was supported by the U.S. Department of Energy, Office of Science, under Contract W-31-109-ENG-38.

7. References

1. Present affiliation: Carl Zeiss, Oberkochen, Germany.

2. P.G. Klemens and M. Gell, Thermal Conductivity of Thermal Barrier Coatings, *Mater. Sci. Engr. A*, **245**, 143-149 (1998).

3. S. Raghavan, H. Wang, R.B. Dinwiddie, W.D. Porter, and M.J. Mayo, The Effect of Grain Size, Porosity and Yttria Content on the Thermal Conductivity of Nanocrystalline Zirconia, *Scripta Mater.*, **39**, no. 8, pp. 1119-1125 (1998).

4. G. Soyez, J.A. Eastman, L.J. Thompson, G.-R. Bai, P.M. Baldo, A.W. McCormick, R.J. DiMelfi, A.A. Elmustafa, M.F. Tambwe, and D.S. Stone, Grain-Size-Dependent Thermal Conductivity of Nanocrystalline Yttria-Stabilized Zirconia Films Grown by Metal-Organic Chemical Vapor Deposition, *Appl. Phys. Lett.*, **77**, no. 8 1155-1157 (2000).

5. G.H. Geiger and D.R. Poirier, Transport Phenomena in Metallurgy, (Addison-Wesley: Reading, MA, 1973), p. 190.

6. V.B. Efimov and L.P. Mezhov-Deglin, Phonon Scattering in Diamond Films, *Physica B*, **263-264**, 745-748 (1999).

7. A. Majumdar, Microscale Energy Transport in Solids, in Microscale Energy Transport, ed. C.-L. Tien, A. Majumdar, and F.M. Gerner, (Taylor and Francis: Washington, D.C., 1998), Ch. 1.

8. K. Lu, R. Lück, and B. Predel, Investigation of the Heat Capacities of Ni-20 at.% P in Different States, *Z. Metallkd.*, **84**, 740-743 (1993).

9. C. Moelle, M. Werner, F. Szücs, D. Wittorf, M. Sellschopp, J. von Borany, H.-J. Fecht, and C. Johnston, Specific Heat of Single-, Poly- and Nanocrystalline Diamond, *Diamond and Related Materials*, **7**, 499-503 (1998).

10. M.J. Mayo, R.W. Siegel, A. Narayanasamy, and W.D. Nix, Mechanical Properties of Nanophase TiO_2 as Determined by Nanoindentation, *J. Mater. Res.*, **5**, no. 5, 1073-1082 (1990).

11. G.E. Fougere, L. Riester, M. Ferber, J.R. Weertman, and R.W. Siegel, Young's Modulus of Nanocrystalline Fe Measured by Nanoindentation, *Mater. Sci. Engr. A*, **204**, 1-6 (1995).

12. F.J. Walker and A.C. Anderson, Low-Energy Excitations in Yttria-Stabilized Zirconia, *Phys. Rev. B*, **29**, no. 10, 5881-5889 (1984).

13. J.S. Blakemore, Solid State Physics. 2nd Edition, (Cambridge University Press: Cambridge, 1985), p. 138.

14. D.G. Cahill, Heat Transport in Dielectric Thin Films and at Solid-Solid Interfaces, in Microscale Energy Transport, ed. C.-L. Tien, A. Majumdar, and F.M. Gerner, (Taylor and Francis: Washington, D.C., 1998), Ch. 2.

15. D.G. Cahill, S.K. Watson, and R.O. Pohl, Lower Limit to the Thermal Conductivity of Disordered Crystals, *Phys. Rev. B*, **46**, no. 10, 6131-6140 (1992).

16. J.E. Graebner, S. Jin, G.W. Kammlott, J.A. Herb, and C.F. Gardinier, Unusually High Thermal Conductivity in Diamond Films, *Appl. Phys. Lett.*, **60**, no. 13, 1576-1578 (1992).

17. J.E. Graebner, J.A. Mucha, L. Seibles, and G.W. Kammlott, The Thermal Conductivity of Chemical-Vapor-Deposited Diamond Films on Silicon, *J. Appl. Phys.*, **71**, no. 7, 3143-3146 (1992).

18. J.E. Graebner, S. Jin, G.W. Kammlott, Y.-H. Wong, J.A. Herb, and C.F. Gardinier, Thermal Conductivity and the Microstructure of state-of-the-art Chemical-Vapor-Deposited (CVD) Diamond, *Diamond and Related Materials*, **2**, 1059-1063 (1993).

19. D.P.H. Hasselman, L.F. Johnson, L.D. Bentsen, R. Syed, and H.L. Lee, Thermal Diffusivity and Conductivity of Dense Polycrystalline ZrO_2 Ceramics: A Survey, *Am. Ceram. Soc. Bull.*, **66**, no. 5, 799-806 (1987).

20. T.R. Anthony, W.F. Banholzer, J.F. Fleischer, L. Wei, P.K. Kuo, R.L. Thomas, and R.W. Pryor, *Phys. Rev. B*, Thermal Diffusivity of Isotopically Enriched ^{12}C Diamond, **42**, 1104-1111 (1990).

21. B. H. Kear, *Nanostr. Mater.*, Chemical Processing and Applications for Nanostructured Materials, **6**, 227-236 (1995).

22. J. Karthikeyan, C. C. Berndt, J. Tikkanen, S. Reddy, J. Y. Wang, and H. Herman, Nanomaterial Powders and Deposits Prepared by Flame Spray Processing of Liquid Precursors, *Nanostr. Mater.*, **8**, no. 1, 61-74 (1997).

23. U. Erb, Electrodeposited Nanocrystals: Synthesis, Properties, and Industrial Applications, *Nanostr. Mater.*, **6**, 533-538 (1995).

24. H. Hahn and R.S. Averback, The Production of Nanocrystalline Powders by Magnetron Sputtering, *J. Appl. Phys.*, **67**, no. 2, 1113-1115 (1990).

25. B. M. Gallois, R. Mathur, . R. Lee, and J. Y. Yoo, Chemical Vapor Deposition of Ultrafine Ceramic Structures, in Multicomponent Ultrafine Microstructures, ed. L.E. McCandlish, D.E. Polk, R.W. Siegel, and B.H. Kear, *Mater. Res. Soc. Proc.*, **132**, 49-60 (1989).

26. Y. J. Liu, H. J. Kim, Y. Egashira, H. Kimura, and H. Komiyama, Using Simultaneous Deposition and Rapid Growth to Produce Nanostructured Composite Films of AlN/TiN by Chemical Vapor Deposition, *J. Am. Ceram. Soc.*, **79**, no. 5, 1335-1342 (1996).

27. Inorgtech, distributed by First Reaction, Hampton Falls, NH.

CONVENTIONAL AND NEW MATERIALS FOR THERMAL BARRIER COATINGS

R. VAßEN AND D. STÖVER
Institut für Werkstoffe und Verfahren der Energietechnik 1,
Forschungszentrum Jülich GmbH, 52425 Jülich, Germany

1. Abstract

Processing, microstructures and properties of standard thermal barrier coating (TBC) systems based on 7 - 8 wt. % yttria stabilised zirconia and prepared by electron beam physical vapour deposition (EB-PVD) or atmospheric plasma spraying (APS) will be reviewed in the first part of the paper. The limited temperature capability of zirconia at increased surface temperatures above 1200 °C due to sintering and insufficient phase stability will be discussed.

The second part of the paper will be focused on new materials and concepts for TBC systems. Our own work on zirconates and metal glass composites and also results from literature on oxides with high melting points will be presented. Additionally, the potential of graded and nanophase structures will be discussed.

2. Introduction

The area of thermal barrier coating development is attracting interest from a lot of different international research groups. This can be illustrated by the large number of review articles which exist on this topic. Some recent articles are listed in the references [1, 2, 3, 4, 5, 6, 7]. The number of articles is certainly also promoted by the high complexity of the subject. A TBC system includes in most cases the substrate, the so-called bond coat and finally the insulative – in most cases ceramic – top coat. All these components react with each other or the environment to a more or less extent and/or they undergo detrimental changes due to thermo-mechanical treatments during operation. In the present paper a brief description of typical TBC systems and their main short comings and failure mechanisms will be given. The second part of the paper will then be focused on recent developments to improve the ceramic part of the TBC system.

The use of thermal barrier coating systems in gas turbines or diesel engines lead to improved performance. The insulative layer can provide a reduction of the temperature of the metallic substrate which results in an improved component durability. Alternatively, an increase of efficiency can be achieved by allowing an increase of the turbine inlet temperatures [8]. A typical example of the temperature profile within a

M.-I. Baraton and I. Uvarova (eds.),
Functional Gradient Materials and Surface Layers Prepared by Fine Particles Technology, 199–216.

leading edge of a vane for an aero-engine application is given in [4]. During a typical flight mission the metal surface temperature is reduced by the use of a 200 μm TBC with a thermal conductivity of 1.9 W/m/K from 1017 °C to 966 °C. A low thermal conductivity coating (1.1 W/m/K) leads to a further reduction of the substrate temperature to 933 °C. Simultaneously, the surface temperature of the coating rises from 1190 °C to 1260 °C. This example illustrates the reasons for major directions of development in the area of TBCs. First, it is tried to improve the insulative capability of the coatings by the reduction of the thermal conductivity. This can either be achieved by looking for new materials with a low thermal conductivity or by a modification of the microstructure of the coating. Details on both directions of development will be given in the present paper. A second important area of research is related to the improvement of the temperature capability of the TBC. The state-of-the-art coating for TBCs is nowadays 7 –8 wt. % yttria stabilised zirconia (YSZ). This material has been identified during the last 2 and a half decade of extensive work on TBCs as a favourite material for this application, e.g. in the fundamental early work of Stecura [9]. However, the maximum surface temperatures of these coatings is limited to about 1200 °C for long-term operation. The reason for this limitation will be discussed. Hence, an ideal new thermal barrier coating material should combine a reduced thermal conductivity and an enhanced temperature capability compared to YSZ while the other critical properties (thermal expansion coefficient, corrosion resistance, etc.) should be comparable to the properties of YSZ.

3. Conventional Thermal Barrier Coating Systems

3.1 SUBSTRATES

In modern gas turbines Ni or Co-based alloys are used in most cases in highly thermal and thermo-mechanical loaded parts as combustor segments and blades [10]. Chemical compositions of these alloys were optimised with respect to strength, creep, fatigue and crack growth behaviour, and to a lower extent to oxidation and corrosive resistance [11, 12]. Material strength could mainly be increased by developing new casting processes. These developments led to directional solidification (DS) and, finally, to single-crystal (SC) solidification. In directionally solidified materials, grain boundaries perpendicular to the main stress directions are avoided and in the single crystal version fully eliminated. As a result, creep strength is increased [10].

A significant further increase of the maximum operation temperature of the Ni- or Co based superalloys cannot be expected because the operation temperature is approaching the melting temperature of the alloys. New materials concepts, e.g. ceramic matrix composites, are certainly a long term development task [10].

In parallel to the alloy development a considerable increase of the turbine inlet temperature was possible by improved cooling techniques starting from uncooled structures to those with cooling holes, convection cooling, film cooling and forced convection [10]. Some new ideas, like transpiration cooling, are under development [13].

A disadvantage of extensive cooling is the reduction of the efficiency of the gas turbine. Thermal barrier coatings can improve this situation by allowing an increased turbine inlet temperature without significant increase of substrate temperature and gas cooling. In the following the intermediate layer, which is in most cases necessary to produce a reliable insulative thermal barrier coating system, is described.

3.2 BOND COATS

Commonly used types of bond coats are MCrAlY (M=Co, Ni) or Pt modified diffusion aluminide coatings. The first type of coating is in most cases deposited by vacuum plasma spraying (VPS). At the beginning of the TBC system development the cheaper atmospheric plasma spraying (APS) has been used, however, meanwhile it is known that the oxygen pick up during spraying under air leads to a degradation of the oxidation resistance of the coating [5, 6]. Alternatively, also electron beam physical vapour deposition (EB-PVD) can be used to produced MCrAlY bond coats [14] although the process is more expensive than VPS [6]. The most important phases present in these bond coats in the as-prepared state are β-NiAl and γ-Ni. Additionally, depending on the composition and the thermal treatment also γ′ (Ni₃Al-based), α-Cr and others are found.

The second commonly used bond coat for challenging application (e.g. in the aeronautical field) in combination with EB-PVD top coats is the Pt-aluminide coating. First, an about 10 μm thick Pt-layer is deposited e.g. by galvanic technique and then an aluminide coating is produced, e.g. via pack cementation. In this process the substrate is packed in a powder bed of aluminium containing powders. At the deposition temperature an activator (e.g. AlF) leads to a gas phase transport from the powders to the substrate surface [15].

Typically the outer zone of the aluminide bond coat contains β-NiAl with Pt in solution and as a second phase PtAl₂ [1, 15].

During the high temperature operation a thermally grown oxide (TGO), in most cases an alumina scale, is formed on the bond coat. This leads to an aluminium depletion in the region beneath the surface and to a reduction of the amount of β-phase and an increase portion of γ′ and γ phases.

3.3 CERAMIC TOP COAT

As stated above the standard material for the ceramic insulative top coat is 7-8 wt. % yttria stabilised zirconia (YSZ). Two methods of coating deposition are widely used: atmospheric plasma spraying (APS) or electron beam physical vapour deposition (EB-PVD). Both methods lead to the formation of a non-equilibrium t′ phase in the coating [16, 17] which does not undergo phase transformation during thermal cycling (non-transformable) and leads to attractive thermo-mechanical properties of the coating. Microstructure and, to a certain degree, properties of APS and PVD coatings are different and will be discussed briefly below.

3.3.1   EB-PVD TBCs

In the EB-PVD process an ingot – made of YSZ for conventional TBCs - is heated by the electron beam to temperatures above the melting point which leads to the evaporation of the material [18]. The evaporated atoms then condense on the substrates and form after a short movement via surface diffusion nuclei which begin to grow. Grain growth along favourable crystallographic orientations leads to a columnar grain structure. A typical microstructure of a PVD YSZ coating produced by the DLR Köln, Germany is shown in Figure 1. The orientation of the columns is affected by the processing conditions e.g. substrate temperature, substrate rotation speed, and orientation of the surface towards the evaporation source [19]. It should be noted that a bond coat surface with low roughness values (typically about 1 micrometer) is advantageous for the formation of uniformly dense coatings.

The columnar structure of the PVD TBCs is generally excepted to lead to a certain amount of strain tolerance as long as the individual columns are not chemically bond to each other. This close contact might be formed due to sintering during high temperature operation (see below).

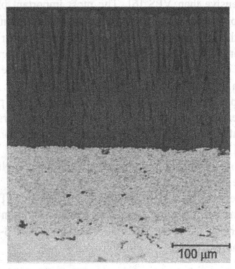

Figure 1. Micrograph of a PVD coating produced by the DLR Köln, Germany, on a nickel based superalloy (IN738) with a plasma sprayed NiCoCrAlY bond coat (produced by FZ Jülich).

For demanding applications in gas turbine as the high-pressure turbines airfoils in advanced aircraft engines in most cases PVD coatings are used [1]. The mentioned good strain tolerance is one reason for this. Additionally, the PVD process has compared to the APS process a reduced tendency to close the small cooling holes in the airfoils. On the other hand the PVD process is more expensive, especially for large parts, than the APS process due to significantly higher capital costs [6]. As a result, both processes find applications in gas turbine engines.

The specific microstructure of the PVD coatings with rather dense columns and a low amount of porosity or microcracks perpendicular to the direction of heat flow leads to relatively high values of the thermal conductivity of PVD coatings of about 1.5 W/m/K or more at 1000°C [4]. This value is close to the value of sintered, dense YSZ (2.2 W/m/K, [20]). On the other hand, this value does not increase as fast during high temperature operation as it is found for plasma sprayed TBCs (see below).

Typical thickness values of 250 μm or below are applied to airfoils for gas turbines, which seems to be a compromise between thermal insulation, additional weight of the TBC and cost effectiveness [21].

3.3.2 APS TBCs

In the plasma spraying process particles are introduced with a carrier gas into a plasma torch. In the hot plasma flame the particle will be partially or totally melted and accelerated by the plasma gases. The molten particles are then deposited on a substrate where they deform and cool down in short times scales (in the microsecond range). The process typically produces a lamellar structure with large pores with pore diameters above 1 μm. Additionally, small microcracks with diameters of about 200 nm are observed. The origin of these cracks are thermal stresses which arise from the rapid cooling during the spray process. A micrograph of a typical TBC system is shown in Fig. 2. It consists of a Ni based substrate, a VPS NiCoCrAlY bond coat, and an APS YSZ-TBC. In the ceramic top coat the two types of porosity are visible. A bimodal porosity size distribution is also found in mercury porosity measurements [22].

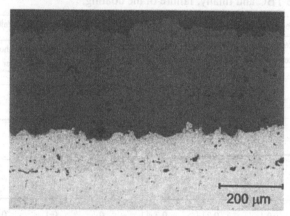

Fig. 2 Microstructure of plasma-sprayed TBC system consisting of a Ni based superalloy substrate (In738), a VPS NiCoCrAlY bond coat, and an APS 8 wt. % YSZ top coat.

The typical microcracked microstucture leads to relatively low thermal conductivity values in APS TBCs between about 0.5 and 1.0 W/m/K, which is about half the value found for PVD TBCs. Due to sintering effects (see below) these low values might increase quite fast during high-temperature operation.

Thickness values of APS TBCs are in general larger than those of PVD coatings. For airfoils values of 300 μm are often used while in the combustion chamber thicker coatings with different porosity values are envisaged [2]. Optimised porosity values are strongly depending on the special operational conditions. In thermal cycling tests Stecura [9] identified optimal porosity values of about 13 %.

Thick APS TBCs with thickness values up to the millimeter range are also frequently used in diesel engines to improve the combustion process [23 24].

3.4 FAILURE MECHANISMS OF YSZ THERMAL BARRIER COATINGS

Several different failure mechanisms have been identified to contribute to the failure of TBCs depending on the conditions of mechanical and thermal loading and the system itself [7, 6]. Some of the most important ones will briefly be reviewed in the following.

3.4.1 Bond coat oxidation

One generally excepted important factor, which influences and often determines the failure lifetime of the TBC systems, is the oxide scale formation at the bond coat - TBC interface [25]. This oxide scale, typically α-alumina, introduces high stress levels into the system and promotes spallation at the interface (often found in PVD-TBCs, [26]) or close to it in the TBC (typical for APS-TBCs, [27]). Fig. 3 shows the stress profile as a function of the distance from the rough bond coat / TBC interface in a plasma-sprayed system. The system was assumed to be stress-free at high temperatures and cooled down to room temperature [27]. It is found that the growth of the TGO leads to a stress conversion at the valley locations of the bond coat. This conversion promotes crack growth within the TBC and finally, failure of the coating.

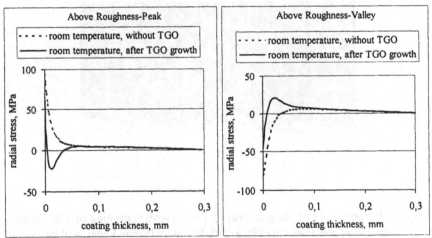

Fig. 3 Radial stress distribution in a TBC system with a rough bond coat / TBC interface at room temperature [27].

Under certain conditions also the formation of spinells instead of alumina is observed, which reduces the strength level of the interface and promotes spallation. These conditions can be a long term exposure with considerable aluminium depletion in the bond coat [1] or also an oxidation of certain bond coat phases [28].

3.4.2 Corrosion and erosion
Engines with TBC systems are often operating in corrosive atmospheres. These atmospheres might contain elements like vanadium, phosphorus, sulphur, and others. Under these corrosive conditions the zirconia top coats are often destabilised due to a reaction of the stabilising oxide with these impurities [29]. Standard YSZ seems to be stable against sodium and sulphur corrodents. However, destabilisation is observed in vanadium containing atmospheres. Sc_2O_3 stabilisers have been proposed to improve the corrosive resistance.

Another important degradation mechanism is particulate erosion leading to wear of the coating on several parts of the engines like turbine blades and vanes [30]. The erosion rate of plasma sprayed coatings are often higher than these of PVD-coatings. The absolute values are strongly depending on the velocity of the particles and the impact angle [31].

3.4.3 Degradation due to the limited high temperature capability of YSZ
At higher temperatures (> 1200 °C) additionally to the above mentioned mechanisms the degradation of the YSZ top coat becomes increasingly important. One degradation mechanism of the standard YSZ is the accelerated phase transformation from the t´-phase to the equilibrium tetragonal and cubic (t + c) phases above 1200 C [32]. During cooling, the equilibrium tetragonal phase transforms to the monoclinic (m) phase, which is accompanied with a volume increase of about 4 %. This volume increase might result in a disintegration of the coating especially with increasing amounts of transformed t´-phase. Furthermore, enhanced sintering of the coating, i.e. the reduction of the porosity and an increase of Young's modulus [33] will take place at elevated temperatures. These changes of microstructure and mechanical properties will result in higher thermally induced stresses and a decrease in thermal fatigue life time of the coating [6]. Additionally, the sintering leads to an increase of the thermal conductivity and hence, to a reduced insulative protection of the substrate. In [34] it is shown that aging at 1038 °C for 1000 h led to a 60 % increase of thermal conductivity in APS coatings. PVD-coatings showed a significant lower effect of aging on thermal conductivity.

4.0 New concepts and materials for improved thermal barrier coatings

In the following the development of new, mainly ceramic thermal barrier coatings is classified into four categories. Although the classification is somewhat arbitrary and some developments might belong to several categories, it is helpful in structuring the large amount of work performed in this area.

4.1 MODIFIED CHEMICAL COMPOSITIONS
Several investigations on the reduction of the sinterability of YSZ are found in literature. It is shown that the amount of sintering in YSZ coatings can be reduced by a lower amount of impurity phases, especially SiO_2 [35]. However, there seems to be a lower limit of sintering, which can not be further reduced even with lower impurity levels [36]. A different approach to reduce sintering, the addition of oxides with limited solubility in YSZ or CeO_2 [37] have not been successful. Al_2O_3 additions in sintered specimens gave

no significant effect on density, it only reduced grain size. 10 wt. % TiO_2 addition in YSZ increased the sinterability of plasma-sprayed coatings considerably. Also plasma-sprayed CeO_2 coatings doped with Al_2O_3 sintered much faster than pure YSZ coatings.

The limited phase stability at elevated temperatures is addressed by the change of the oxides used for stabilisation of the ZrO_2. In this paper only recent developments are mentioned. Frequently in other applications used stabilisers for zirconia like CaO or MgO have already been shown to give a lower performance than standard yttria doped zirconia [38].

In [39] and [40] the effect of scandia instead of yttria addition on phase stability is investigated. An improved thermal stability and also hot - corrosion stability against $NaVO_3$ was found. However, the phase transition from t′ to t and c during annealing at 1400°C was not completely depressed.

Also ceria doped YSZ was investigated with respect to an application as material for thermal barrier coatings. Results in [41] indicate an improved thermal cycling behaviour for an optimised ceria content compared to YSZ. On the other hand, recent investigations reveal some disadvantages of the ceria doped zirconia system. Disintegration of the ceria doped YSZ occurred, especially under reducing atmospheres, at temperatures above 1100 °C [42]. Additionally, high sintering rates are found at elevated temperatures in the system [43].

A very attractive new class of stabilizer for zirconia seem to be oxides with pentavalent cations, i.e. Ta_2O_5 and Nb_2O_5 [44]. Certain compositions show high thermal stability even during aging at 1400 °C [45]. Currently, thermal cycling tests of these materials are under progress.

4.2 NEW CHEMICAL COMPOSITIONS

A selection of thermal barrier coating materials based on new chemical compositions will be given in this section. Certainly, the list of materials cannot be complete due to the fact that only part of the work is published.

One approach is to substitute the zirconia by the chemically quite similar hafnia. Results of investigations on an Y_2O_3 doped hafnia are given in [46, 47], their performance was promising, but an improvement compared to the best YSZ materials could not be found.

Padture and Clemens measured thermal conductivity in garnets with respect to an application as thermal barrier coating [48]. They found high temperature values between 2.4 and 3.1 W/m/K for the investigated compositions. Garnets with a thermal conductivity clearly below that of YSZ (about 2.0-2.7 W/m/K for stabilised zirconia [20]) have not been found.

Within the last years Japanese researchers investigated the potential of materials based on the CaO-SiO_2-ZrO_2 system for TBC applications [49]. They found superior oxidation behaviour up to 1100 °C compared to YSZ, on the other hand the new coatings were unstable under corrosive environment.

Relatively thick mullite coatings have been compared with YSZ coatings at temperatures typical for diesel engine applications (<1000 °C). In this application the mullite has advantages [50]. On the other hand, the cyclic behaviour at higher temperatures seems to be not sufficient for an application in advanced gas turbines [51].

At the university in Stuttgart, Germany, Lanthane Aluminate, a material with magnetoplumbite structure, has been investigated as new thermal barrier coating material [52]. The material shows very interesting properties, which are comparable (thermal conductivity), slightly worse (thermal expansion coefficient) or even better (density, Young's modulus) than these of YSZ. The performance of the new material in a thermal barrier coating system, e.g. under cyclic loading, has to be demonstrated in the future.

In Germany also several groups are working on TBCs based on a perovskite structure [53, 37]. Of course, only a few compositions of this large materials class are of interest, especially those with high melting points (> 2000°C). In our institute two perovskites, namely $BaZrO_3$ and $SrZrO_3$, have been investigated. With respect to thermal expansion coefficient and thermal conductivity $SrZrO_3$ is comparable to YSZ, while $BaZrO_3$ is worse. However, it turned out, that both materials do not have sufficient thermal stability for a TBC application [54].

Within the Advanced Turbine System (ATS) program the development of new TBCs have been funded [55]. Also no information on the chemical composition of the new materials are available for the public, it was demonstrated that TBCs made of the new candidate materials could have an improved cyclic behaviour compared to YSZ. This was only achieved after an optimisation of the processing conditions. Hence, the thermo-physical properties of a new composition are not sufficient to identify their complete potential.

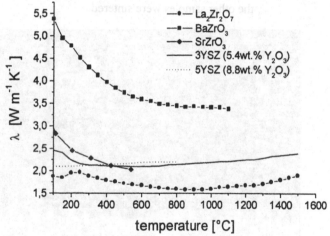

Fig. 4 Thermal conductivity of new thermal barrier coating materials compared with standard YSZ.

Another very interesting candidates for new TBCs are materials with pyrochlore structure [56, 54]. Especially the lanthanum zirconate, $La_2Zr_2O_7$, shows favourable properties for this type of application, e.g. low thermal conductivity (Fig. 4). Its thermal stability has been demonstrated in annealing experiments up to 1400°C [57]. Thermal cycling test of plasma-sprayed coatings revealed promising results even without optimisation of the coating. Further improvements of these materials are now

208

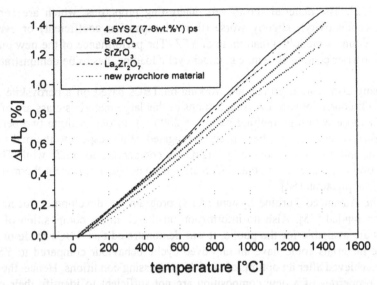

Fig. 5 Thermal expansion coefficient of further improved new TBC material compared to BaZrO$_3$, SrZrO$_3$, La$_2$Zr$_2$O$_7$ and YSZ. The YSZ sample was a plasma-sprayed coating, the other samples were sintered.

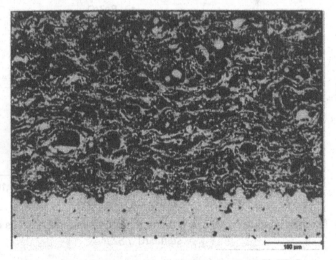

Fig. 6 Microstructure of Metal-Glass Composite (MGC) TBCs prepared by plasma-spraying.

focused on an increase of the thermal expansion coefficient and a further improved chemical stability. First results are very promising with respect to this demands (see Fig. 5). Additionally, a reduction of the thermal conductivity is envisaged.

Finally, a new thermal barrier coating, which is different from the previously mentioned materials, will be described. It is a composite material consisting of a metal and a glass (Metal-Glass-Composite (MGC)) with a low porosity [58, 59]. A typical MGC microstructure is shown in Fig. 6. The thermal expansion coefficient of the MGC is adjusted very closely to the one of the substrate. As a result of the low thermal stresses thermal cycling life times of plasma-sprayed MGC coatings are comparable or even better than these of YSZ. Additionally, the dense coatings enables a protection against oxidation, which results in longer life times in isothermal oxidation tests.

4.3 NANOSTRUCTURED MATERIALS

Since about two decades nanostructured materials become a fast growing field of research [60,61, 62]. Recently, these materials have been suggested as candidates for thermal barrier coatings [63]. An interesting property of these materials is the possible reduction of thermal conductivity due to phonon scattering at grain boundaries. For SiC indeed a dramatic decrease of room temperature thermal conductivity has been observed [64]. An about 30 % decrease was also predicted in nanocrystalline 7 wt.% YSZ of 100 nm grain size in a theoretical work by Klemens and Gell [65]. However, experimental results did not show any significant reduction down to grain sizes of 70 nm in 5.8 wt. % YSZ [66]. These results might be qualitatively understood by the high amount of oxygen vacancies in YSZ, which are effectively scattering phonons and hence, reducing thermal conductivity. A further reduction, e.g. by a high density of grain boundaries, seems to be difficult and only possible at very small grain sizes. A factor of two reduction of thermal conductivity was indeed observed in 10 nm grain-sized YSZ films produced by metal-organic chemical vapour deposition (MOCVD, [67]). Little grain growth occurred below 900 °C. Above this temperature a stabilisation of the structure is required, e.g. by the use of composite structures like YSZ / Al$_2$O$_3$.

It was also proposed to use multilayer nanostructured ceramic TBCs to reduce thermal conductivity [68]. This invention also includes the use of a reflective nanostructured multilayer system, which reflects a part of the incident radiant energy. The radiative contribution to thermal conductivity becomes increasingly important at higher operation temperatures and hence, an effective reduction would be of high significance.

The effectiveness of a so-called phonon scattering zone consisting also of nanostructured system of ceramic multilayers seems to be more critical. Experimental work on multilayers made of YSZ and Al$_2$O$_3$ indicate that the interfaces do no contribute to the thermal resistance [69]. However, the minimum thickness of these layers was 120 nm. Approaching the "real" nanophase regime (<100 nm) might lead to significant effects also for these multilayer systems. On the other hand further decreasing sizes of the structures would increase the problem of stabilisation of these structures.

As a summary of the section on nanostructured materials two points are of special importance:

1) Further work is necessary to clearly demonstrate the effect of nanophase structures especially on thermal conductivity. The investigated structures are probably much finer than 100 nm.

2) It has to be demonstrated that these structures can be stabilised at typical operation temperatures of advanced gas turbines (>1200 °C) for time scales similar to the life times of the engines (> 1000 h).

A further interesting development is the application of quasi-crystalline materials as thermal barrier coating material [70]. The investigated alloy showed a value of thermal conductivity of 2.3 W/m/K which is similar to the value of YSZ. Cyclic oxidation and hot corrosion studies gave promising results. Future work has to be focused on the avoidance of reaction between the substrate and the quasi-crystalline materials by the use of appropriate diffusion barriers. Also sufficient thermal stability of the material has to be demonstrated.

4.4 MULTILAYERS AND GRADED COATINGS

Stresses in ceramic coatings on metals can effectively be reduced by the use of a graded interlayer between the metallic substrate and the ceramic layer [71]. This approach was also applied to thermal barrier coatings applying a graded coating between bond coat and top coat. However, due to the higher surface area of a graded bond coat compared to a step wise compositional change fast oxidation of the bond coat was observed [72].

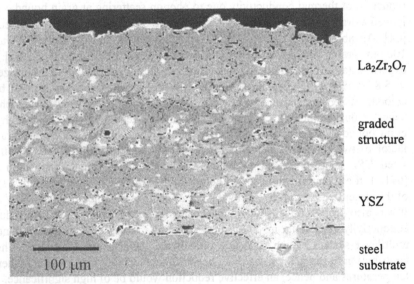

La$_2$Zr$_2$O$_7$

graded structure

YSZ

steel substrate

100 μm

Fig. 7 SEM micrograph of a graded thermal barrier coating consisting of YSZ and La$_2$Zr$_2$O$_7$ ceramic top coats.

Very promising are approaches which use graded or multilayer structures within the ceramic top coat. One possibility is the use of a two layer top coat, in which the ceramic layer in contact with the hot gases is especially resistant against corrosive or erosive attack [73]. Alternatively, it is possible to use newly developed ceramic materials with improved high temperature capability. Often these materials have a lower coefficient of thermal expansion (TEC) or a reduced mechanical strength, which might lead to early failure in thermal cycling tests. Using a standard YSZ layer with a high TEC and good mechanical properties will reduce crack growth in the most critical region close to the bond coat. An even further improvement can be expected from a graded layer between the YSZ and the new ceramic. Both systems have been produced by APS using 8 wt.% YSZ and La$_2$Zr$_2$O$_2$ in our institute. In Fig. 7 the micrographs of these coatings are shown. Thermal cycling tests of these coatings are now under progress. The first results indicate an further increase of thermal cycle life time in the graded structures.

5. Summary

A description of conventional thermal barrier coating systems was given. Special emphasis was put on the limited high temperature capability of these systems. New concepts designed to improve the shortcomings of the standard YSZ material have been described. The presented options included a modification of the stabilising additives in zirconia and the use of completely new compositions. Furthermore, new concepts like nanostructured materials, multilayered or graded coatings were discussed. The variety of new, promising solutions in the field of thermal barrier coatings is very encouraging. Although the list of demands for a new TBC is large, it is expected that improved coatings with higher temperature capability will be developed in the near future.

6. Acknowledgement

The authors would like to thank all members of our group working on new TBCs, partially supplying results for the present paper: Dr. X. Cao, Dr. M. Dietrich, V. Verlotski, and H. Lehmann. Special thanks to the technicians K.H. Rauwald and R. Laufs for performing the plasma-spraying experiments and to S. Schwartz for the careful characterization of samples by different techniques. Furthermore, the dilatometer measurements of Dr. F. Tietz, the laser flash measurements of Dr. Schenk and D. Spitzer and the contributions in the field of finite element calculations of G. Kerkhoff are gratefully acknowledged.

7. References

1 Stiger, M.J., Yanar, N.M., Topping, M.G., Pettit, F.S., and Meier, G.H. (1999) *Zeitschrift für Metallkunde* **90**, 1069-1078.
2 Nelson, W.A. and Orenstein R.M. (1997) TBC experience in land-based gas turbines, *Journal of Thermal Spray Technology* **6 (2)**, 176-180.
3 Beardsley M.B. (1997) Thick thermal barrier coatings for diesel engines, *Journal of Thermal Spray Technology* **6 (2)**, 181-186.
4 Arnault, V, Mévrel R. Alpérine, and S. Jaslier, Y. (1999) Thermal barrier coatings for aircraft turbine airfoils: thermal challenge and materials, *La Revue Métallurgie – CIT/Science et Génie des Matériaux* **96 (5)**, 585-597.
5 Miller, R.A. (1987) Current status of thermal barrier coatings – an overview, *Surface Coating Technology* **30**, 1-11.
6 Thornton J. (1998) Thermal barrier coatings, *Materials Forum* **22**, 159-181.
7 Stöver D. and Funke C. (1999) Directions of developments of thermal barrier coatings in energy applications, *Materials Processing Technology* **92-93**, 195-202.

212

8 Hancock, P., and Malik M. (1994) Materials for Advanced Power Engineering Part 1, D. Coutsouradis et al. (eds.), Kluwer Academic Publishers, Dordrecht, p. 658.

9 Stecura, S. (1986) Optimization of the Ni-Cr-Al-Y/ZrO$_2$-Y$_2$O$_3$ thermal barrier system, *Advanced Ceramic Materials* **1 [1]** 68-76.

10 Kool, G.A. (1996) Current and future materials in advanced gas turbine engines, *Journal of Thermal Spray Technology* **5 [1]** 31-34.

11 Kern, T.-U., Schmitz, F., and Stamm, W. (1998) *Stainless Steel World* **10 [8]** 19-27.

12 Bormfield, R.W., Ford, D.A., Bhanu, J.K., Thomas, M.C., Frasier, D.J., Burkholder, P.S., Harris, K., Erickson, G.L., Wahl, J.B. (1998) Development and turbine engine performance of three advanced rhenium containing superalloys for single crystal and directionally solidified blades and vanes, *Journal of Engineering for Gas Turbines and Power* **120** 595-608.

13 Sonderforschungsbereich 561, "Thermisch hochbelastete, offenporige und gekühlte Mehrschichtsysteme für Kombi-Kraftwerke", SFB261, RWTH Aachen, Germany.

14 Fritscher, K., and Leyens, C. (1997) Grenzschichtproblematik und Haftung von EB-PVD-Wärmedämmschichtsysstemen, *Materialwissenschaften und Werkstofftechnik* **28**, 384-391.

15 Bürgel, R., (1998) Handbuch Hochtemperatur-Werkstofftechnik, Vieweg & Sohn Verlagsgesellschaft mbH, Braunschweig, Germany.

16 Levit, M., Berger, S., Grimberg, I., and Weiss, B.-Z. (1994) Microstructure, Phase Composition, and Interface Phenomena in ZrO$_2$-7Y$_2$O$_3$ Plasma-Sprayed Thermal Barrier Coatings, *Journal of Materials Synthesis and Processing* **2 [1]** 11-27.

17 Schulz, U., Fritscher, K., Leyens, C., Peters, M., and Kaysser, W.A. (1997) Thermocyclic Behaviour of Differently Stabilized and Structured EB-PVD Thermal Barrier Coatings *Materialwissenschaft und Werkstofftechnik* **28**, 370-376.

18 Rigney, D.V., Viguie, R., Wortman, D.J., and Skelly, D.W. (1997) PVD thermal barrier coating applications and process development for aircraft engines, *J. of Thermal Spray Technology* **6 [2]**, 167-175.

19 Schulz, U, and Schmücker, M.(2000) Microstructure of ZrO$_2$ thermal barrier coatings applied by Eb-PVD, *Materials Science and Engineering* **A 276**, 1-8.

20 Raghavan, S., Wang, H., Dinwiddie, R.B., Porter, W. D., and Mayo, M. (1998) The effect of grain size, porosity and yttria content on the thermal conductivity of nanocrystalline zirconia, *Scripta Materialia* **39 [8]** 1119-25.

21 Bose, S., and DeMasi-Marcin, J. (1997) Thermal barrier coating experience in gas turbine engines at Pratt&Whitney, *J. of Thermal Spray Technology* **6 [1]** 99-104.

22 Siebert, B., Funke, C., Vaßen, R., Stöver, D. (1999) Changes in porosity and Young's modulus due to sintering of plasma sprayed thermal barrier coatings *Journal of Materials Processing Technology* **92-93** 217-223.

23 Yonushonis, T.M. (1997) Overview of thermal barrier coatings in diesel engines, *Journal of Thermal Spray Technology* **6 [1]** 50-56.

24 Winkler, M.F., and Parker, D.W. (1992) *Advances in Materials Processes* **5** 17-22.

25 R. A. Miller (1984) Oxidation-Based Model for Thermal Barrier Coating Life ; *J. of the American Ceramic Society* **67 [8]** 517-521 .

26 Sergo, V., and Clarke, D.R. (1998) Observation of subcritical spall propagation of thermal barrier coating, *J. of the American Ceramic Society* **81 [12]** 3237-42.

27 Kerkhoff, G., Vaßen, R., Stöver, D. (1999) Numerically calculated oxidation induced stresses in thermal barrier coatings on cylindrical substrates, *European Federation of Corrosion Publications* **27** 373-382.

28 Quadakkers, W.J., Tyagi, A.K., Clemens, D., Anton, R., and Singheiser, L., (1999) The significance of bond coat oxidation for the life of TBC coatings, *Elevated Temperature Coatings: Science and Technology*, TMS, Warrendale, PA, eds. J.M. Hampikian, N.B. Dahotre, p. 119.

29 Jones, R.L. (1997) Some Aspects of the Hot Corrosion of Thermal Barrier Coatings, *Journal of Thermal Spray Technology* **6, 1** 77-84.

30 Janos, B.Z., Lugscheider, E., and Remer, P. (1999) Effect of thermal aging on the erosion resistance of air plasma sprayed zirconia thermal barrier coating *Surface and Coatings Technology* **113** 278-285.

31 Nicholls, J.R., Deakin, M.J., and Rickerby, D.S. (1999) A comparison between the erosion behaviour of thermal spray and electron beam physical vapour deposition thermal barrier coatings, *Wear* **233-235** 352-361.

32 Miller, R. A., Smialek, J.L., and Garlick, R.G. (1981) Phase stability in plasma-sprayed partially stabilized zirconia-yttria, *Science and Technology of Zirconia*, Advances in Ceramics, Vol. 3, A.H. Heuer and L.W. Hobbs (eds.), The American Ceramic Society, Columbus, OH, USA, pp. 241-251.

33 Funke, C., Siebert, B., Vaßen, R., and Stöver, D. (1998) Properties of $ZrO_2 - 7$ wt. % Y_2O_3 thermal barrier coatings in relation to plasma spraying conditions in Thermal Spray: A United Forum for Scientific and Technological Advances, Proceedings of the United Thermal Spray Conference (15.-19. September 1997, Indianapolis, Indiana), C.C. Berndt (ed.), ASM International, Materials Park, OH, pp. 277-284.

34 Dinwiddie, R.B., Beecher, S.C., Porter, W.D., and Nagaraj, B.A. (1996) The Effect of Thermal Aging on the Thermal Conductivity of Plasma Sprayed and EB-PVD Thermal Barrier Coatings (1996) *The American Society of Mechanical Engineers*, ASME-96-GT-282, New York, USA, 1-8.

35 Goedjen, J.G., Brindley, W.J., and Miller, R.A. (1995) Sintering of plasma-sprayed sol gel zirconia-yttria as a function of silicon content, *Advances in Thermal Spray Science and Technology*, Proceedings of the 8th National Thermal Spray Conference (11-15 September 1995, Houston, Texas), C.C. Berndt, S. Sampath (eds.), ASM International, Materials Park, OH, pp. 73-77.

36 Vaßen, R., Czech, N., Malléner, W., Stamm, W., Stöver, D. Influence of impurity content and porosity of plasma sprayed yttria stabilised zirconia layers on the sintering behaviour, to be published in *Surface and Coatings Technology*.

37 Vaßen, R.,Tietz, F., Kerkhoff, G., Wilkenhöner, R., Stöver, D. (1998) New materials for advanced thermal barrier coatings, Proceedings of the 6th Liège

Conference, Part III, Materials for Advanced Power Engineering, eds J. Lecomte-Beckers, F. Schubert, P.J. Ennis, Forschungszentrum Jülich GmbH, Jülich, Germany, 1627-1635.

38 Bratton, R.J., and Lau, S.K. (1981) Zirconia thermal barrier coatings, in Advances in Ceramics, Vol.3, Science and Technology of Zirconia, eds. A.H. Heuer, L.W. Hobbs, pp. 226-253.

39 Jones, R.L., Reidy, R.F., and Mess, D. (1996) Scandia, yttria stabilized zirconia for thermal barrier coatings Surface Coating and Technolgy 82 70-76.

40 R.L. Jones, and D. Mess (1996) Improved tetragonal phase stability at 1400°C with scandia, yttria-stabilized zirconia, Surface and Coating Technologies 86- 87 94-101 .

41 Harmsworth, P.D., and Stevens, R. (1991) Microstructure and phase composition of ZrO_2-CeO_2 thermal barrier coatings, Journal of Materials Science 26 3991-3995.

42 Thornton, J., Majumdar, A., and McAdam, G. (1997) Enhanced cerium migration in ceria-stabilised zirconia, Surface Coatings and Technology 94-95 112-17.

43 Wilden, J., Wank, M., Steffens, H.D., and Brune, M. (1998) New thermal barrier coating system for high temperature applications, Proc. of the 15th International Thermal Spray Conference, 25-29 May, 1998, Nice, France.

44 Kim, D.-J. (1990) Effect of Ta_2O_5, Nb_2O_5, and HfO_2 alloying on the transformability of Y_2O_3-stabilized tetragonal ZrO_2, J. of the American Ceramic Society 73, 1 115-120.

45 R.G. Rhangavan, private communication.

46 R.A. Miller, "Thermal barrier coatings for aircraft engines: history and directions" Journal of Thermal Spray Technology, 6 [1] 35-42 (1997).

47 D. Zhu, and R.A. Miller (1998) Sintering and creep behaviour of plasma-sprayed zirconia and hafnia-based thermal barrier coatings, Surface Coatings and Technologies 108-109 114-20 .

48 Padture, N.P., and Klemens, P.G. (1997) Low thermal conductivity in garnets, J. of the American Ceramic Society 80 [4] 1018-20.

49 M. Yoshiba, K. Abe, T. Aranami, Y. Harada (1996) High-temperature oxidation and hot corrosion behavior of two kinds of thermal barrier coating systems for advanced gas turbines, Journal of Thermal Spray Technology 5 [3] 259-268 .

50 Pierz, P.M. (1993) Surface Coatings Technology 61 60-66.

51 Ramaswamy, P., Seetharamu, S., Varma, K.B.R., Rao, K.J. (1998) Thermal shock characteristics of plasma sprayed mullite coatings, Journal of Thermal Spray Technology 7 [4] 497-504.

52 Friederich, C.J., Gadow, R., Schirmer, T. (2000) Lanthane Aluminate – a new material for atmospheric plasma spraying of advanced thermal barrier coatings, to be published in the Proceedings of the International Thermal Spray Conference And Exibition, 8-11 May 2000, Montréal, Québec, Canada.

53 MaTech-Programm "Wärmedämmschichten für Oberflächentemperaturen > 1300 °C".

54 Vassen, R., Cao, X., Tietz, F.,Basu, D.,Stöver, D. Zirconates as New Materials for Thermal Barrier Coatings, to be published in the *J. of the American Ceramic Society*.

55 Subramanian, R., Sabol, S.M., Goedjen, J., and Arana, M. (1999) Advanced thermal barrier coating systems for the ATS engine, 1999 ATS Review Meeting, Nov. 8-10, 1999.
see http://www.fetc.doe.gov/publications/proceedings/99/99ats/4-6.pdf

56 Vaßen, R., Cao, X., Tietz, F., Kerkhoff, G., and Stöver, D. (1999) $La_2Zr_2O_7$ - a new candidate for thermal barrier coatings, Proc. of the United Thermal Spray Conference, 17.-19.3.99, Düsseldorf, Hrsg. E. Lugscheider, P.A. Kammer, Verlag für Schweißen und Verwandte Verfahren, Düsseldorf, 1999, p. 830-834.

57 Cao, X.Q., Vassen, R., Schwartz, S., Jungen, W., Tietz, F., and Stöver, D. Chemical and thermal stability of lanthanum zirconate plasma-sprayed coatings, submitted to the J. of the American Ceramic Society.

58 Vaßen, R., Cao, X., Verlotski, V., Lehmann, H., Dietrich, M., and Stöver, D., Two new candidates for thermal barrier coatings, to be published in *Surface and Coatings Technology*.

59 Verlotski, V., Stöver, D., Buchkremer, H.P., and Vaßen, R. Wärmedämmende Glas-Metall/Keramik-Schichten, German patent 198 52 285, Date of Patent: May 3th, 2000.

60 Siegel, R.W. (1994) Synthesis, characterization, and properties of nanophase ceramics, *Encyclopedia of Applied Physics* **11** 173.

61 Mayo, M.J., Hague, D.C., and Chen D.-J. (1993) Processing naoncrystalline ceramics for applications in superplasticity, *Materials Science and Engineering* **A 166** 145.

62 R. Vaßen, D. Stöver (2000) Processing and properties of nanophase non-oxide ceramics, to be published in *Scripta Materialica*.

63 see e.g. Melissa Stewart´s (Renssellaer Polytechnic Institute in Try, NY) webpage entitled "Using nanostructured materials in thermal barrier coatings for aircraft gas turbines engines":
http://www.eng.rpi.edu/dept/materials/courses/nano/stewart/indey.html

64 Vaßen, R., and Stöver, D. (1999) Manufacture and Properties of Nanophase SiC *J. of the American Ceramic Society* **82 [10]** 2583-93.

65 Klemens, P.G., and Gell, M. (1998) Thermal conductivity of thermal barrier coatings, *Materials Science and Engineering* **A245** 143-149.

66 Raghavan, S., Wang, H., Dindiddie, R.B., Porter, W.D., and Mayo, M.J. (1998) The effect of grain size, porosity and yttria content on the thermal conductivity of nanocrystalline zirconia, *Scripta Materialia*, **39, 8** 1119-1125.

67 Eastman, J.A., Choi, U.S., Soyez, G., Thompson, L.J., and Dimelfi, R.J. (1999) Novel thermal properties of nanostructured materials, *Materials Science Forum* Vols. 2-6 629-634.

68 Courtright, E.L., Johnson, R.N., and Bakker, W.T. (1999) Multilayer nanostructured ceramic thermal barrier coatings, United States Patent 5998003, date of patent: 7.12.1999.

69 An, K., Ravichandran, K.S., Dutton, R.E., and Semiatin, S.L. (1999) Microstructure, texture, and thermal conductivity of single-layer and

multilayer thermal barrier coatings of Y_2O_3-stabilized ZrO_2 and Al_2O_3 made by physical vapor deposition, *J. of the American Ceramic Society* **82, 2** 399-406.

70 Sánchez, A., Garcia de Blas, F.J., Algaba, J.M., Alvarez, J., Vallés, P., Garcia-Pogio, M.C., and Guero, A. (1999) Application of quasi-crystalline materials as thermal barriers in aeronautics and future perspectives of use for these materials *Materials Research Society Proceedings* **553** 447-458.

71 Teixeira, V., Andritschky, M., Stöver, D. (1999) Modelling of thermal residual stresses in graded ceramic coatings *Materials Science Forum Vols.* **308-311** 930-935.

72 Alaya, M., Oberacker, R., and Diegle, E. (1995) Aspekte zum thermozyklischen und oxidativen Verhalten von plasmagespritzten Duplex- und zusammensetzungsgradierten ZrO_2 - Wärmedämmssystemen, *Fortschrittsberichte der deutschen keramischen Gesellschaft* **10** 30-40.

73 Bornstein, N.S., and Zatorski, R.F. (1999) Method of applying an overcoat to a thermal barrier coating and coated article European Patent EP 0937 787 A1, date of publication: 25.8.1999.

THERMAL EXPANSION AND WEAR PROPERTIES OF CENTRIFUGALLY PRODUCED Al-Si/SiC MMCs

H. AKBULUT and M. DURMAN
Sakarya University, Metallurgy and Materials Engineering,
Esentepe Campus, 54187, ADAPAZARI-TURKEY

Abstract

Centrifugal casting was used to establish a gradient profile of reinforcement particle densities in a carrier matrix of an Al-Si alloy. Cast hollow cylinders of different volumes and different sizes of SiC particle-reinforced Al-Si/SiC MMCs have been produced. They show a gradient in SiC concentration increasing from a denuded region adjacent to the axis of rotation of the cylinder to a maximum density at the periphery of the cylinder. The CTEs (Coefficient of thermal expansions) decrease in the particle rich sections compared with particle free matrix alloy. An increase of the introduced particle volume results in a decrease of the CTEs. Wear tests show a gradual increase in the mass-based wear rates from the outer section to the denuded region of centrifugally produced MMCs.

Keywords: Centrifugal casting, gradient MMCs, CTEs, wear

1. Introduction

Within the past five years, a family of foundry grade particle reinforced metal matrix composites (MMCs) produced by dispersion process have become increasingly available. Volume loading of reinforcement ranges typically between 5 and 20 percent. At the higher levels of particle loading, the tensile properties and coefficient of thermal expansion can approach those of cast irons [1]. Particle reinforced MMCs that have variable or selective reinforcement have been produced by means of powder metallurgy, spray deposition, and a variety of preform infiltration processes. Each type of processes can yield a controlled distribution of particle reinforcement; but they are not as economically attractive or as readily accessible as the more traditional foundry particles that can be used to produce cast MMC parts having a homogeneous particle distribution [2]. Accordingly, the present investigation fulfills a requirement for a process enabling control over the distribution of reinforcement within casting, which is also applicable to a standard environment.

217

M.-I. Baraton and I. Uvarova (eds.),
Functional Gradient Materials and Surface Layers Prepared by Fine Particles Technology, 217–223.
© 2001 *Kluwer Academic Publishers. Printed in the Netherlands.*

2. Experimental

2.1. PRODUCTION OF MATERIALS

An aluminum-Silicon piston alloy (Al-12 wt % Si, 1.16 wt % Cu, 1.21 wt % Mg, 0.90 wt % Ni) was chosen as a matrix material. Matrix alloy was melted in an electrically heated furnace. The Al-Si melt that superheated to 600 °C, stirred (1000 rpm/min.) with a mechanical stirrer to obtain a vortex in the melt. SiC powders with average grain sizes of 36 μm, 17 μm and 9 μm were added to the vortex at the volume fractions of 5, 10 and 15 % under the protective nitrogen gas atmosphere. The mixture was then overheated to 850 °C. All the mixtures were poured into a cast iron mould, which was rotated centrifugally at a rate of 1600 rpm/min. The cylindrical cast iron mould was heated to 400 °C prior to pouring the composite mixtures. The produced cylindrical samples are 100 mm in diameter and 15 mm thick.

2.2. METALLOGRAPHY AND MICROHARDNESS

The thickness of SiC particle rich layers formed during centrifugal acceleration was observed by an optical microscope. The microhardness of the MMCs samples were determined by means of a LECO M-400 microhardness tester. A Vickers microhardness indenter was used to provide indentation depths from the periphery to the core of the particle free regions. A 500 g load was applied for 15 seconds.

2.3. CTE MEASUREMENTS

For CTE (Coefficient of thermal expansions) measurements the samples were cut by a diamond disk from the outer region of the cylindrical cast parts and then slightly grounded to 25x2x3mm rectangular small pieces. The volume fraction of the SiC particles in the samples whose thermal expansion was studied, was measured via archimedian procedure. Densities used for Al-Si alloy and SiC particles were 2.72 g/cm^3 and 3.22 g/cm^3, respectively. Thermal expansion measurements were made using an Orton dilatometer (Model 1000D). The dilatometer was interfaced with a computer, which was used to calculate and record the experimental data. The thermal expansion of the samples was detected by a LVDT having resolution of ±0.25 % of the LVDT range ±0.318 cm and a repeatability of 65 x 10^{-4} cm. All the thermal expansion measurements were carried out in the temperature interval of 20°C to 500 °C. The coefficient of thermal expansion was calculated from these results.

2.4. WEAR TESTS

Dry abrasion tests at room temperature were conducted on metalographically polished rectangular (3x3 mm) specimens that were cut from the cylindrical cast parts. The flat end of rectangular samples were pressed against the plane surface of a rotating disc, carrying bonded abrasive SiC paper whose mean abrasive particle size was 38 μm. The applied loads were chosen as 30 N, 60 N and 90 N. The specimens were allowed to move to-and-fro against the abrasive contained wheel that rotated at a speed of 0.41

ms^{-1}. The wear tests were performed in two cycles corresponding a total sliding distance of 100 m for each run. An electronic balance having an accuracy level of 0.01 mg was used for weighing samples.

3. Results

Figure 1 is a photomicrograph showing cross-sectional view of the centrifugally cast composites. As can be seen from the Figure, SiC particles having more density than the liquid aluminum-silicon alloy were moved from the inner region of the cylindrical part

TABLE 1. Real SiC particle volume fraction, experimental and theoretical CTEs (10^{-6}°C^{-1}) obtained from the centrifugally produced MMCs.

Material	SiC Vol.%	Exp. 25-250°C	Exp. 250-500°C	Exp. 25-500°C	ROM	Kerners' Model	Turners' Model
Al-Si	-	21.73	24.96	23.42	-	-	
Al-Si(36 μm SiC)	18	18.97	20.58	19.92	20.03	19.09	13.96
Al-Si(36 μm SiC)	32	15.64	17.15	16.42	17.42	16.09	10.57
Al-Si(36 μm SiC)	42	13.51	15.67	14.71	15.55	14.11	08.98
Al-Si(17 μm SiC)	19	18.17	19.89	18.99	19.85	18.87	13.65
Al-Si(17 μm SiC)	31	15.24	16.87	16.15	17.60	16.30	10.76
Al-Si(17 μm SiC)	40	13.86	15.82	14.91	15.92	14.50	09.27
Al-Si(9 μm SiC)	20	17.91	19.78	18.82	19.66	18.65	13.35
Al-Si(9 μm SiC)	33	15.47	16.70	16.04	17.23	15.89	10.39
Al-Si(9 μm SiC)	44	13.42	14.17	13.83	15.17	13.72	08.72

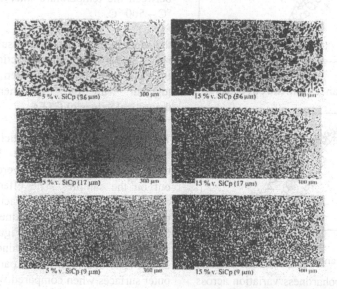

Figure 1. The cross-sectional view of centrifugally produced MMCs showing the gradient particle distribution in the 5 vol. % and 15 vol. %SiC reinforced materials.

220

to the outer section. Micrographs show that Al-Si/SiC particle reinforced MMCs have been produced with a gradient of SiC increasing from denuded region adjacent to the axis of rotation of the cylindrical cast part to a maximum density at the periphery of the cylinder. Figure 1 also reveals a homogeneous particle distribution and no important particle segregation and porosity even in the MMCs containing 9 μm particle. As it can be clearly seen from the micrograph, all the centrifugally produced MMCs present a very high amount of SiC content at the outer regions. Table 1 shows that the particle volume percentage were definitely higher than the introduced percentages. For example, the 36 μm size and 5 vol. % SiC added material had 18 vol. % of the particle density at the periphery zone. It becomes 42 vol. % in the 15 vol. % SiC introduced material with

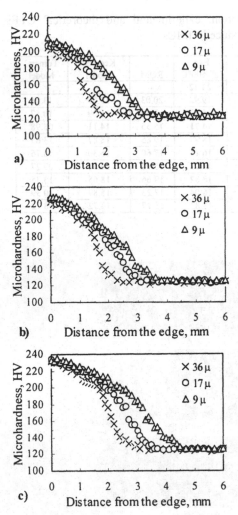

a)

b)

c)

Figure 2. Microhardness variation across the cross-sections; a) 5 vol.% SiC, b) 10 vol.% SiC and c) 15 vol. % SiC addition.

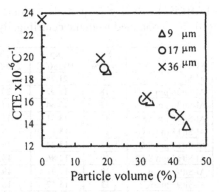

Figure 3. The relationship between particle volume and CTEs measured between the temperature intervals of 25 °C – 500 °C.

the same particle size. The results of the real SiC particle volume fraction at the SiC rich region are presented Table 1 together with the experimental and theoretical values of CTEs.

To evaluate the effects of particle volume fraction and size on the hardness, microhardness transverses were carried out on the samples. The effect of the added particle volume fraction and particle size on the microhardness of Al-Si alloy is illustrated in Figure 2. A significant increase in the hardness, up to 234 HV, is observed in the particle rich outer surfaces when compared with about 125 HV in the unreinforced alloy. The introduction of a higher volume fraction of SiC particles and smaller SiC particle size also resulted in high hardness values. Even at the lowest percentage of particle addition (5 vol. % becoming 18-20 vol. % because of centrifugal acceleration), hardness values as high as 205 HV were obtained. Addition of 15 vol. % of 9 μm SiC particles produced 40-44 vol % particle volume in the surface and maximum surface hardness values around 230 HV. It can be clearly seen from the Figure 3 that increasing volume fraction of SiC particles resulted in the thickness increase of the particle rich section.

A comparison of the experimental and the theoretical CTE values obtained by using the rule of mixtures (ROM), Turner's and Kerner's equations are presented in Table 1. The experimental values agree better with Kerner's model than with the Turner's and rule of mixture models. This is not unexpected since Kerner's equation is closer to a rule of mixture approximation, and the constraint term is small.

Figure 4 presents results obtained during the wear tests in the case of the unreinforced Al-Si alloy and centrifugally formed Al-Si-SiC composite samples against 38 μm SiC abrasive paper at loads of 30 N, 60 N and 90 N. For brevity sake, only the results on 17 μm sized SiC particle reinforced materials are given as example. In the Figure 4, the relationship between particle size, particle volume and applied normal load as a function of sliding distance is presented. As shown in Figure 4a, the mass-based wear rates of unreinforced Al-Si alloy showed almost linear change with sliding distance and a slight decrease is observed in the wear rates with sliding distance. It can be seen from the Figure 4 that the wear rates decrease with increasing particle volume percent. Since the SiC particle volume decreases through to inner part of the samples, the wear rates increase with increasing sliding distance. Parallel to the microhardness results, the concentration increase of introduced SiC particles into Al-Si matrix resulted in an increase of the thickness of the reinforced zone. This caused a decrease of the wear

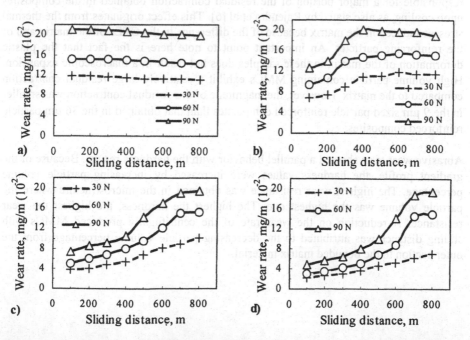

Figure 4. The mass-based wear rates with sliding distance: a) Al-Si unreinforced alloy,

rates. Since the particle volume was gradually decreased from the periphery to the inner part of the samples in the centrifugally produced MMCs, the wear rates of the samples were gradually increased moving from the outer (periphery) section to the inner regions of the materials.

4. Discussion

In typical static castings, the development of a dendritic structure during solidification rejects the SiC particles to the interdendritic regions and result in a non-uniform local distribution of reinforcement particles [3, 4]. The centrifugal cast MMC cylinders showed a markedly different structure at increasing particle densities with a suppressed dendritic structure and increasing particle incorporation in α-Al regions. The cylinders that were centrifugally cast to develop a functionally gradient profile of particle loading showed the progression of dendritic suppression with increasing particle density. Towards the center of the cylinders, the SiC particles were segregated to the interdendritic regions. As the particle density increases, the arm spacing was gradually suppressed. The gradient structure and the uniform distribution in the highly densified regions of the centrifugally cast MMC cylinders suggests a greater resistance to crack propagation than that observed in static castings [5].

The CTE tests that were carried out in unreinforced alloy and MMCs showed that gradient composites showed a decreased CTE values compared with unreinforced alloy. For determining matrix related effects the unreinforced alloy was tested. The expansion curve for the unreinforced matrix alloy is compared with those of the particle reinforced composites. Results showed that unreinforced alloy did not start deforming plastically until a temperature of 500 °C was reached. The soft matrix was thought to be responsible for a major portion of the residual contraction obtained in the composites upon cooling, as also stated by Rajendra et al [6]. This effect originates from the thermal stresses induced in the matrix because of the difference in the CTEs of the matrix and of the reinforcing particles. An important point to note here is the fact that the plastic deformation of the matrix in these samples does not result in a macroscopic expansion. High volume particle containing MMCs exhibit a relatively large residual contraction compared to the matrix. However, the magnitude of this residual contraction was smaller in the 9 μm sized particle reinforced composites than that obtained in the 36 μm particle reinforced composites.

Abrasive wear tests showed a parallel behavior with the hardness results. Because of the gradient profile, the hardness values were increased by increasing particle volume percentage. The highest wear resistance was obtained in the microstructure where the particle volume was the highest one. The highest the hardness, the highest the wear resistance. A reduction on the wear rate of the centrifugally produced MMCs with sliding distance was attributed to the decreasing particle volume percentage from the outer section to the denuded matrix material.

5. Conclusions

1. A gradient profile of SiC particles has been obtained in the Al-Si matrix alloy by applying 1600-rpm centrifugal acceleration. The settling profile upon solidification of the matrix shows a ring of high SiC particle density at the outer edge.
2. Uniform distributions of SiC particles have been obtained in the particle rich region without macro-porosity and particle agglomeration.
3. The gradient profile of SiC particles results in high hardness at the periphery of the cast cylinders and the hardness decreases as the particle volume decreases through the interior of the cast MMCs
4. The incorporation of ceramic particles into an Al-Si matrix results in lowering of the CTEs in the ring of SiC rich outer region. This is expected to improve the dimensional and relaxation stability.
5. The plastic deformation in the matrix must be taken into account in any application involving exposure to elevated temperatures as it can affect the properties of the composite.
6. Kerner's model appears to predict the thermal expansion of the particle reinforced composites quite reasonably
7. The wear resistance in the outer section of the cylinder parts has been increased compared with matrix alloy. The wear rates decrease apart from the outer edge of the centrifugally produced MMCs.
8. Results show that centrifugal casting of MMCs leads to hard, thermally stable SiC-rich coating layer without an interface with the substrate.

References

1. Chamle, T.C. (1999) Particulate Field Distribution in centrifugally cast composites, *USA Patent No: 5,980,792*, Nov. 9.
2. Akbulut, H., Üstel, F., and Yilmaz, F. (1999) Thermal plasma coating of metal matrix Composites, Proc. Of the 14th Int. Symposium on Plasma Chemistry, August 2-6 Prague. Czech Republic, Ed. M. Hrabosky, M. Konrad and V. Kopecky, Vol: **4**, 2007-2012.
3. Stefenascu, D.M., Dhindaw, R.K., Kacar I. A., and Moire, A. (1988) Behavior ceramic particles at the solid-liquid metal interface in metal matrix composites, Metall. and Mater. Trans. **19A**, 2847-2855.
4. Rohatgi, P.K., Asthana, R and Das, S. (1992) Solidification, structures, and properties of cast metal-ceramic particle composites, Int. Mater. Rev., **31**, 85-96.
5. Mortensen, A. and Jin, I. (1992) Solidification processing of metal matrix composites, Int. Mater. Rev., **37**, 101-128.
6. Rajendra, U. V., and Chawla, K.K. (1994) Thermal expansion of metal matrix composites. Comp. Sci. & Technol., **50**, 13-22.

TOPICAL PROBLEMS IN THE THEORY OF TECHNOLOGICAL PROCESSES FOR THE NANOSTRUCTURED MATERIALS PREPARATION

V. V. SKOROKHOD
Frantsevich Institute of Problems in Materials Science NAS of Ukraine
3, Krzhizhanovsky St., 03142 Kiev, Ukraine

1. Introduction

In the great diversity of technological processes suitable for manufacturing of nanostructured materials, the technology of small particles plays the most important role. This technology is close to conventional powder metallurgy (including the technology of high-temperature oxide and non-oxide ceramics), however when nanosized particles are used they exert a great effect on the main consistencies in the evolution of a disperse system on a way from the ensemble of free particles to the monolithic nanostructured material [1].

The following three main stages in the technology of small particles might be highlighted:

- initial packing of particles by precipitation in liquid or gaseous medium under action of gravitation, electrostatic and other fields aimed at obtaining a green body;
- preliminary heating aimed at removing any technological dopants (binder, lubricant, residual liquids etc.) and initial sintering; initial formation of necks between particles makes the green body stronger and more rigid;
- final sintering at sufficiently high temperature, which results in densification of a material and formation of its microstructure in accordance with functions demanded.

The mentioned stages do not differ markedly from those commonly inherent to the technology of powder materials (both metallic and ceramic) especially if slip casting or injection molding are used [2].

Meanwhile, as it was stated above, new phenomena are revealed in ultradisperse powders consisting of nanosized particles. These phenomena inherent to the mentioned stages must be taken into consideration when optimizing the technological process as a whole. When small particles precipitate, the Van-der-Vaals interaction forces between particles essentially result in the particles adhesion and the decrease of the packing density. Such a spontaneous aggregation of small particles plays an important role in formation of mesostructure within the green body.

Heating the ensemble of small particles leads not only to the fast growth of interparticle necks, but also to coagulation and coalescence processes in the ultradisperse system, that can decisively affect the densification kinetics and structure evolution at high-temperature sintering.

225

M.-I. Baraton and I. Uvarova (eds.),
Functional Gradient Materials and Surface Layers Prepared by Fine Particles Technology, 225–237.
© 2001 *Kluwer Academic Publishers. Printed in the Netherlands.*

Finally, high-temperature sintering stage of nanograined and nano-porous body coming from the intermediate stage, has a lot of specific features, such as distinct tendency to local (differential) shrinkage and formation of large pores. In this case the mesostructure formed on the first technological stage becomes important subsequently. At the final sintering stage and at high density of a nanograined body, there is an abrupt dependence of mean grain size on both absolute residual porosity and geometry and topology of pores.

Present paper aims at theoretical treatment of a number of features inherent to the processes taking place in the technology of small particles.

2. Micro-and mesostructure of randomly packed spherical particles.

At the stage of compacting, the structure of porous body can be presented as a system of spatially randomly packed small particles. The simplest model of such a system represents an ensemble of identical spherical particles, the centers of which are randomly arranged in space [3]. The variation of the volume concentration of material in such system corresponds to the variation of the ratio between particle radius and mean interparticle distance (taken between centers). In the framework of this model, each value of the mean relative density corresponds to the definite particle coordination number distribution, herewith the average coordination number becomes close to its most probable magnitude (Fig. 1, [3]). In case of very low relative densities (low volume concentration of particles), this model allows the coordination numbers to be equal to 1 and even 0. The real powder system should be coherent at any density. For the last decade a theory of structure suitable for such kind of systems has been proposed on the basis of general principles of statistical physics [4].

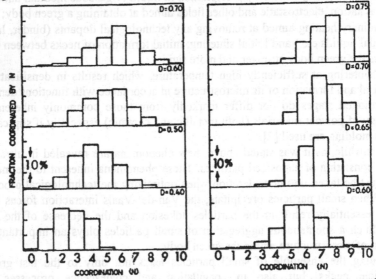

Figure 1. The variation of coordination number distribution with the change of density for the systems with the differential initial density (computer-simulated results).

It should be pointed out that this theory is fair for small particles demonstrating substantial mobility in the particulate medium. As a consequence, the structure of a deposited porous layer can be considered as a result of dynamic equilibrium.

The main result of the system statistical analysis applicable to the ensemble of adjacent contacting small particles consists in the following. The medium of randomly packed particles may as a rule be mesoscopically inhomogeneous. Mesoelements represent clusters with quite high coordination number and high density of particle packing surrounded by a loose shell. The coordination within this loose shell is substantially minor. When changing the macroscopically mean density of the system, the volume fractions (statistical weight) of clusters and loose shells vary as well, whereas average density and coordination number within each of the subsystems varies slowly (Fig. 2, [4]). As it will be shown below, the described mesostructure evolves inhomogeneously in the course of a technological process and can result in undesirable phenomena of local densification and pore coarsening.

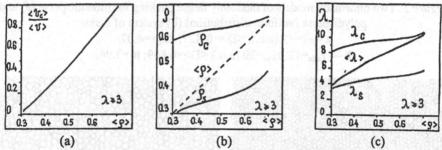

Figure 2. Dependence of relative cluster volume (a), relative packing density in clusters ρ_c and in loose spaces ρ_s (b) and coordination numbers in clusters λ_c and in loose spaces λ_s (c) on the average density of randomly packed spheres.

The real structure of the system of precipitated or deposited nanosized particles substantially depends on forces of interparticle cohesion, which can be interpreted as forces of friction. Experimental study of the micro- and mesostructures in the ensemble of small particles and especially the dynamics of structure formation faces insurmountable problems.

Therefore, researchers more often use a computer modeling of multiparticle system, taking into account as many real parameters as possible. Computer modeling gives a qualitative prediction of substantial features in the behavior of particulate ensemble during packing and further densification of the powder body (layer). Presently, novel results have been obtained in computer modeling of packing of particles with different size distributions (including monodisperse and polymodal systems) taking into account friction between particles [5]. The latter allows one to model the formation of the most dense random packing as well as a more realistic loose packing by varying the parameter of interparticle friction [5]. Today's computer code considers primarily two-dimensional models, although there is first success in modeling of 3D packing [6]. Some results of computer modeling of particulate packing are presented in Figs. 3,4 [5].

In Fig. 3, the most dense random packing of the monodisperse system is compared with the system of particles distributed uniformly in the definite range of sizes. One can

228

see that in case of monodisperse systems the packing is fragmented by clusters with regular structure (the coordination number is equal to 6), divided by boundaries where the coordination number is markedly reduced. The average coordination number in the packing is approximately equal to 4, and average relative density is around 0.827.

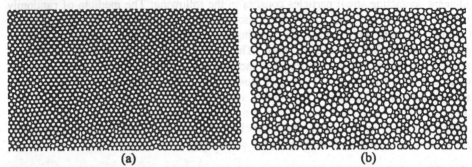

(a)   (b)

Figure 3. Two dimension models of randomly dense packing for monodisperse (a) and polydisperse (uniform distribution) (b) system of disks:

a — r=12 (a.u.), <D>=0,827; <n_c>=4.07.

b — r_{min}=12; r_{max}=20 (a.u.) <D>=0,809; n_c=3.94.

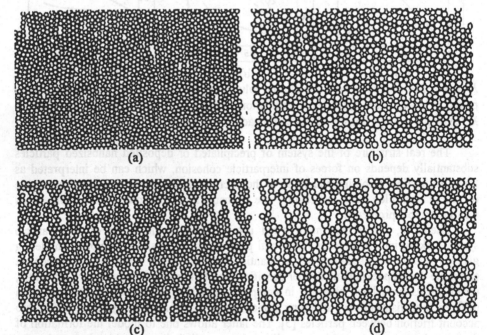

(a)   (b)

(c)   (d)

Figure 4. Two-dimension models of packing of monodisperse (a,b) and polydisperse (c,d) systems with different value of the interparticle friction angle:

a,c — ∠α=0.2: a — <D>=0.738, <n_c>=3.24; c — <D>=0.735, <n_c>=3.28;

b,d — ∠α=0.6: b — <D>=0.594, <n_c>=2.44; d — <D>=0.575, <n_c>=2.43.

In polydisperse system, the structure of random packing is more homogeneous and there are no clusters formed, hereby the average density of packing and the coordination number differs a little compared to monodisperse system. This confirms the validity of well known technological prescriptions. The best homogeneity and density is attainable during sintering of ceramics if narrow particle size distribution is used rather than monodisperse one [7]. Fig. 4 demonstrates the structures of both mono- and polydisperse systems taking into account increase of interparticle friction (cohesion). These structures are quite similar to the real powder fillings or precipitated powder layers.

Thus, the statistical analysis and computer modeling of spatial packing of small particles on the first technological stage might serve instrument for prognosis and optimization of regimes in technological processes and particle size distribution of initial powder.

3. Reduction of surface area in the system of small particles without volume densification.

When heated, the green body compacted from small particles expires the relaxation processes, which substantially reduce free surface energy of particles and proceed without marked variation in density of the particulate system as a whole. These processes are thermally activated and enable to proceed at temperatures sufficient for surface and grain boundary diffusion [8]. The surface area relaxation begins by forming the contact necks and their subsequent growth. R.German proposed that the surface area decrease should be used as a measure characterizing the particulate system at the initial stage of sintering [9]. Let us consider several of the simplest relationships defining the reduction of the surface area in the system of identical spherical particles due to formation and growth of the interparticle necks.

The specific surface area in such system is equal to:

$$S_v = \frac{4\pi R^2}{4/3 \pi R^3} = \frac{3}{R} \left[\frac{cm^2}{cm^3} \right], \tag{1}$$

where R is a radius of a particle.

Formation of interparticle contacts of radius R results in the decrease of the specific surface area:

$$S_v^{(x)} = S_v^0 \left(1 - \frac{n_c}{8} \frac{x^2}{R^2} \right), \tag{2}$$

where n_c is the average coordination number for a given packing of particles, x is the neck radius [8]. The regular and dense packing corresponds to $n_c=12$. In real porous structures, however, n_c can be arbitrary but less than 12 depending on relative density. In very loose packing $n_c=3\div4$. Taking into consideration the fact that the maximum value of relative neck size x/R is close to 0.7 (spherical particles being in dense packing transform to polyhedrons and loose chain-like structure, the chains of particles transform to the cylinder shaped fibers), one can estimate the maximum value of relative decrease in specific surface area due to growth of the necks.

In accordance with the formula (2) one can deduce for the most dense packing (n_c=12) $\Delta S_v^{max} / S_v^{(0)} = 0{,}74$ and for very loose packing (n_c=3) $\Delta S_v^{max} / S_v^{(0)} = 0{,}2$. The kinetics of the surface area decrease due to neck growth is defined by well known relationship like

$$\left(\frac{x}{R}\right)^n = K_1 t, ,$$ (3)

where n=7 for surface diffusion and n=6 for grain boundary diffusion, and kinetic constant K_1 is inversely proportional to R^t [9]. Since

$$\left[\frac{\Delta S_v}{S_v^{(0)}}\right]^\gamma \approx \frac{x^2}{R^2},$$

than taking into consideration the equation (3), the kinetics of surface area decrease will follow the relationship:

$$\left[\frac{\Delta S_v}{S_v^{(0)}}\right]^\gamma = K_2 t,$$ (4)

where $\gamma = n/2 = 3 \div 3{,}5$.

Being elaborated in [9], the analysis of experimental data used in calculation of the specific surface area reduction during isothermal firing of ultrafine oxide systems in oxidizing media shows the validity of the equation (4) in many cases. However, a substantial number of experiments does not correlate with this simple theory: the exponent is very often higher than 3.5 (4 and higher) and the general surface area decrease largely exceeds the value estimated from the above equations for loose packing ($\Delta S_v^{max} / S_v^0 \approx 0.2$).

Such deviations originate from particle coarsening due to coalescence and coagulation. To describe the kinetics of surface diffusion controlled coalescence in the system of small adjacent particles, the author proposed the equation [10]:

$$R^4 = R_0^4 (1 + Bt),$$ (5)

where B is a kinetic constant, which is close to K_1 in (3) for surface self diffusion. It was assumed that the initial particle size distribution following the formula

$$\chi_0 = \frac{R_0^{max} - R_0^{min}}{\overline{R}_0}$$

is quite wide (around 1). Equation (5) is in good correlation with many experimental data obtained for oxide and metallic powders [8]. It appears from this that γ can be equal to 4 in the equation (4).

If the initial particle size distribution is narrow ($\chi_0 \ll 1$), the kinetics of coalescence would be substantially changed. The calculations performed with some simplifying assumptions showed that the kinetics of coalescence is very complex taking into account the initial width of particle size distribution. The general case gives:

$$\frac{1}{4}\left[\left(\frac{R}{R_0}\right)^4 - 1\right] + \frac{1}{3}(1-\chi_0)\left[\left(\frac{R}{R_0}\right)^3 - 1\right] + \frac{1}{2}(1-\chi_0)^2\left[\left(\frac{R}{R_0}\right)^2 - 1\right]$$

$$+ (1-\chi_0)^3\left[\left(\frac{R}{R_0}\right) - 1\right] + (1-\chi_0)^4 \ln\left(\frac{R/R_0 + \chi_0}{\chi_0}\right) = \frac{B}{4}t \tag{6}$$

When χ_0 is small, the logarithmic term predominates and the kinetics of mean particle size growth can be described by S-type curve at the early stages of the process (curve 2 in Fig. 5). As the coalescence is in progress and χ increases, the curve 2 tends to curve 1, corresponding to equation (5), which is a particular case derived from (6) at $\chi_0=1$. Thus, in monodisperse system the initial stage of the coalescence has a behavior similar to that of the chemical reactions with an explicit incubation period [8].

The coalescence in the particulate system does not mean any mobility of the particles and the mass transfer proceeds through the diffusion mechanism i.e. atomically.

In the colloidal systems, the particle coarsening and their corresponding surface area decrease might occur through coagulation. This mechanism assumes substantial mobility of the particles in the surrounding medium, their collisions, and further formation of agglomerates, which gradually transform to coarser particles.

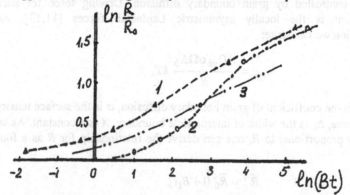

Figure 5. The coalescence kinetics under the different value of χ_0 parameter: 1 — $0.5<\chi_0<1$; 2 — $\chi_0<<1$ (~0.01); 3 — kinetics of coagulation ($B_1<<B$).

The classic equation of kinetic coagulation derived by M. Smolukhovsky may be written here:

$$\frac{dN}{dt} = -KN^2, \tag{7}$$

where N is the number of particles per unit volume;
K is a kinetic constant of coagulation. If particles motion follows the Brawn law

$$K = \frac{4}{3}\frac{kT}{\eta}, \tag{8}$$

where η is the viscosity of the particulate system, k is the Boltzmann's constant, T is the absolute temperature. Integration results in hyperbolic dependence N versus t:

$$N = N_0\left(1 + KN_0 t\right)^{-1}, \tag{9}$$

and the mean particle radius R increases with time as a cubic parabola:

$$\overline{R}^3 = \overline{R}_0^3 + K_1 t \tag{10}$$

Where the new constant K_1 is defined proportional to the volume concentration of particles in the colloidal medium $C_{rol} = \dfrac{4}{3}\pi\overline{R}^3 N$.

If particles are quite mobile, the processes like coagulation should proceed in the system of contiguous small particles. Indeed, these processes are the local sintering, which results in agglomerates consisting of several particles. In this case the reduction of free surface energy occurs due to reorganization of particle external surface to the internal surface of grain boundaries. Nevertheless we can apply the formalism of coagulation kinetics to the description of such a local sintering.

To do that it appears sufficient to substitute the coagulation constant in the equation (7) by the kinetic constant defining the rate of interparticle sliding or rotation, which are controlled by grain boundary diffusion. Driving force for such particle rearrangement is the locally asymmetric Laplacian forces [11,12]. As a first approximation we can write:

$$K_{eff} = \frac{AD_{gb}\sigma\Omega\Delta_b}{\overline{R}} kT, \tag{11}$$

where G_{gb} is the coefficient of grain boundary diffusion, σ is the surface tension, Ω is an atomic volume, Δ_b is the width of interparticle boundary, A is a constant. As soon as K_{eff} is inversely proportional to R, one can derive the relationship for R as a fourth power parabola:

$$\overline{R}^4 = \overline{R}_0^4\left(1 + B_1 t\right), \tag{12}$$

where B_1 is the summarized kinetic constant equal to

$$B_1 = \frac{A_1 D_{gb}\sigma\Omega\Delta_b}{kT} \tag{13}$$

and A_1 is the new numerical constant.

This equation is fair for stationary coagulation because on the earliest stage the rate of interparticle sliding strongly depends on interparticle neck size. In other words we suggest that the specific time necessary to double R is substantially larger than that of neck growth up to the value $x/R \approx 1$. Equation (12) coincides with presented above equation of coalescence kinetics, the kinetic constant B_1 in (12), however, is substantially less than the kinetic constant in (5), as their ratio is proportional to the ratio of grain boundary to the surface diffusion coefficients at the same temperature.

Despite of this fact, the coagulation in the system of ultrafine particles of very narrow size distribution is able to prevail over coalescence for the reasons stated above. It is necessary to note that the surface diffusion controlled coalescence is not accompanied by the variation of mean density of a system (no change of distance between particles), the coagulation like local sintering can lead to some densification of the system as a whole.

Experimental study of the competitive role of both coalescence and coagulation on the second stage of technology is of interest because of surface impurities inevitably inherent to ultrafine media. These impurities behave differently in the two processes. During coalescence it is necessary to expect a substantial refinement of surface. When particles coagulated, the impurities should pinch off and remain on the internal interfaces or grain boundaries.

4. Sintering of ultradisperse systems: densification, evolution of porous and grain structure.

One of the characteristic features of sintering in the ultrafine media is a tendency to local densification or differential shrinkage [8]. Local densification usually becomes apparent as an increase of mean pore size on a stage of open porosity during sintering of micron- and submicron-size particles. On the late sintering stages the homogeneity of densification can be restored and high density in sintered materials can be achieved. Local densification, however is usually irreversible and pores in size of tens or hundreds times larger than the size of particles are formed and cannot be eliminated even if temperature would increase up to near fusion temperatures.

As stated above, in the system of small particles there is a tendency to an inhomogeneous spatial distribution of pores as soon as green packing is prepared. Such inhomogeneity can intensify due to coagulation at the early sintering stages. In addition to formation of mesostructure, the inhomogeneity of density of much larger size than that of mesoelement appears to become possible. Finally, in the systems of polydisperse powders and especially in those with a two-modal particle size distribution, the local inhomogeneity of mean pore size as well as the inhomogemeity of Laplacian pressure become possible [13,14]. On sintering, such inhomogeneity is able to provoke appearance of zones with tensile stresses, resulting in macro-cracking inside the sintered body [7,8]. Thus, the characteristics of initial powder, the methods of green sample compacting, the degree of particle coagulation on heating play a determinant role in reaching the high density material while preserving homogeneous and fine grained structure.

The main principles of choice of the optimal particle size distribution and the best method of particle condensation on compacting have been stated above. Achievement of maximal coordination number at given packing density allows preventing the coagulation process by means of rapid heating in the temperature range where the surface diffusion driving coalescence predominates. This way, it appears to be possible to retain nanosized particles unchanged up to temperatures of intensive densification [14]. All these arrangements, however, are rather necessary than sufficient to obtain fully dense nanocrystalline structure on sintering.

Computer modeling of two-dimensional sintering accomplished by means of particle approach [15] demonstrates that at the same and quite homogeneous packing the local densification can be either negligible (so called "coherent" sintering) or strongly developed. Such behavior is a function of kinetic parameters of the process. Fig. 6 represents a dynamics of densification in the same system and starting from the same initial state by "coherent" path compared with strong localization of densification in the framework of two-dimensional model [15]. In this case the variable parameter is the relative rate of particle approach divided by time of retardation. This is similar to the variation of the time spent by one particle to activate new contacts, which form while density and average coordination number increase. Recently, similar results have been obtained for 3-D computer models [16].

Figure 6. Stricture evolution under sintering (two-dimension computer modeling): a — coherent sintering, b— local densification and formation of macropores.

Let us go back to the problem of optimization of the temperature-time path for sintering. The results of numerous experiments show that the isothermal sintering regime is not optimal for ultrafine systems. It inevitably results in coarsening of microstructure and substantial remnant porosity as large pores. Physical and mechanical properties of such sintered material are markedly decreased. The non-isothermal sintering regime, for instance constant heating rate regime, becomes much better. The main features of non-isothermal sintering have been analyzed in [1,14]. One of them is the independence of densification degree on heating rate at given maximal temperature of heating [1]. Constant heating rate sintering, however, does not lead to a material with nanocrystalline structure. The systematic mercury porosimetry investigations of porous structure evolving in the course of non-isothermal sintering show visible increase of mean pore size and shift the pore size distribution curve to the range of larger radius [17]. The best combination of densification and grain growth gives non-isothermal and non-linear heating rate regime, so called Rate-Controlled Sintering (RCS), which had been first elaborated by H. Palmour III in [18].

Both theoretical basis and experimental results of RCS applications in nanocrystalline bulk materials from ultrafine powders are stated in [1]. In practice the RCS is the sintering regime with variable heating rate depending upon instantaneous density of a material. The optimization of the temperature-time path based on series of specially organized experiments allows the solution of the three main problems of the third technological stage, which is high-temperature sintering: to prevent coarsening of microstructure on the initial heating; to impede undesirable local densification and formation of large pores on the intermediate stage of sintering; to obtain fully dense nanostructured material [1].

The most difficult problem is the suppression of grain growth at the last stage of sintering when porosity becomes less than 10 %. The general scheme of microstructure evolution on sintering is presented in Fig. 7 [19] and shows that the residual pores effectively prevent grain growth while pores have channel-like shape (capillary). When isolated pores formed, the grain growth is sharply intensified. Apparently, under RCS, some favorable conditions appear for retaining intercommunicating pores even at low overall porosity.

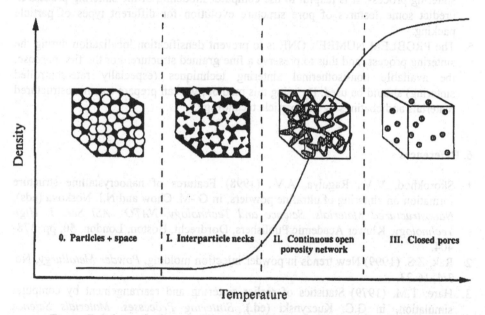

Figure 7. Schematic diagram illustrating the three stages of sintering [19].

Thus, application of RCS allows fully dense nanocrystalline materials to be obtained without external pressure. This technology was successfully used for manufacturing nanocrystalline bulk materials based on nickel, titanium nitride, barium titanate and yttria-doped zirconia [20]. It should be considered as the most prospective for manufacturing surface layers and graded nanostructured materials if ultrafine powders are used as initial product.

236

5. Conclusion

1. Non-uniform packing of particles is typical for highly porous compacts. One of the characteristic features of the random packing of identical particles is its meso-structure that consists of closely packed clusters and loosely packed interclusters space.
2. Computer modeling of the packing process predicts the main features including the effect of interparticle friction and particle size distribution on the compact structure and packed density.
3. Decrease in the surface area of a system of fine particles can be achieved at the preliminary heating stage without its bulk densification, through the processes of coalescence and coagulation. Coagulation dominates at the early stage when packing is loose, and the contact necks are small, especially, for the systems with narrow particle size distribution. The kinetic equations for both processes are similar.
4. The initial structure of green compacts, and the degree of surface area relaxation at the preliminary heating stage plays an important role for the final high-temperature sintering process. It is helpful to use computer modeling of the sintering process to predict some features of pore structure evolution for different types of particle packing.
5. The PROBLEM NUMBER ONE is to prevent densification localization during the sintering process, and thus to preserve a fine grained structure. For the first purpose, the available non-isothermal sintering techniques (especially rate-controlled sintering) should be used. By using this method, we can prepare true nanostructured materials and coatings by fine particle technology.

6. References

1. Skorokhod, V.V., Ragulya, A.V., (1998) Features of nanocrystalline structure formation on sintering of ultrafine powders, in G.-M. Chow and N.I. Noskova (eds), *Nanostructured Materials. Science and Technology, NATO ASI Ser. 3. High Technology*, Kluwer Academic Publishers, Dordrecht, Boston, London, 50, pp. 378-404.
2. Rak, Z.S. (1999) New trends in powder injection molding, *Powder Metallurgy*, No. 3/4, 16-23.
3. Hare, T.M. (1979) Statistics of earlier sintering and rearrangement by computer simulation, in G.C. Kuczynski (ed.), *Sintering Processes. Materials Science Research*, Plenum Press, New York, pp. 77-88.
4. Kovaltchenko, M.S., Nikolenko, A.M. (1994) Statistical Analysis of Structure Features in Powder Composites in Terms of Random Particles Packing, in *Congress Mondial de Metallurgie de Poudres*, Les Editions de Physique, Paris, 3, pp. 2295-2298.
5. Kartuzov, V.V., Kartuzov, E.V., Krassikov, I.V. (1999) Computer generation of two-and three dimensional packing as a background for numerical modeling of sintering processes, *Science of Sintering* 31, No. 3, pp. 157-162.

6. Nurkanov, E.Yu., Kadushnikov, R.M., Kamenin, I.G., Alievski, V.M., Alievski, D.M., Kartashev, V.V. (2000) Study of density properties of three-dimensional random packing of spherical particles using computer modeling, *Powder Metallurgy*, in print.

7. Handwerker, C.A., Blendell, J.E., and Coble, R.L. (1989) Sintering of Ceramics, in D.P. Uskokovic, H. Palmour III., R.M. Spriggs (eds.), *Science of Sintering. New Direction for Materials Processing and Microstructural Control*, Plenum Press, New York, London, pp. 3-24.

8. Skorokhod, V.V., Solonin, Yu.V., and Uvarova, I.V. (1990) Chemical, Diffusion and Rheological Processes in Powder Materials Technology [in Russian], Naukova. Dumka, Kiev.

9. German, R.M. (1978) Sintering Parameter for Sub-micron Powders, *Science of Sintering* 10, No. 1, 11-25.

10. Skorokhod, V.V. (1987) Surface relaxation, dynamics of geometrical structure and macro-kinetics of densification during sintering of ultrafine powders, in G.C. Kuczynski, D.P. Uskokovic, H. Palmour III, and M.M. Ristic (eds.), *Sintering '85*, Plenum Press, New York, London, pp. 81-88.

11. Ashby, M.F., and Verrall, R. (1973) Non-uniform viscous flow of polycrystalline bodies and superplasticity, *Acta Metall.*, 21, No. 2, 53-61.

12. Exner, H.E. (1979) Principles of single phase sintering, *Rev.of Powder Metallurgy and Physical Ceramics* 1, Nos.1-4, 7-251

13. Skorokhod, V.V., Panichkina, V.V., Ragulya, A.V. (1999) Effect of size distribution and heating rate on initial and intermediate sintering stages, *Functional Materials*, 6, No. 2, 215-220.

14. Skorokhod, V.V. (1999) Rapid Rate Sintering of Disperse Systems: Theory, Processing and Problems, *Powder Metallurgy and Metal Ceramics* 38, No7-8, 350.

15. Kadushnikov, R.M., Skorokhod, V.V. (1991) Research of the zonal separation during sintering of powder bodies by computer modeling methods [in Russian], *Poroshk. Metall.*, No.7, 31-37.

16. Kadushnikov, R.M., Skorokhod, V.V., Kamenin, I.G., Alievski V.M., Nurkanov, E.Yu., Alievski, D.M. (2000) Three dimension modeling of sintering of spherical particles, *Powder Metallurgy*, in print

17. Ragulya, A.V., Skorokhod, V.V. (1997) Validity of Rate Controlled Sintering Method for Consolidation of Dense Nanocrystalline Materials, *Proc. of the 14-th Plansee Seminar* 2, pp. 735-744.

18. Palmour III, H. (1989) Rate controlled sintering of ceramics and selected powder materials, in D.P. Uskokovic, H. Palmour III., R.M. Spriggs (eds.), *Science of Sintering. New Direction for Materials Processing and Microstructural Control*, Plenum Press, New York, London, pp. 337-356.

19. Mayo, M.J. (1996) Processing of nanocrystalline ceramics from ultrafine particles, *Int. Materials Reviews* 41, No. 3, 85-115.

20. Andrievski, R.A. (2000) New Superhard Materials Based on Nanostructured High-Melting Compounds: Achievements and Perspectives, in M.-I. Baraton (ed.), *Functional Gradient Materials and Surface Layers Prepared by Fine Particle Technology*, *NATO ASI Ser. 3. High Technology*, Kluwer Academic Publishers, Dordrecht, Boston, London, present issue.

NANOSTRUCTURED CARBON COATINGS ON SILICON CARBIDE: EXPERIMENTAL AND THEORETICAL STUDY

YURY GOGOTSI,[1] VALENTIN KAMYSHENKO,[2]
VLADIMIR SHEVCHENKO,[2] SASCHA WELZ,[3]
DANIEL A. ERSOY,[3] MICHAEL J. MCNALLAN[3]

[1]University of Illinois at Chicago, Department of
Mechanical Engineering, Chicago, IL 60607-7022.
Current Address: Drexel University, Department of
Materials Engineering, Philadelphia, PA 19104,
E-mail: gogotsi@drexel.edu

[2]Institute for Problems of Materials Science,
3 Krzhizhanovskogo St., Kiev, 252134 Ukraine

[3]University of Illinois at Chicago, Department of Civil and
Materials Engineering, 842 West Taylor St., Chicago, IL
60607

Abstract

Nanotechnology has been recognized as an emerging technology of the new century. Control over the structure of materials on nanoscale can open opportunities for the development of nanostructured materials with controlled properties, if the structure/property relations are known. This paper describes a technique that can produce a broad range of potentially important carbon nanostructures that may be used in future technologies. Nanostructured carbon coatings can be obtained either by deposition from the gas phase onto a substrate, or by surface treatment of a carbon-containing substrate. The method presented in this paper is accomplished through the extraction of metals from carbides (SiC and TiC) using chlorine or chlorine-hydrogen mixtures. This is a versatile technology because a variety of carbon structures can be obtained on the surface of carbides in the same reactor. Not only simple shapes, but also fibers, powders and components with complex shapes and surface morphologies can be coated. This technology allows the control of coating growth on the atomic level, monolayer by monolayer, with high accuracy and controlled structures.

In this work, the structure and properties of carbon coatings obtained on the surface of carbides have been investigated using transmission electron microscopy (TEM). Molecular dynamics (MD) was used to model the growth of carbon coatings on SiC and the simulation data are compared with TEM results. The simulation can provide guidance to understand the growth mechanisms and potentially possible carbon structures. Ordered and disordered graphite, nanoporous carbon (specific surface area of >1000 m^2/g) and hard carbon films can be formed on SiC surfaces depending on the

239

M.-I. Baraton and I. Uvarova (eds.),
Functional Gradient Materials and Surface Layers Prepared by Fine Particles Technology, 239–255.
© 2001 Kluwer Academic Publishers. Printed in the Netherlands.

temperature and gas composition. These carbon coatings can be used as tribological coatings having a low-friction coefficient for a variety of applications, from heavy-load bearings to micro-electro-mechanical systems (MEMS); protective coating for sensors and tools, intermediate thin films for further chemical vapor deposition (CVD) deposition of diamond, weak coatings on SiC reinforcements for composite materials, coatings on SiC powders for improved sinterability, catalyst supports, and molecular membranes.

1. Introduction

Carbide ceramics such as SiC, B_4C, WC and TiC find a number of applications because of their high hardness and excellent wear resistance[1]. Since their dry friction coefficient is about 0.6-0.7, better tribological performance can be expected if the ceramic is coated with a carbon film to reduce the friction coefficient. Tungsten carbide tools are often coated with diamond films to reduce the wear. Graphite is widely used as solid lubricant for carbides and other materials or as a counterbody in dynamic seals and other tribological applications[2].

Either CVD or physical vapor deposition (PVD) is used to obtain a variety of carbon coatings ranging from diamond to graphite and amorphous carbon. However, adhesion of vapor deposited coatings is often a problem[3]. Additionally, the size of the coated component increases, which is not always desirable, especially when these objects are thin, e.g., SiC fibers or MEMS devices. It is also difficult to deposit coatings with the thickness in excess of several microns. There is no current deposition technology that allows the producion of uniform carbon coatings within pores in a porous ceramic body, or on whiskers/platelets/fiber preforms without bridging the particles and fibers. Since carbides contain carbon in their structure, it is possible to produce a carbon coating by leaching the metal and leaving the carbon layer behind[4]. Such coating is not deposited onto the surface, but, rather, the surface is transformed into carbon. Therefore, the thickness of the coated component remains the same or decreases slightly. Since the structure and composition of the coating can be controlled with high precision, leaching metal atoms layer by layer is particularly useful for nanostructured coatings[5].

Carbon films can be produced on SiC surfaces by high temperature chlorination[5]. $SiCl_4$ is more thermodynamically stable than CCl_4 at elevated temperatures; thus, chlorine reacts selectively with the Si at SiC surfaces leaving carbon behind:

$$SiC + 2\,Cl_2 = SiCl_4 + C \qquad (1)$$

$$SiC + 2/3\,Cl_2 = SiCl_3 + C \qquad (2)$$

The structure of the carbon layer depends on the temperature and composition of the chlorinating gas mixture. Carbon films have been produced on β-SiC powders[5], as well as SiC based fibers[6], monolithic CVD, and sintered ceramics[7] exposed to Ar-H_2-Cl_2 gas mixtures at atmospheric pressure and temperatures between 600°C and 1000°C.

Previous work showed that after a treatment of β-SiC powder in 3.5% Cl_2-Ar at 600°C, the X-ray diffraction (XRD) pattern was similar to that of as-received powder. After reacting with 3.5% Cl_2-Ar at 800°C and 1000°C, the intensity of β-SiC reflections diminished and they disappeared with increasing reaction time. However, no other peaks were detected. Only long-term treatment of bulk ceramics resulted in some well-crystallized graphite that could be detected by XRD.

The powder treated at 1000°C in Cl_2 had a specific surface area in excess of 1000 m^2/g [5], which is a typical value for nanoporous carbons. Values of 1400 and 1200 m^2/g were reported for carbon produced by chlorination of TiC and SiC respectively [8]. The pore size of 0.8 nm to 2 nm was reported for SiC treated in Cl_2 at 900°C [8].

Carbon films have also been formed on the surfaces of commercially available monolithic SiC specimens by high temperature chlorination at atmospheric pressure in Ar-Cl_2 and Ar-H_2-Cl_2 gas mixtures [7]. The carbon film formed in two layers. The outer layer is a loosely adherent, coarse graphitic carbon. The lower layer, which is strongly adherent to the SiC, is amorphous or nanocrystalline. The kinetics of its growth at 1000°C was linear with time up to thickness in excess of 50 μm. This suggests the chemical reaction controlled process and not a diffusion controlled one [7]. Thus, the carbon film allows penetration of CCl_4 molecules through the layer. Raman spectroscopy and XRD identified it as a highly disordered carbon with characteristics of nanocrystalline graphite. However, the composition of the layer depends on the treatment conditions, such as temperature and presence of hydrogen in the environment. Pin-on-disk tribology testing using a silicon nitride ball showed that the presence of this layer reduced the friction coefficient by a factor of approximately six and the wear rate by more than an order of magnitude [9]. TiC and other carbides can be coated as thermodynamic modeling and experiments show [10,8].

In this work, we present results of transmission electron microscopy (TEM) studies of carbon films produced on α-SiC ceramics and single crystals, as well as a TiC powder, and compare them with the molecular dynamics simulation of carbon growth on α-SiC (6H).

2. TEM Study

The TEM work was performed using a JEOL 3010 microscope with the accelerating voltage of 300 kV, providing a lattice resolution of 0.14 nm (point resolution 0.17 nm). Samples were prepared by dispersing the treated TiC powder (H.C. Stark) in acetone with ultrasound and subsequent deposition on copper grids coated with lacey carbon. The carbon coating on α-SiC 6H single crystals (Cree Research, Inc., n-type, (1000) orientation, 3.5° off axis) peeled off after treatment and was collected on the grid. In case of highly adherent coatings on sintered α-SiC ceramics (Hexoloy™ SA, Carborundum), carbon was scraped from the surface using a sharp tungsten carbide tool.

TEM analysis of carbon films that delaminated from the SiC single crystal after chlorination at 1000°C (Fig. 1) shows their high degree of graphitization. They are thin graphite layers that peel off after cooling to room temperature, probably due to a thermal expansion mismatch between graphite and SiC. As can be seen from Fig. 1, graphite grew parallel to the surface of the crystal. Since bonding between graphene layers in c direction is very weak [11], spallation of the coating is possible.

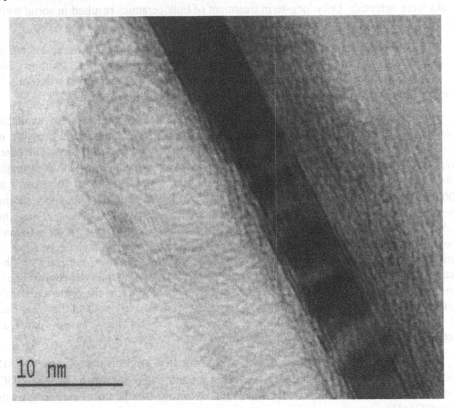

Figure 1. Carbon film that delaminated from the surface of a SiC single crystal treated in Ar+3%Cl$_2$ at 1000°C. The film thickness is less than 20 graphene layers.

TEM analysis of coatings produced on the surface of sintered SiC (Fig. 2) shows the presence of amorphous and mesoporous carbon, as well as ordered planar graphite similar to that produced on the single crystal (compare Fig. 1 and Fig. 2a). The amorphous carbon is built of basic structure units of graphite with very little order in c direction. This metastable amorphous material may crystallize to form the crystalline graphite. The graphitization of amorphous carbon occurs by ordering and stacking of graphite basic structure units due to the sustained heat treatment. However, since self-diffusion rate in carbon is slow, very high temperatures (close to 3000°C) are required for solid-state graphitization. As can be seen in Fig. 2, graphite fringes (0.34 nm spacing) were always accompanied by amorphous carbon. There are many faults in graphite (marked with arrows in Fig 2a). Amorphous carbon structure seen in the lower part of Fig 2b is called mesoporous carbon, which is a suitable material for membranes and electrodes [11].

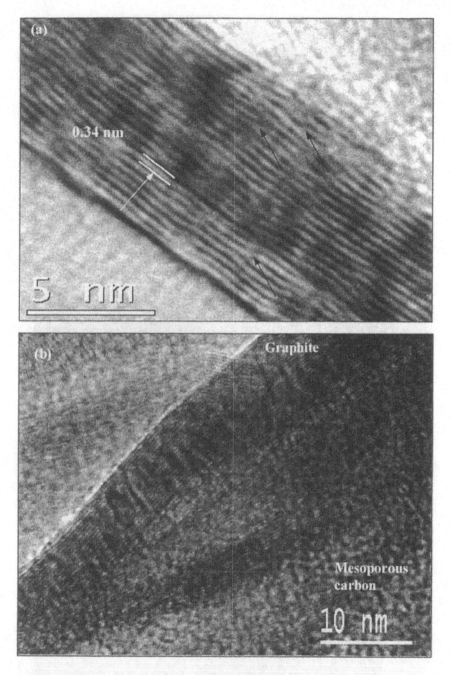

Figure 2. TEM micrographs of carbon coatings produced on sintered α-SiC treated in Ar-3.5%Cl$_2$ at 1000°C, showing graphite fringes and amorphous carbon. Faults in graphite can be seen in (a). Well-ordered graphite and mesoporous carbon are shown in (b).

carbon along with graphite fringes which form a nodule-like cut and onion-like structures (b). The sample was treated with an Ar-Cl$_2$ gas mixture at 1000°C. The carbon onion (marked by arrows) has a diameter of 13 nm.

244

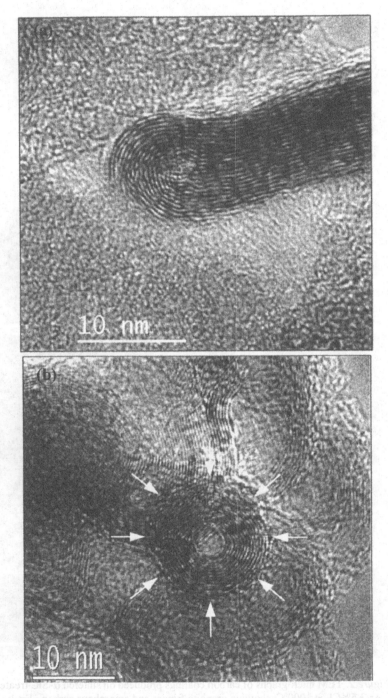

Figure 3. TEM images of carbon coatings on sintered α-SiC after treatment in Ar-3.5%Cl$_2$ for 20 hrs (a) and 24 hrs (b) at 1000°C showing mesoporous carbon along with graphite fringes which form nanotube-like (a) and onion-like structures (b). The sample was treated with an Ar-Cl$_2$ gas mixture at 1000°C. The carbon onion (marked by arrows) has a diameter of 13 nm.

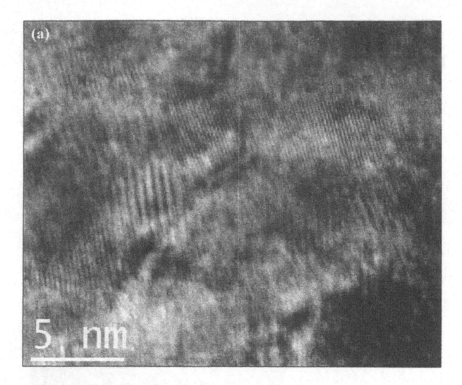

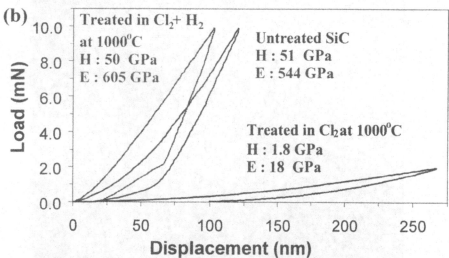

Figure 4. (a) TEM image of nanocrystalline structures with a varying crystal size. (b) Indentation load-displacement curves illustrating the difference between hardness (H) values and Young's modulus (E) of different carbon coatings in comparison with untreated SiC. The sample was treated with an Ar-Cl$_2$-H$_2$ gas mixture for 24 hrs at 1000°C.

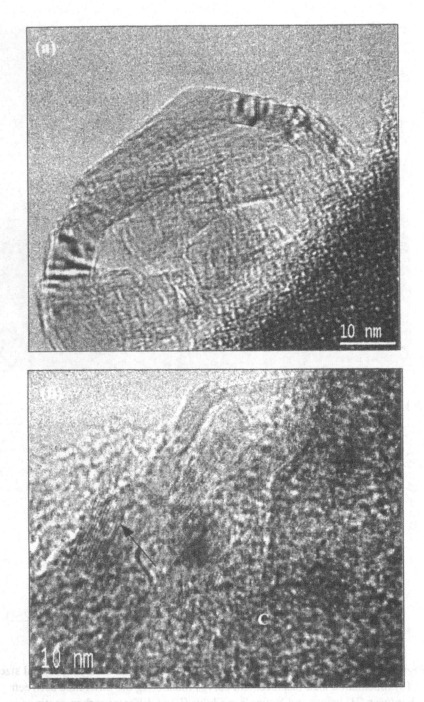

Figure 5. TEM images of carbon produced by treatment of TiC powder for 2 hrs at
800°C in an Ar-Cl₂ gas mixture. Arch-like graphitic structures on the surface and
amorphous carbon can be seen.

Along with planar graphite, bent graphite fringes were observed, which can ultimately result in nanotube-like shapes (Fig. 3a) or carbon onions and ribbons (Fig. 3b). The mechanism of the formation of those structures is still under discussion, but onion-like structures were also observed in carbon produced by chlorination of TiC [8]. Carbon onions were embedded in mesoporous material or present on the surface of the carbon coating. The distance between the successive graphene layers in the onion is ~0.34 nm. They vary in shape from nearly spherical ones to barrel-like. The size of the onion-like structures varies between ~10 nm and ~30 nm, and it appears to be larger on the surface (no constraint for growth) than within the mesoporous carbon layer. The carbon onions are sensitive to the electron irradiation; thus, graphite fringing could be damaged during the TEM observation, if the electron beam intensity was high.

Nanocrystalline carbon regions (Fig. 4a) were found in samples treated with Cl_2-H_2 gas mixtures throughout the layer, and in the lower, strongly adherent layer close to the SiC interface in samples treated with Cl_2. Their structure is composed of small crystallites of 5 to 10 nm in size. It is well known that the presence of hydrogen can stabilize dangling carbon bonds and lead to the formation of diamond or amorphous sp^3 carbon [12]. The sp^3 hybridization allows the material to develop diamond-like properties such as high hardness and chemical inertness. Lattice fringes at d=0.206-0.208 nm that can be assigned to diamond were observed along with other, unidentified fringes. The nanocrystalline material seen in Fig. 4a might be responsible for very high hardness values and Young's modulus determined in indentation tests (Fig. 4b). As can be seen in Fig. 4b, the hardness of the chlorine/hydrogen treated sample was about 50 GPa (and could reach 60 GPa) in contrast to the chlorine treated sample which shows the hardness of 1.8 GPa, close to that of glassy carbon. Values of the Young's modulus are significantly higher for the hydrogen treated material (605 GPa for the given curve and up to 800 GPa in some points) compared to the Young's modulus of the chlorine treated material (18 GPa). The hardness and modulus values of these films are equal to or exceed that of SiC and can be explained by presence of diamond-structured carbon in the film.

TEM analysis of TiC powder treated in an Ar - 3% Cl_2 gas mixture shows that carbon structures similar to that on SiC are formed (Fig. 5). Graphite was present in planar or complex-shape structures. Amorphous and mesoporous carbon was also observed, similar to SiC-derived carbon. There were no carbon onions found in TiC powder treated under these conditions, however treatment at higher temperatures was reported to produce onions in TiC-derived carbon [8].

3. Molecular Dynamics Simulation

Development of coatings with a controlled microstructure requires understanding of factors that control bonding of carbon (sp^2 or sp^3) and properties of the coating. It is also important to understand to what extent the structure of the coating and bonding between the coating and the carbide substrate can be controlled by changing the process variables such as temperature. Recently developed methods of modeling the atomic structure of covalent materials [13] may help to answer these questions. However, these methods have not been previously used to simulate growth of carbon coatings on the surface of carbides.

In this work, we used molecular dynamics to investigate the behavior of carbon layers and clusters on SiC. The interactions between the atoms in the system were described using Tersoff potentials [13,14], in which the potential energy E of the system of interacting carbon and silicon atoms is described as a sum of pairs of atoms

$$E = \sum_i E_i = \sum_{i<j} V_{ij}, \qquad V_{ij} = f_C(r_{ij})\left[f_R(r_{ij}) + b_{ij} f_A(r_{ij})\right]$$

where r_{ij} is bond length between i and j, $f_C(r_{ij})$ is cut-off function which limits the radius of the potential by two nearest neighbors. Functions $f_A(r_{ij})$ and $f_R(r_{ij})$ are repulsive and attracting components of the Morse potential:

$$f_R(r_{ij}) = A_{ij} \exp(-\lambda_{ij} r_{ij}), \qquad f_A = -B_{ij} \exp(-\mu_{ij} r_{ij}).$$

The repulsive component of the potential has a coupled character, while the attracting component is modified by a multi-component function b_{ij}, which is introduced to account for the local surrounding of atom i. Values of parameters of the Tersoff model for the system with Si-C, Si-Si and C-C interactions were taken from [4]. Chlorine and hydrogen were not included in this simulation, because no reliable potentials are available.

α-SiC 6H substrate was taken as a plate with (1000) orientation and thickness equal to the lattice parameter in [1000] direction. The periodical boundary conditions along the (1000) plane corresponded to 5-6 lattice parameters. The same conditions were applied to additional carbon atoms placed on the surface of the substrate. The reactive surface of the 6H-SiC substrate consisted of silicon atoms (silicon-terminated surface). Each atom was bonded to three carbon atoms underneath. The total number of atoms in each of the calculated systems was about 500. Carbon layers with different crystallographic structures were placed at the substrate, and the system was allowed to evolve at different temperatures until no visible changes in the structure were observed (this usually takes from 10^{-11} to 10^{-9} seconds, with the characteristic MD step being 10^{-15} seconds).

To verify the selected approach, a layer of carbon atoms on the Si-surface of (1000) 6H-SiC was calculated such that every carbon atom has only one bond with a Si atom underneath. The results of modeling for T=1000 K and T=2000 K are shown in Fig. 6. Bond length and energy, and bond angles of the atoms marked with arrows are given in Table 1. The obtained results are in good agreement with the published experimental and simulation data[15,16]. In addition, they show a tendency to formation of chains of carbon atoms, which increases with temperature. Formation of chains instead of hexagonal rings or sp^3-bonded clusters may lead to the growth of an amorphous network instead of well-ordered carbon structures.

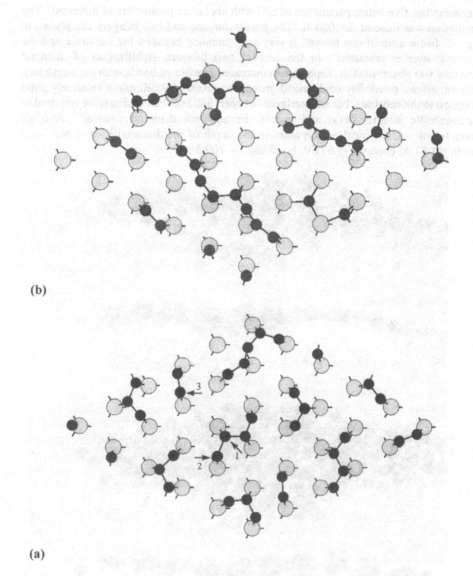

Figure 6. Chains of carbon atoms (black circles) on the Si-surface of 6H-SiC at 1000 K
(a) and 2000 K (b). The figure surface is parallel to (1000) SiC surface. Only the upper
(Si, light circles) layer of the SiC substrate is shown.

For evaluation of the possibility of diamond growth on SiC, flat diamond
layers were placed onto (1000) Si-surface of 6H-SiC by such a way that the orientation
of the diamond crystal coincided with that of SiC (epitaxial growth). Under a "bilayer",
we understand a pair of atomic layers in which each carbon atom has only one free
bond. Diamond layers were compressed in the layer direction for about 2% to

accommodate five lattice parameters of SiC with six lattice parameters of diamond. The simulation was done at T=1000 K. The results for one and two bilayers are shown in Fig. 7. In the case of one bilayer, a very poor bonding between the substrate and the growing layer is predicted. In the case of two bilayers, stabilization of diamond structure was observed (flat shape and an increased number of bonds with the substrate). This suggests a possibility of diamond growth on (1000) 6H-SiC and a relatively good adhesion to the substrate, but an interlayer between SiC and diamond may be required to accommodate internal stress and enable the adherent diamond coating. Average characteristics of a typical carbon atom in the depth of the diamond bilayer are: bond length – 1.51 Å, energy – -3.63 eV, bond angle – 109.5°.

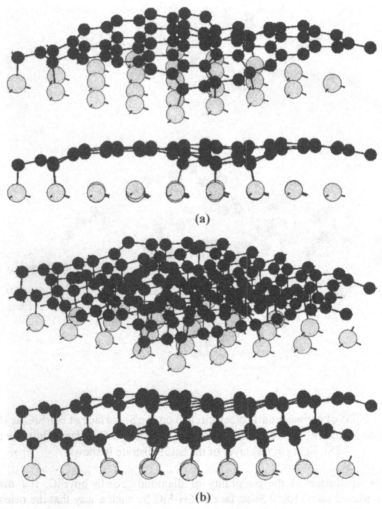

(a)

(b)

Figure 7. One bilayer- (a) and two bilayers (b) of diamond on the Si-surface of (1000) 6H-SiC after annealing at 1000 K (shown in two projections). Only the upper layer of the SiC substrate is shown. Dark circles show carbon atoms and light circles are Si.

TABLE 1. Bond length, energy and bond angles for atoms marked in Fig. 6.

C Atom No.	Bonds				Bond angles	
	No.	Atom	Length, Å	Energy, eV	Bonds	Angle, degrees
1	a	C	1.60	-4.12	a-b	102.4
	b	Si	1.99	-2.69	a-c	110.6
	c	C	1.53	-3.83	b-c	101.2
2	a	Si	1.89	-3.37	a-b	117.3
	b	C	1.60	-4.12		
3	a	Si	1.80	-3.40	a-b	119.8
	b	C	1.41	-4.70		

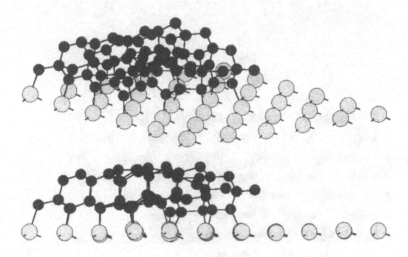

Figure 8. Diamond cluster on the (1000) Si-surface of 6H-SiC annealed at 1000K (shown in two projections). Only the upper layer of the SiC substrate is shown. Dark circles show carbon atoms and light circles are Si.

TABLE 2. Bond length, energy and bond angles for atoms marked in Fig. 9.

C Atom No.	Bonds				Bond angles	
	No.	Atom	Length, Å	Energy, eV	Bonds	Angle, degrees
1	a	Si	1.98	-4.12	a-b	95.4
	b	C	1.80	-2.69	a-c	94.7
	c	C	1.81	-3.83	a-d	96.6
	d	C	1.74		b-c	116.0
					b-d	122.0
					c-d	119.3
2	a	C	1.73	-3.61	a-b	114.4
	b	C	1.80	-3.39	a-c	121.4
	c	C	1.81	-3.30	b-c	114.5

252

Similar to diamond films, diamond clusters on SiC were simulated. Flat diamond layers were placed onto (1000) Si-surface of 6H-SiC by such a way that the orientation of the diamond crystal coincided with that of SiC (epitaxial growth). The simulation was done at T=1000 K. The results for two bilayers (Fig. 8) demonstrate a good adhesion to the substrate and maintained sp^3 coordination of carbon atoms in the cluster.

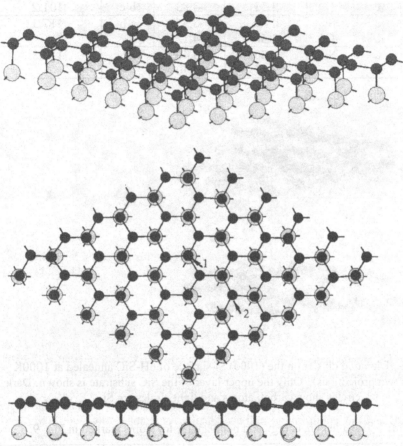

Figure 9. A layer of hexagonal packing of carbon atoms on the Si-terminated (1000) surface of 6H-SiC shown in three different projections. Only the upper layer of the SiC substrate is shown. Dark circles show carbon atoms and light circles are Si.

We have also considered formation of a hexagonal carbon network as a possible route for growing graphite films on the Si-surface of (1000) 6H-SiC. The results of this simulation confirm the possibility of the stabilization of such a film (Fig. 9 and Table 2) with a following change in properties of the SiC surface. This is in agreement with the experimentally observed growth of graphite parallel to the crystal surface (Fig. 1). However, strong bonding between the carbon layers may be lost already starting from the third layer (Fig. 10). This may lead to delamination of continuous graphite layers grown on SiC single crystals that we observed in our experiments with 6H-SiC.

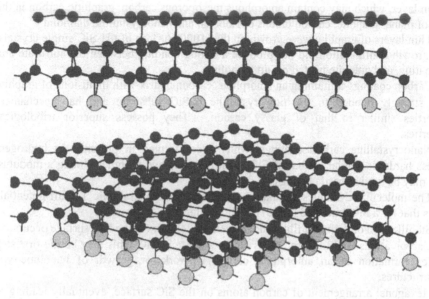

Figure 10. Three layers of hexagonally arranged carbon atoms above the Si-terminated (1000) surface of 6H-SiC. Only the upper layer of the SiC substrate is shown. Dark circles show carbon atoms and light circles are Si.

4. Discussion

Comparison between the experimental data and molecular dynamics simulation shows that experimental data are, in general, in agreement with the simulation. Both sp^2-bonded carbon and sp^3-bonded carbon can grow on the surface of SiC. This can result in the formation of carbon films that may maintain the sp^3-structure which carbon has in carbide, or in the growth of sp^2-bonded graphite or amorphous carbon. Hardness tests and TEM analysis confirm possibility of all these outcomes. However, hydrogen presence was required to promote the formation of hard sp^3-bonded carbon. Diamond is not a thermodynamically stable phase of carbon under ambient pressure, however the initial sp^3-bond configuration in SiC and presence of hydrogen can stabilize sp^3-carbon under the conditions of our experiments. These effects may be taken into account within the same MD framework and will be a subject of our future investigations. Alternatively, SiC can be transformed to the graphite, which is thermodynamically stable under these pressures, or form various amorphous or disordered structures intermediate to diamond and graphite. By changing the experimental conditions, it is possible to tune the structure of the coating. For example, high temperatures will favor the graphitization, while amorphous coatings will be formed at low temperatures, when surface reconstruction and diffusion rates are slow. Addition of hydrogen to chlorine will stabilize sp³ bonding leading to the formation of sp^3-bonded carbon [5,17].

254

5. Conclusions

High-temperature chlorination of silicon carbide and titanium carbide produces a carbon layer, which may contain amorphous mesoporous carbon, graphitic carbon in the form of planar graphite, carbon onions or ribbons, and nanocrystalline diamond.

Thin layers of graphite were grown on the (1000) surface of 6H-SiC single crystals. This growth configuration did not produce a strong bonding between the substrate and the coating, and graphite peeled off after cooling.

Carbon coatings containing an amorphous carbon matrix with inclusions of graphite were strongly bonded to the polycrystalline α-SiC substrate and had mechanical properties similar to that of glassy carbon. They possess superior tribological properties.

Nanocrystalline carbon coatings, which were formed in presence of hydrogen, possess hardness similar to that of SiC substrate and have a higher Young's modulus. They may be used as hard coatings.

The molecular dynamics simulation using empirical interatomic Tersoff potentials shows that for a Si-terminated (1000) 6H-SiC surface:

- Stabilization of diamond films and, preferably, clusters at the SiC surface occurs.
- Carbon atoms tend to form chains on the surface of SiC. This may be the first step for formation of an amorphous carbon network, or growth of nanotube-type structures.
- Hexagonal arrangement of carbon atoms on the SiC surface, eventually leading to graphite growth, is possible.

Acknowledgments

This research was supported in part by the National Science Foundation under Grant # CMS-9813400. The electron microscopes used in this work are operated by the Research Resources Center at UIC. Visit of V. Kamyshenko to UIC was supported by a COBASE grant from the National Research Council.

References

1. Gogotsi, Y. G. & Andrievski, R. A. (Eds.) *Material Science of Carbides, Nitrides and Borides* (Kluwer, Dordrecht, NL, 1999).

2. Holmberg, K., Matthews, A. *Coatings Tribology: Properties, Techniques and Applications in Surface Engineering* (ed. Dowson, D.) (Elsevier, New York, 1994).

3. Rickerby, D. S. & Matthews, A. *Advanced surface coatings: A handbook of surface engineering* (Blackie, Glasgow, UK, 1991).

4. Gogotsi, Y. G. & Yoshimura, M. Formation of Carbon Films on Carbides under Hydrothermal Conditions. *Nature* **367**, 628-630 (1994).

5.   Gogotsi, Y. G., Jeon, J. D. & McNallan, M. J. Carbon Coatings on Silicon Carbide by Reaction with Chlorine-Containing Gases. *J. Mater. Chem.* **7**, 1841-1848 (1997).

6.   Gogotsi, Y. *et al.* Formation of Carbon Coatings on SiC Fibers by Selective Etching in Halogens and Supercritical Water. *Ceram. Eng. Sci. Proc.* **19**, 87-94 (1998).

7.   Ersoy, D. A., McNallan, M. J. & Gogotsi, Y. High Temperature Chlorination of SiC for Preparation of Tribological Carbon Films. In *High Temperature Corrosion and Materials Chemistry, vol. 98-9,* Edited by P.Y. Hou, M.J. McNallan, R. Oltra, E.J. Opila, and D.A. Shores, 324-333, (The Electrochemical Society, 1998).

8.   Zheng, J., Ekström, T. C., Gordeev, S. K. & Jacobs, M. Carbon with an onion-like structure obtained by chlorinating titanium carbide. *J. Mater. Chem.* **10**, 1039-1041 (2000).

9.   Ersoy, D.A., McNallan, M., Gogotsi, Y. & Erdemir, A., Tribological Properties of Carbon Coatings Produced by High Temperature Chlorination of Silicon Carbide, *STLE Tribology Transactions*, in press (2000).

10.   McNallan, M., Gogotsi, Y. & Jeon, I. D. Formation of Carbon Films on Ceramic Carbides by High Temperature Chlorination. In *Tribology Issues and Opportunities in MEMS* (ed. Bhushan, B.) 559-565 (Kluwer, 1998).

11.   Pierson, H. O. *Handbook of Carbon, Graphite, Diamond and Fullerenes: Properties, Processing and Applications* (Noyes Publications, Park Ridge, NJ, 1993).

12.   Spear, K. E. Diamond - Ceramic coating of the future. *J. Am. Cer. Soc.* **72**, 171-191 (1989).

13.   Tersoff, J. Empirical interatomic potential for silicon with improved elastic properties. *Phys. Rev. B* **38**, 9902 (1988).

14.   Tersoff, J. Modeling solid-state chemistry: Interatomic potentials for multicomponent systems. *Phys. Rev. B* **39**, 5566 (1989).

15.   Pizzagalli, G. F. L., Catellani, A. & Baratoff, A. Theoretical study of the (3x2) reconstruction of beta-SiC(001). *Phys. Rev. B* **60**, 5129R (1999).

16.   Sabisch, M., Kruger, P. & Pollmann, J. *Ab initio* calculations of structural and electronic properties of 6H-SiC(0001) surfaces. *Phys. Rev. B* **55**, 10561 (1997).

17.   Gogotsi, Y. G., Kofstad, P., Nickel, K. G. & Yoshimura, M. Formation of sp^3-Bonded Carbon upon Hydrothermal Treatment of SiC. *Diamond and Relat. Mater.* **5**, 151-162 (1996).

REINFORCED n-AlN-CERAMICS

I. P. FESENKO*, M. O. KUZENKOVA*, G. S. OLEYNIK**
* V.N.Bakul Institute for Superhard Materials, Avtozavodska Str. 2, 04074
Kyiv, Ukraine
** Institute of Materials Science Problems, Krzhizhanovsky Str. 3, 03142
Kyiv, Ukraine

1. Introduction

Earlier studies have found that in oxygen-containing AlN materials, which were obtained under the conditions of polytype formation, two kinds of microstructures typical for self-strengthened ceramics can form [1-4]. The first form is due to separation of individual plate-like layers in grains, the second form is the result of growth of plate-like grains. In this case, both intragranular layers and grains of the similar morphology contain multilayer polytypes (MP) of AlN combined with a 2H AlN-based solid solution.

It has been shown in [5], that polytypes in AlN is formed only during the formation of 2H AlN-O solid solution, though other impurities may be contained in it (e.g., Ti in AlN material with TiN and TiC additives [2], Si in (AlN-SiO$_2$) material [3]). The decisive role of oxygen during polytype formation in AlN is defined by two factors [5]: (1) The nitrogen-oxygen pair meets all the requirements of isomorphous heterovalence substitution; (2) The formation of the 2H AlN-O solid solution is accompanied by the formation of aluminum vacancies. It is because of the presence of the vacancies in the solid solution that such transformations as stratification, periodic separation of three-layered oxygen-containing defects and development of Wardsley shifts in defects are possible. The final stage of the above-mentioned sequence of transformations is a periodic (in local volumes) separation in the 2H AlN structure of two-dimensional layers of Al$_2$O$_3$ (formed due to a shear reconstruction of defects), which is responsible for the formation of concentration polytypes in AlN.

The mechanism of formation of the second type self-strengthened microstructure has been partially discussed in [1, 6].

In the present paper, the formation mechanism of the second type self-strengthened microstructure has been proposed, which is based on the analysis of the evolution of the

M.-I. Baraton and I. Uvarova (eds.),
Functional Gradient Materials and Surface Layers Prepared by Fine Particles Technology, 257–264.
© 2001 Kluwer Academic Publishers. Printed in the Netherlands.

grain structure in AlN materials (beginning with the stage of the solid solution separation and ending with the growth of plate-grains).

2. Experimental procedure

The AlN samples have been prepared from porous compacts of fine (0.05 - 0.1 μm) plasmochemically prepared AlN powder by sintering in the 1700-2000 °C temperature range at 100-degrees intervals. The oxygen content of AlN was 3 wt%, unbound aluminum content was about 0.1 wt%. The duration of sintering at each temperature was 90 min.

Microstructural investigations of samples were conducted using a combination of transmission electron microscopy (TEM) of thin foils and replicates from natural fracture surfaces, and X-ray diffraction microanalysis. The typical electron fractographs of samples, that illustrate the grain structure geometry and histograms of grain distribution in samples, are presented in Fig. 1a-d. Hydrostatic weighing showed that all the samples were pore-free.

We will describe morphological and substructural characteristics of grain structure of the samples with increasing sintering temperatures.

3. Results and discussion

The samples, prepared at 1700 °C, consisted of equiaxial grains. There were basal stacking faults in some grains, that intersected a grain from one boundary to another. Following the procedure described in [7] it has been found that stacking faults are attributable to defects of growth or deformation, and not due to defects of interstitial type. The intergranular boundaries were well formed. Observations in characteristic radiation of oxygen have revealed that oxygen is not localised in boundaries or triple grain junctions. It is distributed uniformly over the grains, and this points to the formation of AlN-O solid solution.

Five successive stages can be distinguished in the development of grain structure of samples sintered at 1700 °C and above.

1. Separation of polytype layers in AlN-O grains at T = 1800 °C (mostly in combination with the oxygen-enriched solid solution) in the form of bulk plates (Fig. 2a). This process defines the transition of grains from a single-phase to the composite (2H AlN + MP) state. In a single grain, only one plate is formed and it intersects the grain from one boundary to another. The plane-parallel surfaces of plates are faceted only by basal planes and mate with the matrix along these planes. The thickness of the plates is equal to or less than 0.5 μm. The X-ray diffraction analysis shows that such

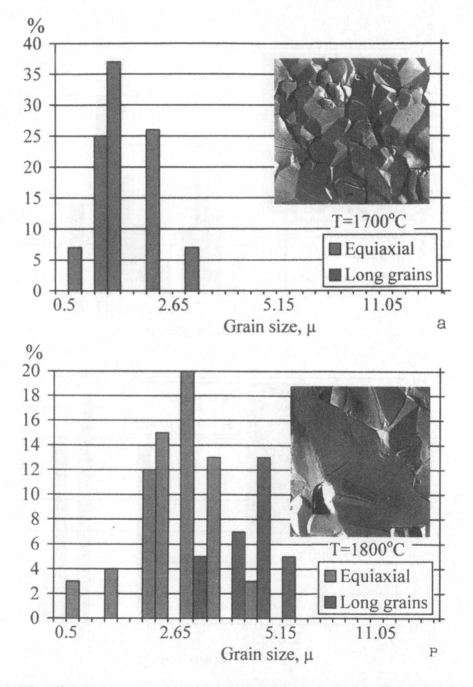

Figure 1. Electron micrograph of fracture from AlN samples, prepared in the 1700 – 2000 °C temperature range, and its grain size distributions: a) at 1700 °C – equiaxial grains; b) at 1800 °C – with polytype layers;

260

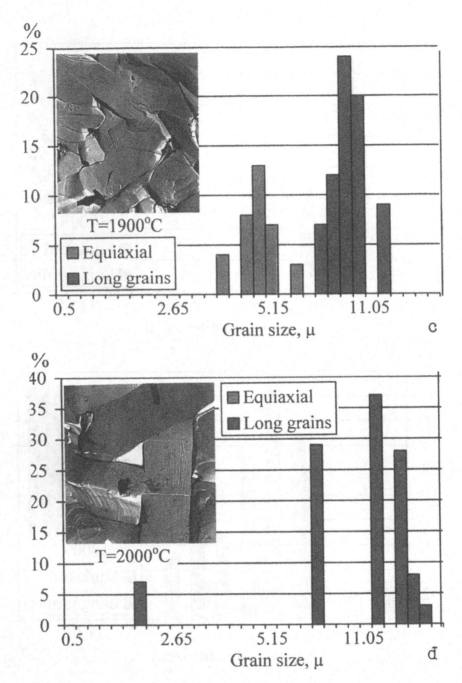

Figure 1. Electron micrograph of fracture from AlN samples, prepared in the 1700 – 2000 °C temperature range, and its grain size distributions: c) at 1900 °C – connection of layered grains and plate-grains; d) at 2000 °C – plate-grains.

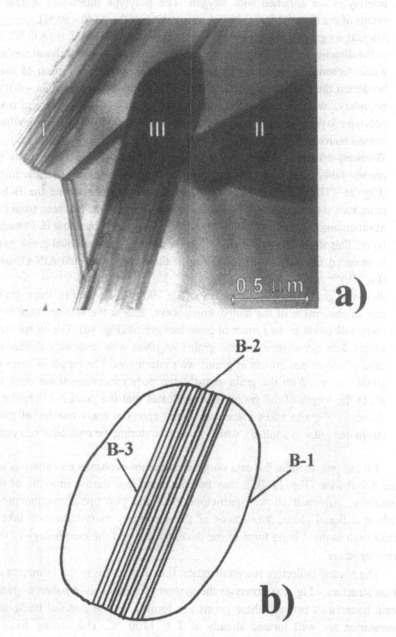

Figure 2. (a) Electron microscope image of three grains with layers; and (b) the schematics of an individual grain with the layer: I – the grain, in which uniformity of boundary matrix-layer is maintained; II – the grain, in which the joint (B-1 – B-2 – B-3) is observed at boundary of this type; and III – the grain with the alienation of boundary, bordering the layer, from the common boundary. The sample was sintered at 1800 °C.

interlayers are enriched with oxygen. The polytype interlayers appear not in all grains of the material, but only in some portion of them (40 - 50 %).

2. Anomalous growth of grains with polytypes layers arises at T > 1800 °C. It proceeds in the direction of plate lengthening, i.e. the development of its basal surfaces without change of thickness (Fig. 1b). As grain grows, a section of the boundary bordering the plate is bulging and the triple joint appears (Fig. 2a - II): (B-1 grain boundary, which is common for the whole grain) - (B-2 boundary of mating of the polytype layer with the adjacent grain) - (B-3 intragranular crystallographically caused boundary) (Fig. 2b).

3. Breaking-off the B-2 from the common B-1 boundary of a composite grain. This process takes place only after formation of the well-pronounced triple joint of grains (Fig. 2a - III). The B-2 boundary contains dislocations unlike the B-1 boundary; moreover, the radius of B-2 curvature is much smaller (as it is seen from Fig. 2a).

4. Autonomous migration of the B-2 boundary with the immovable B-3 boundary of the layer. This stage defines the growth of the plate as an individual grain, part of which is included into the matrix consisting of the oxygen-depleted AlN-O solid solution (Fig. 1c).

5. Plate-like grains grow in two stages at T ≥ 1900 °C. At the first stage grains increase due to absorption of the matrix component, and at the second stage by collective recrystallization in the system of plate-like grains (Fig. 1d). The morphology of such grains does not change and the grains lengthen with invariable thickness as in the case of anomalous growth of grains with interlayers. The length of these grains may attain 20 μm. With the grain growth, the polytype composition does not change along the length of the grains. This indicates that the process of regular solid-state growing of grains takes place due to the epitaxial mass transfer of the substance (from one grain to another), which is realised during the collective recrystallization.

Taking into account the data on grain structure evolution and analysis of the grain size distribution (Fig. 1a-d), it may be concluded that during sintering of the oxygen-containing AlN-material with participation of the polytype formation processes and without a liquid phase, four stages of the secondary recrystallization take place that differ both in the driving force of the development and the morphology of the resulting grain structure.

The normal collective recrystallization (i.e. that results in the formation of isotropic grain structure - Fig. 1a) occurs in the system of AlN-O solid solution grains once the grain boundaries between these grains are formed. In the material being studied such boundaries are well formed already at T = 1700 °C. The driving force of such a recrystallization is defined by the gradient of the grain boundary energy. An additional contribution to the driving force is made by the gradient of the elastic energy because of the different concentrations of oxygen dissolved in the grains. It has been mentioned above that the formation of the AlN-O solid solution is accompanied by the formation of vacancies that leads to a generation of compression stresses in grains.

Once plate-like interlayers, which contain MP, are formed in a portion of grains of the solid solution, the anomalous grain growth takes place, i.e., the secondary collective recrystallization proceeds. It develops in two stages. At the first stage, the growth centers are grains of a complex composition containing polytype interlayers (Fig. 1b). The driving force of such recrystallization is defined by a combination of the grain boundary energy and bulk elastic energy. The bulk elastic energy is caused by the fact that the formation of polytypes is a relaxation process of reconstruction of solid solution grains, and is related, as it was mentioned above, to the stratification of the solid solution. The development of this process promotes a decrease of grain bulk energy, and, thus, an appearance of an additional driving force for grain growth. The morphology of grains, forming at this recrystallization stage (i.e., grain lengthening in the direction of the development of the basal plate surfaces) is dictated, first of all, by a high migration rate of B-2.

The second stage of the secondary recrystallization is growth of plate-like grains in the grain system of: (i) the initial solid solution 2H AlN-O, in which interlayers have not been formed, and (ii) fragments of the oxygen impoverished matrix, connected with the plate-like grains (Fig. 1c). Growth centers of such grains appear as a result of a breaking away of B-2 from the common B-1 boundary (Fig. 2a - III). The driving force for the grain growth is defined by a surface energy gradient in a combination with the additional driving force, caused by the morphology of a migrating B-2 boundary (i.e., by a very small radius of curvature) and its high nonequilibrium.

After disappearance of grains consisting of the AlN-O solid solution, the normal collective recrystallization in the material takes place in the system of plate-like grains (see Fig. 1d) on retention of grain crystal geometry, while the aspect ratio (l/d is the ratio of size to thickness in a developed surface) increases due to an increase of l.

4. Summary and concluding remarks

The analysis of the above data allows a conclusion that in AlN materials under the conditions of sintering, the formation of individual plate-like grains based on the initial equiaxial grains, containing the polytype layers, is caused by a combination of a fragment of a highly migratory boundary that borders the layer and plane crystallographically caused boundaries of mating the layer and the matrix with a low surface energy. The separation of such a boundary fragment from the common boundary promotes the formation of a plate-like grain nucleus, the growth of which is controlled by the surface energy gradient.

The peculiarities that characterize the grain structure evolution in AlN during the development of the 2H AlN-O → MP transition are also typical of SiC-materials being formed under the conditions of the 3C → 6H, 3C → 4H, and 6H → 4H polytype transitions. The first type transition is realized during heating without any additives, the second and the third types in the presence of B, B_4C, Al, Al + Y_2O_3 and other additives [4].

An anomalous recrystallization, defined by separation of migrating boundaries, is known both for metals and ceramics [8]. However, the nature of this phenomenon differs from that observed in AlN- and SiC-materials, because its development is related to an increase in the mobility of boundaries due to their separation from boundary precipitations, pores, and impurity atmospheres.

5. References

1.   Kisly, P.S., Kuzenkova, M.A., Oleynik, G.S., and Pilyankievich, A.N. (1985) Recrystallization sintering of ultrafine aluminum nitride, Proceedings of the III-rd International Conference on Powder Metallurgy (Dresden, GDR,) pp. 51-57.

2.   Gunchenko, V.A., Shevchenko, O.A.,. Oleynik G.S., and B.M. Vereshchak (1991) Peculiarities of micro-structure of material on the base of aluminum nitride with silicon dioxide addition, in Urgent Problems of Materials Science (Institute of Problems in Materials Science of the Ukr. Nat. Ac. Sci., Kyiv, Ukraine), pp. 106-115 (in Russian).

3.   Tkachenko, Yu.G., Yurchenko, D.Z., Oleynik, G.S., et al (1992) Self-Reinforced Materials on the Base of Aluminum Nitride, Poroshkovaya Metallurgiya No.9, 69-73.

4.   Oleynik, G.S. (1993) "Selfstrengthened Ceramic Materials" (Proceedings of the Institute of Problems in Materials Science of the Ukr. Nat. Ac. Sci., Kyiv, Ukraine, No. 9) (in Russian).

5.   Oleynik, G.S., and Danilenko, N.V. (1997) Polytype formation in non-metal substances, Uspechy chimii 6, No. 7, 615-640 (in Russian).

6.   Oleynik, G.S., and Shevchenko, O.A. (1992) Mechanism of anomalous grain growth in polycrystalline AlN,   in Modern Achievements in the Field of Physical Material Science (Proceedings of the Institute of Problems in Materials Science of the Ukr. Nat. Ac. Sci., Kyiv, Ukraine), pp. 148-161 (in Russian).

7.   Blank, H., Delavignette, P., Gevers R. and Amelinckx, S. (1964) Fault structures in wurtzite, Phys. Stat. Sol. 7, No.3, 747-760.

8.   Lange, F.F. (1989) Powder processing science and technology for increased reliability, J. Am. Ceram. Soc. 72, No. 1, 3-15.

STRUCTURE AND STRENGTH OF CERAMIC MULTILAYERED COMPOSITES

O.N.GRIGORIEV, A.V.KAROTEEV, A.V.KLIMENKO.
E.V.PRILUTSKY, E.E. MAIBORODA, N.D. BEGA
Institute for Problems of Materials Science,
Krzhizhanovsky str., 3, 03142, Kiev, Ukraine

1. Abstract

The effect of structure and residual stresses on strength of some multilayer composite ceramic materials was studied. The use of β-SiC powders has allowed to obtain SiC layers with porous structure reinforced by prismatic crystals. Such structures possess the relaxation ability of thermal strains that excludes formation of cracks at material production and provide the high strength of SiC/TiB_2 composite. By the thermoshock, the layers may lose mechanical durability with lateral deflection, consecutive fracture and formation of a crater in a central part of a plate.

2. Introduction

Silicon carbide is one of the most prospective ceramic materials for structural applications because of its unique thermomechanical properties and high corrosion resistance. However, many potential applications of silicon carbide are not realized for the reason of low fracture toughness and reliability. It is well known [1-2] that the additions to silicon carbide of titanium and zirconium borides in amount of 15 - 30 % enable to increase the strength and fracture toughness by 50 - 100 %. However, the corrosion resistance of ceramics is sharply decreased, and it is undesirable for the majority of high-temperature applications.

Therefore, the some multilayer compositions may be promising with external layers with high corrosion resistance (SiC without any impurities) and internal SiC/MeB_2 ones with high strength and fracture toughness. Moreover, the external SiC layers having lower coefficient of thermal expansion compared with internal SiC/MeB_2 layers will be under thermal compression stresses. It will increase apparent fracture toughness as well as strength under contact interaction and reliability. The studies of the last few years have shown that the increase in strength and/or fracture toughness of multilayer ceramic composites may provide the tolerance of material to damages. However, the formation

265

M.-I. Baraton and I. Uvarova (eds.),
Functional Gradient Materials and Surface Layers Prepared by Fine Particles Technology, 265–272.
© 2001 *Kluwer Academic Publishers. Printed in the Netherlands.*

of such compositions requires the decision of layer bonds problems. The very important significance belongs to the optimization of composite production conditions and its structure state to ensure elastic strains relaxation .

In this study we explored the effect of structure and residual stresses on strength of multilayer composites. A layer's structure is changed due to the sintering additives, use of various raw material as well as changing of technological conditions. Residual thermal stresses are controlled by change of layers compositions.

3. Materials and procedures

Two kinds of α-SiC powders were used: - 1) technical abrasive powders, M5 grade, produced by the Zaporozhye abrasive plant, Ukraine, and 2) powders of UF05 and UF10 grades from the N.C Starck company, Germany, designed for ceramics production. Both powders were mixtures of polytypes: mainly 6H, 15R and 3C. The β-SiC powder, produced by IPMS, had the content of 3C-polytype up to 100%. TiB_2 powders (TC 6-09-03-7-75) from the Donetsk factory of chemical reagents (Ukraine), and abrasive B_4C powders from the Zaporozhye abrasive plant (GOST 5744-74), were used as sintering additives.

Some powders' properties in the as-received condition are given in Table 1.

TABLE 1. The characteristic of powders

Powder	Size of particles, d_{50}, μm	Content of oxygen, wt.%	Free carbon, wt. %
SiC_{M5}	5	1.5	1 - 2
SiC_{UF05}	1.47	0.55	-
SiC_{UF10}	0.7	1.2	0.17
β-SiC	0.1-0.2	$\leq 0.5*10^{-2}$	-
TiB_2	30	0.3	<0.1
B_4C	20	1.5	2

SiC powders were very different as for their defectiveness and sinterability. The powders UF05 and M5 were characterized with low width of X-ray diffraction peaks and good resolution of K_α-doublets and, therefore, had a high degree of structural perfection. At the same time, the XRD peaks of UF10, as well as of β-SiC powder, were very broad due to high density of defects (stacking faults, polytypes interlayer, and non-homogeneous microstrains) which, apparently, facilitated an increase of their activity during sintering.

Joint milling and blending of charge components were carried out in a planetary ball mill. The particles have a sufficiently small grain size for hot pressing (2.5 and 1.1 μm, respectively), ensuring optimum dispersion. The B_4C additives and, in some cases, TiO_2 ones were introduced into SiC- and TiB_2- based charges. In the presence of TiO_2, the reactionary hot pressing with formation of secondary TiB_2 during reaction $TiO_2 + B_4C \rightarrow TiB_2 + CO$ took place. The additives as pointed out, were introduced with the aim of reducing and matching of hot pressing temperature of various composition layers.

We used a slip casting method for lamina films production. From powders of various compositions the ceramic slips with subsequent slip casting of films of thickness ~50 μm were prepared. With the aim of casting defects removal, the films were folded in rolls and then rolled up to the thickness of 400 μm. From the sheets obtained, the plates of the demanded size were cut out and the packages containing 11-13 pairs of alternating layers were obtained for chosen compositions.

Hot pressing was carried out using a pilot induction heating unit in graphite dies without vacuum chamber. The temperature of isothermal sintering under load was in the range of 1600 - 2150°C, pressure - 26 - 30 MPa, time of isothermal densification - 7 - 20 minutes. The grinding using diamond wheels with a grain size of 125/100 and 80/63 μm was conducted after hot pressing. Bending strength of the samples (4 -12 pieces) per point was measured by three and four-point bending for the spans of 30 mm and 20x40 mm. Vickers hardness was determined under the load of 5 N.

4. Results and discussion

4.1 MONOLITHIC CERAMICS

The bending strength of single-phase and heterogeneous ceramics with composition similar to the layers ones is shown in Table 2. Single-phase silicon carbide ceramics had rather high porosity (5-10 %), average grain size being 5-10 and up to 100 μm for raw powders of α- SiC and β-SiC, respectively. In the latter case the high grain size is stipulated by grain growth during β→α transformation of silicon carbide at hot pressing. Thus, hot pressing of pure silicon carbide results in the production of porous coarse-grained materials with the low strength (110 - 190 MPa).

Introduction of boron carbide allows to reduce porosity of ceramics to 1-3 % with the relevant increasing of strength up to 300-370 MPa. Simultaneous introduction of 5% B_4C and 12 % TiB_2 results in an increase of strength up to 408 MPa (Tab. 2) with increase of strength of 650 MPa at content TiB2 20-25 %. The ceramics TiB_2 - B_4C had the strength of 415 MPa at a grain size of 5-10 μm with practically zero porosity. The results of detailed study of structure and mechanical behaviour of SiC - TiB_2 - B_4C ceramic system are presented in [2].

TABLE 2. Compositions and mechanical properties of monolithic ceramics

№	Composition of ceramics, vol. %	Bending strength (stand. dev.), MPa	Fluctuation factor, %
1	β-SiC	190(40)	21
2	α-SiC$_{M5}$	110(57)	52
3	α-SiC$_{UF05}$	170(48)	28
4	α-SiC$_{M5}$ + 10 % B$_4$C	372(71)	19
5	α-SiC$_{UF05}$ + 10 % B$_4$C	306(56)	18
6	α-SiC$_{UF05}$ + 12% TiB$_2$ + 5% B$_4$C	408(64)	17
7	TiB$_2$ +42% B$_4$C	415(53)	13

4.2 LAMINATED COMPOSITES

At the first stage of work we studied the laminated composites with the maximum thermal expansion misfit between layers. Such systems have the greatest hazard of uncontrollable fracture in fields of thermal stresses. At the same time, just in such materials it is possible to expect the positive effects because of localization of fracture within the single layers and difficulties of cross fracture.

In the temperature range 20-1500^{0}C the effective coefficient of thermal expansion is equal $5.8*10^{-6}$ $/^0$C and $8.9*10^{-6}/$ ^{0}C for SiC and TiB$_2$, correspondingly. Within the framework of Eshelby model, there are the average stresses in a plane of layers $|\sigma_{22}|$ = $|\sigma_{33}| \cong 1.4$ GPa, tension of TiB$_2$ and compression of SiC. According to accepted orientation of axes, the components σ_{22} and σ_{33} of principal stresses are in a plane of layers and the σ_{11}-component is perpendicular to the plane of layers. The values of stresses exceed a possible level of strength and should result in the phenomena of fracture. Actually, the thermal stresses will be essentially lower resulting from the viscoelastic relaxation due to the segregation of impurities on layer's boundaries and also because of a presence of the phase (B$_4$C) with intermediate coefficient of thermal expansion (α=6,05$*10^{-6}/^0$C).

The estimation of thermal stresses and a character of their distribution with the help of finite elements method is more real. The calculations in this work were executed for five-layer ABCBA symmetric configuration , where A, B and C are layers of ceramics: SiC, SiC+20 % TiB$_2$ and TiB$_2$,respectively (Figure 1). The stress distribution is more complex, than it should follow from the Echelby model. There are the edge effects of redistribution of stresses and inhomogeneity of a stress distribution cross the thickness of layers. In particular, near the edges the extensive zone with a tension stresses directed perpendicularly to the layers plane ($\sigma_{11}>$ 0) take place. Within this zone the delamination is possible . Moreover, the various types of fracture and microcracking in composites may take place under the joint effect of both thermal and applied stresses.

These studies have shown that in the composites containing the α-SiC$_{M5}$ powders (the composites 1 and 2, Tab. 3) both types of layers (SiC and TiB$_2$) with low porosity are formed during sintering (Figure 2a).The low relaxation ability of such structures results in a microcracking of composite with realization of main types of fracture, described in [3]. The tension stresses in TiB$_2$ layers form cellular microcracking structure of (Figure 2b). In such layers the transverse cracks approaching to a boundary propagate either into the near-boundary volumes in the case of strong inter-layers boundaries or due on boundaries in the case of weak inter-layers bonds. It was found, that the boundary strength is reduced at formation of TiO$_2$ inter-layers. The delamination cracks were observed at composite butt-ends because of edge redistriburion of thermal stresses as well as in the volume of a material. Described considerable damages of structure result in low strength of composites at bending tests, viz., 20 - 100 MPa.

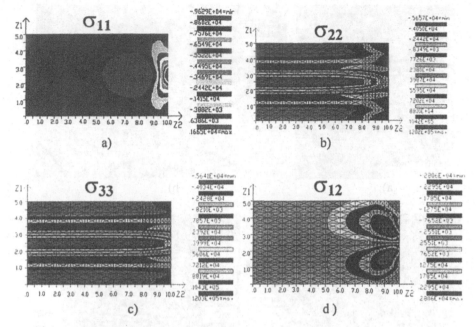

Figure 1. Distribution of principal stresses σ_{11}, σ_{22}, σ_{33} (a - c) as well as σ_{12} (d) in laminated composites SiC/ SiC-TiB$_2$/ TiB$_2$

TABLE. 3. Compositions and mechanical properties of laminated composites

№	Compositions of layers in composite , vol. %	Hardness HV of composite layers, MPa	Bending strength (stand. dev.), MPa	Fluctuation factor, %
1	1) α-SiC$_{MS}$ + 10 % B$_4$C	28	100(14)	14
	2) TiB$_2$ +42% B$_4$C	33		
2*	1) α-SiC$_{MS}$ + 10 % B$_4$C	32	24(7)	29
	2) TiB$_2$ +42% B$_4$C	33.5		
3	1) β-SiC	4	584(560 - 601)**	3
	2) TiB$_2$ +42% B$_4$C	32.4		
4	1) β-SiC + 10% B$_4$C + 4 % TiO$_2$	11	410(34)	8
	2) TiB$_2$ +42% B$_4$C	32.4		

*- Layers TiB$_2$ are of double thickness , interlayer TiO$_2$ on phase boundaries.**- Three samples are tested, in brackets is shown min. max. significances of strength.

The structure and properties of composites obtained with use of β-SiC powders (composites 3-4, Tab. 3) are essentially different from others. The structure of sintered SiC is porous and coarse-grain. The porosity of these layers is exceptionally high - up to 40 %. The prismatic columnar SiC crystals with a size about thickness of a layer (~100 μm) form "engineering" arch structure bonding together the dense strong TiB$_2$ layers (Figure 3). The microscopic studies of composites have not revealed the traces of microcracking.

270

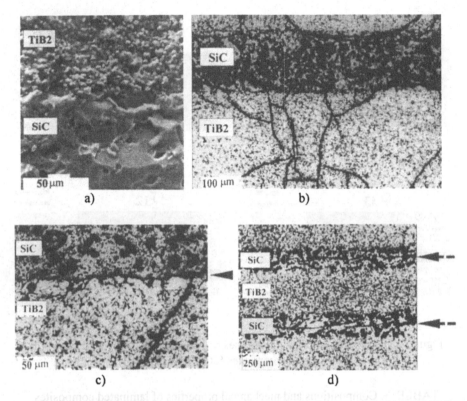

Figure 2. Structure and crack formation in composites with low strength (≤ 100 MPa):
a) microphoto of composite in the field of phase boundaries of layers α-SiC/TiB$_2$;
b) cracks in (TiB$_2$) layer with tension stresses ; c) cracks on the weak phase
boundaries; d) longitudinal cracks along the pivotal zone of compressed SiC layers

The the measurements of bending strength showed its high value (400 - 600 MPa), and
the composites strength exceeds the strength of monolithic ceramics. Thus, there is the
unusual inverse relation between strength of a composite and hardness of SiC layers
(Figure 4). Obviously, the decrease of hardness is determined by the relevant growth of
porosity of silicon carbide. Hence, the increase of composite strength with the growth of
porosity of one of its component is stipulated by a high relaxation ability and,
apparently, high magnitude of critical fracture strains.

The composition and strength of composites prepared using SiC$_{UF05}$ powders with small
amount of impurities and, consequently, perspective for the high-temperature
applications are given in Table 4. As well as in case of SiC$_{M5}$ powders, materials with
large misfit of thermal expansion between layers (the composites 1 and 2) have low
strength - 60-65 MPa. Decrease of thermal expansion misfits between layers
(composites 3 and 4) excludes microcracking and allows to increase the strength of
composites up to 300 - 400MPa, i.e. up to that level when these composites can already
be interesting. It should be noted that for all investigated composites, a standard
deviation and fluctuation factor of strength is much lower, than for monolithic ceramics

(Tab. 2-4). It clearly demonstrates that the laminated structure enables to control effectively a size of defects of material and increase the Weibull module.

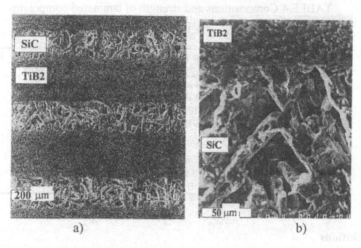

Figure 3. Structure of strong (400-600 MPa) composites with porous SiC layers

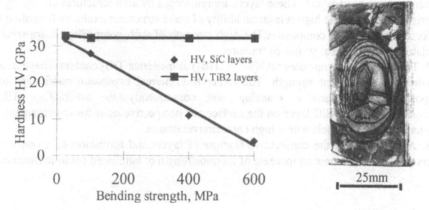

Figure5. Composite plate
after thermoshock test

Figure 4. Relations between strength of composites
and hardness of their layers

The thermoshock experiments were fulfilled under conditions of fast cooling of composites in air from temperature of hot pressing. In this case the extraction of laminated plate from die to air was made at the temperature of ~1500°C. In a quickly cooled plate the tension stresses in external layers and compression stresses in internal ones are risen. At this conditions, the layers can lose mechanical durability with lateral deflection, consecutive fracture and formation of a crater in a central part of a plates (Figure. 5). Thus, cracking and fragmentation of monolithic ceramics at thermoshock is replaced by self-adjustable process of local sequential fracture. Such change of fracture mechanism can ensure the increased reliability of laminates at thermoshock.

TABLE 4.Compositions and strength of laminated composites
based on SiC_{UF05} powders

№	Compositions of layers in composite, vol. %	Bending strength (stand. dev.), MPa	Fluctuation factor, %
1	1) α-SiC$_{UF05}$ 2) TiB$_2$ +42% B$_4$C	65(13)	20
2	1) α-SiC$_{UF05}$ + 12 % TiB$_2$ + 5 % B$_4$C 2) TiB$_2$ +42% B$_4$C	61(6)	10
3	1) β-SiC 2) α-SiC$_{UF05}$ + 12 % TiB$_2$ + 5 % B$_4$C	314(20)	6
4	1) α-SiC$_{UF05}$ + 10 % B$_4$C 2) α-SiC$_{UF05}$ + 12 % TiB$_2$ + 5 % B$_4$C	397(19)	5

5. Conclusions

1. In laminated β-SiC/TiB$_2$ composites, the recrystallization of SiC results in porous structure of its layers. These layers are reinforced by arch structures consisting of the prismatic grains. The high relaxation ability of these structures excludes formation of cracks at production of composites.The high strength of such composites is apparently stipulated by high critical strains of fracture.

2. The laminated composites SiC/(SiC + TiB$_2$) at moderate TiB$_2$ contents have a low porosity and a sufficient strength. The rather low thermal expansion misfits in such composites do not result in cracking, and, consequently, the SiC/(SiC + TiB$_2$) composites with pure SiC layer on the surface are perspective ones for development of high-temperature materials with a high corrosion resistance.

3. At thermoshock the consecutive fracture of layers and formation of a crater are observed that can promote an increase of thermostrength of laminated ceramic materials.

6. References

1. Janney, M.A. (1987) Mechanical properties and oxidation behavior of a hot pressed SiC-15 vol.% TiB$_2$ composite, Amer. Ceram. Soc. Bull. **66**, 322-324.
2. Grigoriev,O.N., Kovalchuk,V.V., Subbotin, V.I., Gogotsi Yu.G., (1999) Structure and properties of SiC-TiB$_2$ ceramics, J. of Mat. Proc. and Manufact. Sci., **7**, No7, 99-110.
3. Cai, P.Z., Green, D.J., Messing, G.L., (1997) Constrained Densification of Alumina/Zirconia Hybrid Laminates, I: Experimental Observations of Processing Defects, J. Amer. Ceram. Soc. **80**, 1929-39.

ANALYSIS OF LAYERED COMPOSITES WITH CRACK DEFLECTION CONTROLLED BY LAYER THICKNESS

M. LUGOVY, N. ORLOVSKAYA, K. BERROTH[*], J. KÜBLER[*]
*Institute for Problems of Materials Science, National Academy of Sciences of Ukraine,
Kiev, 03142, 3 Krzhizhanovsky str., Ukraine
email: lugovoj@materials.kiev.ua
[*] Swiss Federal Laboratories for Materials Testing and Research,
EMPA, Dübendorf, Überlandstrasse 129, CH-8600, Switzerland*

Abstract

A method to analyse the crack deflection behaviour of symmetric two-component layered composites had been developed. With the method it is possible to compute the layer thickness at which cracks bifurcate. The thickness necessary for cracks to bifurcate depends on the elastic constants of the layers, the number of layers, the difference of thermal expansion coefficients and the temperature gradient. Further, the thickness ratios at which crack bifurcation can occur for layers with tensile and compressive residual stresses were computed.

1. Introduction

The strategy of ceramic strengthening is usually associated with the design of ceramic composites to enable different mechanisms of fracture energy absorption or dissipation. In addition, other properties can be modified to eliminate the fracture energy e.g. microstructure design in composites. In functional gradient materials even more complex microstructural architectures have to be developed. Different structural elements such as particles, whiskers, filaments, platelets or laminates are combined to tailor properties to different service requirements.

Composites are complicated systems characterised by a host of interdependent parameters. The theoretical prediction of the mechanical behaviour provides information related to the failure of such materials. The strength of ceramic-matrix layered composites depends on the properties of the separate layers and also the interactions between different layers. Ceramic materials show a lot of outstanding physical and chemical properties, which make them interesting for many engineering purposes. However, their more intensive technical application is restricted by their brittleness. A key feature that imparts good mechanical properties in multilayer

M.-I. Baraton and I. Uvarova (eds.),
Functional Gradient Materials and Surface Layers Prepared by Fine Particles Technology, 273–280.
© 2001 *Kluwer Academic Publishers. Printed in the Netherlands.*

274

systems is the ability to deflect cracks. Cracks that form in one layer are deflected along weak interfaces with adjacent layers or into layers with compressive residual stresses. The formation of weak interfaces with controlled strength is a very difficult technological problem. Furthermore, weak interfaces usually also increase high temperature corrosion due to the high defect density. Therefore, crack bifurcation in layers with compressive residual stresses is the preferred mechanism to improve laminates.

The goal of the present work is the development of an analysis method for two-component ceramic matrix layered composites with crack bifurcation in layers with compressive residual stresses controlled by the layer thickness. The method is necessary for an optimal design of processing parameters for fabrication of such composites.

2. Thermal residual stresses in laminates

In this work two-component layered composites with symmetric macrostructures are considered (Fig.1). The layers alternate, with exposed faces being of the same component. Thus the total number of layers N in the composite sample is odd. In Fig.1 the layers of the first component including the two external (outside) layers are designated with the index 1 (j = 1), and those of the second component (internal) with index 2 (j = 2). The number of layers designated with index 1 is $(N+1)/2$ and the number of layers designated with index 2 is $(N-1)/2$. All layers of each component have the same constant thickness.

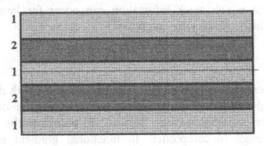

Figure 1. Two-component layered composite: 1- layers of the first component including two external (outside) layers; 2- layers of the second component (internal).

In this work the mechanical behaviour of a layered beam (sample) with a rectangular cross section with the height h and the width b is considered. It is assumed that the two components have different thermal expansion coefficients. Therefore the difference in temperature ΔT between the actual temperature and the temperature at which the layers consituting the material were joined is the important initial parameter which determines the residual thermal stress in the layers.

Characteristics of the individual structural elements such as Young's modulus, strength, fracture toughness, and thermal expansion coefficients affect the composite failure process. The effective elastic modulus of the layers (Fig.1) is determined by the Young's modulus E_j (j=1, 2) and the thickness l_j (j=1, 2) of the layers. Further, the strength of j-th component of layered composite is σ_{cj}, the fracture toughness K_{cj}, and the thermal expansion coefficient α_{Tj}.

During cooling of the sample the deformation difference, due to the different thermal expansion coefficients, is accommodated by creep as long as the temperature is high enough. Below a certain temperature, called the "joining" temperature, the different components become bonded together and internal stresses appear. In each layer, the total deformation after sintering is the sum of an elastic component and of a thermal component [1]. In the case of a perfectly rigid bonding between the layers, the total deformation will be the same for all the layers:

$$e_j = \frac{\sigma_{rj}}{E_j'} + \alpha_{Tj}\Delta T = const \tag{1}$$

where σ_{rj} is the residual stress in the j-th component, $E_j' = E_j / (1 - v_j)$, v_j is Poisson's ratio of the j-th component.

The force balance requires (in normal stresses):

$$\sum_j \sigma_{rj} f_j = 0 \tag{2}$$

where f_j is the volume fraction of j-th component.

For two-component material, f_j is:

$$f_1 = \frac{(N+1)l_1}{2h} \quad \text{and} \quad f_2 = \frac{(N-1)l_2}{2h}.$$

The combination of equation (1) and (2) results in:

$$\sigma_{r1} = \frac{E_1' E_2' f_2 (\alpha_{T2} - \alpha_{T1})\Delta T}{E_1' f_1 + E_2' f_2} \tag{3}$$

and

$$\sigma_{r2} = \frac{E_2' E_1' f_1 (\alpha_{T1} - \alpha_{T2})\Delta T}{E_1' f_1 + E_2' f_2} \tag{4}$$

276

In this paper we will consider components marked with index 1 as the ones with tensile residual stress and those with index 2 as the ones with compressive residual stress.

3. Determination of crack deflection regime

To understand the thinking that led to experiments concerning crack bifurcation, consider a crack propagating through a laminate as shown in Fig.2. When a crack propagates through the laminate, it creates a free surface. When it approaches a layer that is under a residual compressive stress, σ_r, it creates a situation similar to the free edge of a laminate [2,3]. The tensile stress, σ_T, which now arises due to the introduction of the free surface, may cause the propagating crack to bifurcate into the compressive layer. Because this behaviour is similar to the edge-cracking problem, a critical layer thickness is expected below which no crack bifurcation will occur.

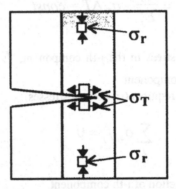

Figure 2. Schematic of a propagating crack in a laminate under residual compressive stress.

Ho *et al.* [3] observed that the occurrence of edge cracks was dependent on the thickness of the compressive layer and the magnitude of the residual compressive stress in this layer. They developed a strain energy release rate function for a crack in this localised tensile stress field. The thickness for which the crack bifurcation occur can be determined from the expression:

$$l_2 \geq \frac{G_{c2}E_2}{0.34(1-v_2^2)\sigma_{r2}^2},$$

(5)

where G_{c2} is the critical strain energy release rate of the second composite component. It should be noted that $G_{c2}E_2 = K_{c2}^2$.

Eqn. 4 can be easily transformed into the following form using the expressions for f_1 and f_2:

$$\sigma_{r2} = \frac{E_2 \Delta\alpha\Delta T}{(1-v_2)\left[1+\dfrac{E_2}{E_1}\left(\dfrac{N-1}{N+1}\right)\left(\dfrac{1-v_1}{1-v_2}\right)\dfrac{l_2}{l_1}\right]} \qquad (6)$$

where $\Delta\alpha = \alpha_{T1} - \alpha_{T2}$.

Criterion (5) is transformed using (6) into:

$$\frac{l_2}{\left[1+\dfrac{E_2}{E_1}\left(\dfrac{N-1}{N+1}\right)\left(\dfrac{1-v_1}{1-v_2}\right)\dfrac{l_2}{l_1}\right]^2} \geq \frac{(1-v_2)^2 K_{c2}^2}{0.34(1-v_2^2)(E_2\Delta\alpha\Delta T)^2},$$

and then

$$l_2 - l_c\left[1+\frac{E_2}{E_1}\left(\frac{N-1}{N+1}\right)\left(\frac{1-v_1}{1-v_2}\right)\frac{l_2}{l_1}\right]^2 \geq 0,$$

where $l_c = \dfrac{(1-v_2)^2 K_{c2}^2}{0.34(1-v_2^2)(E_2\Delta\alpha\Delta T)^2}$ is the characteristic layer thickness. The dependence of characteristic size l_c on the thermal expansion factors difference $\Delta\alpha$ for various temperature differences ΔT is shown in Fig.3. The characteristic size is increased when ΔT and $\Delta\alpha$ are decreased.

Using the substitution $a = \dfrac{E_2}{E_1}\left(\dfrac{N-1}{N+1}\right)\left(\dfrac{1-v_1}{1-v_2}\right)\dfrac{l_c}{l_1}$ we obtain:

$$\frac{a^2}{l_c}l_2^2 - (1-2a)l_2 + l_c \leq 0. \qquad (7)$$

From (7) the upper and lower limits, l_{b1} and l_{b2}, of the crack deflection area can be determined:

$$\frac{(1+\nu_2)}{(1-\nu_2)}\frac{E_2^2}{K_{c2}^2}l_c$$

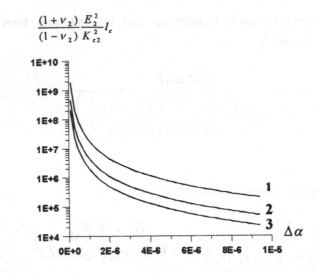

Figure 3. Dependence of characteristic size l_c on the difference of thermal expansion factors $\Delta\alpha$ for various temperature differences: 1 - $\Delta T = 400$ K; 2 - $\Delta T = 800$ K; 3 - $\Delta T = 1200$ K.

$$l_{b(1,2)} = \frac{1 - 2a \mp \sqrt{1-4a}}{2a^2} l_c. \tag{8}$$

The thickness of the second component layer for which crack bifurcation occurs can be determined in this case from the expression:

$$l_{b1}(l_1) \le l_2 \le l_{b2}(l_1). $$

The layer thickness regime where crack bifurcation occurs is shown in Fig. 4(a). These thicknesses depend on the elastic constants of the layers, the number of layers, the difference of thermal expansion factors and the temperature difference. For a given thickness of layers with tensile residual stresses there exists a thickness interval of the layers with compressive residual stresses where the crack bifurcation occurs. The regime boundaries for $E_2 / E_1 = 1$ and for $E_2 / E_1 = 2$ are presented in Fig. 4(b).

Eqn. (8) results in the expression:

$$1 - 4a \ge 0 \quad \text{or} \quad a \le 0.25. \tag{9}$$

From (9) a criterion for the first component layer thickness can be obtained

$$l_1 \ge l_1^*, \tag{10}$$

where $I_1^* = 4\dfrac{E_2}{E_1}\left(\dfrac{N-1}{N+1}\right)\left(\dfrac{1-v_1}{1-v_2}\right)l_c$ is the characteristic thickness of the first

component. No deflection occurs as long as the layer thicknesses of the first component are less than the characteristic thickness (Fig. 4a). The dependence of the first component's characteristic thickness I_1^* on the ratio E_2 / E_1 is presented in Fig.5.

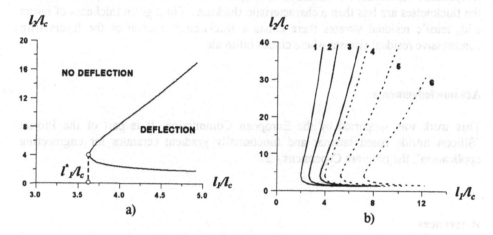

Figure 4. Area of crack deflection (a) and boundaries (b) for $E_2 / E_1 = 1$ and N=3, N=5, N=20 (solid curves 1, 2, and 3), and for $E_2 / E_1 = 2$ and N=3, N=5, N=20 (dashed curves 4, 5, and 6).

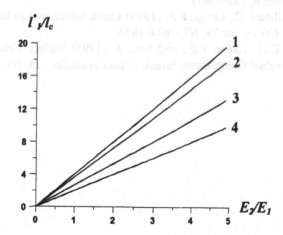

Figure 5. Dependence of first component characteristic thickness I_1^* on ratio E_2 / E_1 and N=∞, N=20, N=5, N=3 (curve 1, 2, 3, and 4).

280

4. Conclusions

A method to analyse the crack deflection behaviour of two component symmetric layered composites has been developed. It could be shown that a layer thickness regime exists where crack bifurcation occurs. This regime depends on the elastic constants of the layers, the number of layers, the thermal expansion factor difference and the temperature difference. No deflection of cracks occurs in the first component layer if the thicknesses are less than a characteristic thickness. For a given thickness of layers with tensile residual stresses there exists a thicknesses interval of the layers with compressive residual stresses where cracks bifurcate.

Acknowledgements

This work was supported by the European Commission. It is part of the Project "Silicon nitride based laminar and functionally gradient ceramics for engineering application", the program Copernicus - 2.

References

1. Lugovy, M., Orlovskaya, N., Berroth, K. and Kübler, J. (1999) Macrostructural Engineering of Ceramic-Matrix Layered Composites, *Composite Science and Technology* 59, Issue 8, 1429-1437.
2. Oechsner, M., Hillman, C., Lange, F.F. (1996) Crack bifurcation in laminar ceramic composites, *J.Am.Ceram.Soc* 79, N7, 1834-1838.
3. Ho, S., Hillman, C.D., Lange, F.F., and Suo, Z. (1995) Surface Cracking in Layer under Biaxial, Residual Compressive Stress, *J.Am.Ceram.Soc.* 78, N9, 2353-2359.

ON SOME PROBLEMS OF BENDING AND VIBRATIONS OF THIN PIEZOCERAMIC PLATES

L.MKRTCHYAN
Senior Researcher
Institute of Mechanics NAS of Armenia,
24B Baghramyan Ave. , Yerevan 375019, Armenia

The interaction between electromagnetic and mechanical fields in deformable solids is widely used in many technological applications. For example, one of the products of this interaction is used to produce elastic vibrations in materials. These vibrations are employed in measuring devices for determining the physical and mechanical properties of structural components
In present work some problems of bending and vibrations of transversely polarized piezoceramic plates are discussed when the justice of Kirchhoff's hypothesis is adopted for the stress-strain state of the plate. Regarding the electric field, no hypothesis will be made, except to assume that the potential of the electric field can be expressed as a sum of symmetric and asymmetric functions with respect to the transverse coordinate. Applicability of known models is discussed.

1. General equations

Let the thin transversely polarised piezoceramic plate of a constant thickness $2h$ be referred to the rectangular coordinate system so that the plane (XOY) coincides with the midplane of the plate.

The simplified constitutive relations [4] in which the normal stress σ_{33} and the transverse shear stresses σ_{31}, σ_{32} are neglected according to the classical theory, are taken in the following form:

$$e_{11} = s_{11}\sigma_{11} + s_{12}\sigma_{22} + d_{31}E_3; \quad e_{33} = 0, \quad e_{31} = e_{32} = 0$$

$$e_{22} = s_{12}\sigma_{11} + s_{11}\sigma_{22} + d_{31}E_3; \quad e_{12} = 2(s_{11} - s_{12})\sigma_{12}$$

$$D_1 = \varepsilon_1 E_1; \quad D_2 = \varepsilon_1 E_2; \quad D_3 = \varepsilon_3 E_3 + d_{31}(\sigma_{11} + \sigma_{22})$$

(1.1)

Here D_i - are the components of the electric displacement vector, E_j - are the components of the electric filed strength:

$$E_j = -\frac{\partial \varphi}{\partial x_j}$$

Then, on the basis of Kirchhoff hypothesis from (1.1) we obtain the stresses in the plate as well as the components of the electric displacement vector [1]
and the components of the electric displacement vector

281

M.-I. Baraton and I. Uvarova (eds.),
Functional Gradient Materials and Surface Layers Prepared by Fine Particles Technology, 281–288.
© 2001 *Kluwer Academic Publishers. Printed in the Netherlands.*

$$D_3 = \frac{d_{31}E}{1-v}\left(\frac{\partial u}{\partial x} + \frac{\partial v}{\partial y} - z\Delta w\right) - \left(\varepsilon_3 - \frac{2Ed_{31}^2}{1-v}\right)\frac{\partial\varphi}{\partial z}$$

(1.2)

$$D_1 = -\varepsilon_1\frac{\partial\varphi}{\partial x} \qquad\qquad D_2 = -\varepsilon_1\frac{\partial\varphi}{\partial y}$$

Here u, v, w - are the displacements of the middle plane of the plate. From the condition of static equivalence, it is convenient to introduce inner forces and moments assigned to a unit length of the middle plane of the plate.

$$T_1 = C\left(\frac{\partial u}{\partial x} + v\frac{\partial v}{\partial y}\right) + A(\varphi_+ - \varphi_-); \quad T_2 = C\left(\frac{\partial v}{\partial y} + v\frac{\partial u}{\partial x}\right) + A(\varphi_+ - \varphi_-)$$

$$S = \frac{1-v}{2}C\left(\frac{\partial u}{\partial y} + \frac{\partial v}{\partial x}\right), \qquad H = -(1-v)D\frac{\partial^2 w}{\partial x\partial y}$$

(1.3)

$$M_1 = -D\left(\frac{\partial^2 w}{\partial x^2} + v\frac{\partial^2 w}{\partial y^2}\right) + hA\left(\varphi_+ + \varphi_- - \frac{1}{h}\int\limits_{-h}^{h}\varphi dz\right)$$

$$M_2 = -D\left(\frac{\partial^2 w}{\partial y^2} + v\frac{\partial^2 w}{\partial x^2}\right) + hA\left(\varphi_+ + \varphi_- - \frac{1}{h}\int\limits_{-h}^{h}\varphi dz\right)$$

Where the following notations are introduced:

$$C = \frac{2Eh}{1-v^2}, \qquad A = \frac{d_{31}E}{1-v}, \qquad D = \frac{2Eh^3}{3(1-v^2)}$$

And φ_+, φ_- denote the values of electrostatic potential on the facial surfaces of the plate $z = \pm h$.

Consider the following surface conditions:

$$\sigma_{33} = 0, \qquad \sigma_{31} = \sigma_{32} = 0$$

(1.4)

a) $\varphi = \varphi^{(e)}$, $\qquad D_3 = D_3^{(e)}$

and also other variants of electric à surface conditions:

b) $\varphi = 0, D_3 = 0$ c) $D_3 = 0$

c) and their combinations on the facial surfaces $z = \pm h$.

Let us average the equations of motion of the plate. Then the expressions of cutting forces as well as equations of plate vibrations are obtained with respect to the displacements of the middle plane and the electrostatic potential.

$$N_1 = -D\frac{\partial}{\partial x}\Delta w + hA\frac{\partial}{\partial x}\left(\varphi_+ + \varphi_- - \frac{1}{h}\int\limits_{-h}^{h}\varphi dz\right)$$

$$N_2 = -D\frac{\partial}{\partial y}\Delta w + hA\frac{\partial}{\partial y}\left(\varphi_+ + \varphi_- - \frac{1}{h}\int\limits_{-h}^{h}\varphi dz\right)$$

(1.5)

$$\Delta u + \frac{1+\nu}{1-\nu}\frac{\partial}{\partial x}\left(\frac{\partial u}{\partial x}+\frac{\partial v}{\partial y}\right) + \frac{d_{31}(1+\nu)}{h}\frac{\partial}{\partial x}(\varphi_+ - \varphi_-) = \frac{2\rho(1+\nu)}{E}\frac{\partial^2 u}{\partial t^2}$$

$$\Delta v + \frac{1+\nu}{1-\nu}\frac{\partial}{\partial y}\left(\frac{\partial u}{\partial x}+\frac{\partial v}{\partial y}\right) + \frac{d_{31}(1+\nu)}{h}\frac{\partial}{\partial y}(\varphi_+ - \varphi_-) = \frac{2\rho(1+\nu)}{E}\frac{\partial^2 v}{\partial t^2}$$

$$D\Delta^2 w + 2\rho h\frac{\partial^2 w}{\partial t^2} - hA\Delta\left(\varphi_+ + \varphi_- - \frac{1}{h}\int_{-h}^{h}\varphi dz\right) = 0$$

$$(1.6)$$

We must add the equation of electrostatics to the obtained equations. It is taken in the three dimensional form

$$\varepsilon_1\Delta\varphi + B\frac{\partial^2\varphi}{\partial z^2} + A\Delta w = 0 \qquad (1.7)$$

Here $B = \varepsilon_3 - 2d_{31}A > 0$. From the system (1.6) it is seen that in the case when constant values φ_+ and φ_- of electrostatic potential are given, then the planar vibrations (vibrations of generalised plane stress state) are separated from the transverse ones.

Let us express the displacements u and v of the middle plane through functions Φ and Ψ.

$$u = \frac{\partial\Phi}{\partial x} + \frac{\partial\Psi}{\partial y}, \qquad v = \frac{\partial\Phi}{\partial y} - \frac{\partial\Psi}{\partial x}$$

Then, after some transformations the system (1.6) yields:

$$c_t^2\Delta\Psi - \frac{\partial^2\Psi}{\partial t^2} = 0; \quad c_l^2\Delta\Phi - \frac{\partial^2\Phi}{\partial t^2} + \chi_1(\varphi_+ - \varphi_-) = 0$$

$$(1.8)$$

$$D\Delta^2 w + 2\rho h\frac{\partial^2 w}{\partial t^2} - \frac{hA}{\varepsilon_1}\left\{\Delta[\varepsilon_2(\varphi_+ + \varphi_-) + 2Aw] + \frac{B}{h}\left[\left(\frac{\partial\varphi}{\partial z}\right)_+ - \left(\frac{\partial\varphi}{\partial z}\right)_-\right]\right\} = 0$$

Here, as usually $\chi_1 = \dfrac{d_{31}E}{2\rho h}$ - is electromechanical connection coefficient, and

$$c_t^2 = \frac{E}{2\rho(1+\nu)}, \quad c_l^2 = \frac{E}{\rho(1-\nu^2)}$$

The functions Φ and Ψ in (1.8) are separated, but they are connected by electric surface conditions, and the problems of planar and transverse vibrations are coupled.

Suppose that it is possible to represent the electrostatic potential as a sum of symmetric and antisymmetric with respect to the transverse coordinate functions :

$$\varphi = \varphi_c + \varphi_a,$$

where $\varphi_c(x,y,h) = \varphi_c(x,y,-h)$, $\varphi_a(x,y,h) = -\varphi_a(x,y,-h)$

Then the system (1.6), and electrostatics equation (1.7) are decomposed into two independent systems:

$$\begin{cases} c_t^2 \Delta \Psi - \dfrac{\partial^2 \Psi}{\partial t^2} = 0 \\[2mm] c_l^2 \Delta \Phi - \dfrac{\partial^2 \Phi}{\partial t^2} + 2\chi_1 \varphi_{a+} = 0 \\[2mm] B\dfrac{\partial^2 \varphi_a}{\partial z^2} + \varepsilon_1 \Delta \varphi_a = 0 \end{cases}$$

(1.9)

$$\begin{cases} D\Delta^2 w + 2\rho h \dfrac{\partial^2 w}{\partial t^2} - \dfrac{2hA}{\varepsilon_1}\left\{ \Delta(\varepsilon_1 \varphi_{c+} + Aw) + \dfrac{B}{h}\left(\dfrac{\partial \varphi_c}{\partial z}\right)_+ \right\} = 0 \\[2mm] B\dfrac{\partial^2 \varphi_c}{\partial z^2} + \Delta(\varepsilon_1 \varphi_c + Aw) = 0 \end{cases}$$

(1.10)

2. Static problems

In [1] some problems of free and forced vibrations of the plates have been discussed, a comparison with known models is brought. Here we would like to consider a static problem, making use of the equations obtained in the first section

Consider the bending problem of a simply supported rectangular plate. On the lateral surfaces the electrostatic potential vanishes:

$$T_1 = 0, \quad W = 0, \quad M_1 = 0, \quad \varphi = 0 \quad \text{for} \quad x = 0, a$$

(2.1)

$$T_2 = 0, \quad W = 0, \quad M_1 = 0, \quad \varphi = 0 \quad \text{for} \quad x = 0, b$$

Consider different electric surface conditions

a)
$$\varphi_+ = \varphi_{0mn} \sin \lambda_m x \sin \mu_n y$$
$$\varphi_+ = -\varphi_{0mn} \sin \lambda_m x \sin \mu_n y$$

(2.2)

b)
$$\varphi_\pm = \varphi_{0mn} \sin \lambda_m x \sin \mu_n y$$

where $\lambda_m = m\pi x/a$, $\mu_n = n\pi y/b$

as well as transversal mechanical loading, acting on the facial surfaces of the plate $z = \pm h$

$$\sigma_{13} = \sigma_{23} = 0, \quad \sigma_{33} = P_{mn} \sin \lambda_m x \sin \mu_n y \quad \text{for} \quad z = \pm h$$

As it is shown in [1], the mechanical boundary conditions in this case are separated. They yield independent systems with respect to w, φ_c and u, v, φ_a.

Consider the second system (1.10) describing transverse vibrations, which in our case takes the form

$$\begin{cases} D\Delta^2 w + 2\rho h \dfrac{\partial^2 w}{\partial t^2} - hA\Delta\left(2\varphi_{\hat{n}+} - \dfrac{1}{h}\int_{-h}^{h}\varphi_{\hat{n}}dz\right) = 0 \\[4mm] \varepsilon_1\Delta\varphi_{\hat{n}} + B\dfrac{\partial^2\varphi}{\partial z^2} + A\Delta w = 0 \end{cases}$$

(2.3)

Seek the solutions of this system in the following form:

$$W = W_{mn} \sin\lambda_m x \sin\mu_n y$$

$$\varphi_c = \varphi_{mn} \sin\lambda_m x \sin\mu_n y$$

(2.4)

Let us substitute (2.4) into the system (2.3). First we solve the differential equation with respect to the electric field potential:

$$B\,\varphi''_{mn}(z) - (\lambda_m^2 + \mu_n^2)(\varepsilon_1\,\varphi_{mn}(z) + AW_{mn}) = 0$$

(2.5)

here

$$B = \varepsilon_3 - 2d_{31}A, \qquad A = \dfrac{d_{31}}{1-\upsilon}$$

The solution of this equation has the following form:

$$\varphi_{mn}(z) = \dfrac{A}{\varepsilon_1}\left(\dfrac{\cosh(\chi Mz)}{\cosh(\chi Mh)} - 1\right)w_0$$

(2.6)

where

$$M^2 = \lambda_m^2 + \mu_n^2 \qquad \chi^2 = \varepsilon_1/B$$

This solution is obtained with the use of electric surface conditions in case a).

Substituting the expression (2.6) into the first equation of (2.3), the following expression is derived for the deflexion W_{mn}

$$W_{mn} = P_{mn}\left[DM^4 + \dfrac{2hM^2A^2}{\varepsilon_1} - \dfrac{2MA^2\tanh(\chi Mh)}{\varepsilon_1\chi}\right]^{-1}$$

(2.7)

In case of absence of mechanical loading: $W = 0$

This solution is obtained with the account of the boundary conditions a). The transversal displacements and the electrostatic potential of the plate have the form (2.4)

Let's consider the first system of (1.9), which in our case has the form:

$$\begin{cases} \Delta u + \dfrac{1+\nu}{1-\nu}\dfrac{\partial}{\partial x}\left(\dfrac{\partial u}{\partial x}+\dfrac{\partial v}{\partial y}\right)+\dfrac{d_{31}(1+\nu)}{h}\dfrac{\partial}{\partial x}(2\varphi_{\delta+})=0 \\[2ex] \Delta v + \dfrac{1+\nu}{1-\nu}\dfrac{\partial}{\partial y}\left(\dfrac{\partial u}{\partial x}+\dfrac{\partial v}{\partial y}\right)+\dfrac{d_{31}(1+\nu)}{h}\dfrac{\partial}{\partial y}(2\varphi_{\delta+})=0 \\[2ex] B\dfrac{\partial^2 \varphi_a}{\partial z^2}+\varepsilon_1\Delta\varphi_a = 0 \end{cases} \qquad (2.8)$$

The solution of the electrostatics equation has the form:

$$\varphi_a(z) = \frac{\varphi_0\sinh(\chi Mz)}{\sinh(\chi Mh)} \qquad (2.9)$$

Here φ_0 is the maximal value of the electrostatic potential on the facial surfaces of the plate, under the electric surface conditions a)

Let us seek the displacement u, v of the middle plane in the form

$$u = \lambda_m\, u_{mn}\, \sin\lambda_m x \sin\mu_n y$$
$$v = \mu_n\, v_{mn}\, \sin\lambda_m x \sin\mu_n y \qquad (2.10)$$

Substituting the expressions (2.10) into the system (2.8), we obtain for the displacements:

$$V_0 = \frac{1+\nu}{1-\nu}\left(\frac{m\pi}{a}\right)^2 \frac{2d_{31}(1+\nu)\chi Mh\cosh(\chi Mh)}{h^2\left[M^2+\left(\dfrac{m\pi}{a}\right)^2\dfrac{1+\nu}{1-\nu}\right]}\varphi_0 \times$$

$$\times\left\{\left(\frac{1+\nu}{1-\nu}\right)^2\frac{\left(\dfrac{m\pi}{a}\right)^2\left(\dfrac{n\pi}{b}\right)^2}{M^2+\left(\dfrac{m\pi}{a}\right)^2\dfrac{1+\nu}{1-\nu}}-\left[M^2+\left(\frac{n\pi}{b}\right)^2\frac{1+\nu}{1-\nu}\right]\right\}^{-1} \qquad (2.11)$$

$$U_0 = \left\{2d_{31}(1+\nu)\chi Mh\,\mathrm{ctanh}(\chi Mh)\varphi_0 - \frac{1+\nu}{1-\nu}\left(\frac{hn\pi}{b}\right)^2 V_0\right\}\times$$

$$\times\left\{(hM)^2+\left(\frac{hm\pi}{a}\right)^2\frac{1+\nu}{1-\nu}\right\}^{-1} \qquad (2.12)$$

Consider the electric surface conditions in case b).

Then, the following expressions are derived for the symmetric part of the electrostatic potential and the maximal deflexion:

$$\varphi_{mnc}(z) = \frac{\dfrac{A}{\varepsilon_1}W_{0mn} + \varphi_{0mn}}{\cosh(\chi Mh)}\cosh(\chi Mz) - \frac{A}{\varepsilon_1}W_{0mn}$$

$$W_{0mn} = \frac{\left[P_b - 2AhM^2\left(1 - \dfrac{\tanh(\chi Mh)}{\chi Mh}\right)\varphi_{0mn}\right]}{DM^4 + \dfrac{2hA^2M^2}{\varepsilon_1}\left(1 - \dfrac{\tanh(\chi Mh)}{\chi Mh}\right)} \tag{2.13}$$

Let us study the system describing the generalised plane stress state. We obtain that the antisymmetric part of the electrostatic potential is equal to zero. The system (2.8) does not depend on piezoeffect. It yields a system describing the generalised plane stress state of transversely isotropic body.

In the case when no mechanical loading acts on the plate, it bends under the influence of the electric field with the following maximal deflexion:

$$W_0 = \frac{2Ah\left(1 - \dfrac{\tanh(\chi Mh)}{\chi Mh}\right)\varphi_{0mn}}{DM^4 + \dfrac{2hA^2}{\varepsilon_1}\left(1 - \dfrac{\tanh(\chi Mh)}{\chi Mh}\right)} \tag{2.14}$$

Consider the problem of bending of the plate under the influence of the electric field,
when the following boundary conditions are given:

$$\varphi_{\pm} = \varphi_{0mn}\sin\lambda_m x \sin\mu_n y, \qquad P_{mn} = 0 \quad \text{for} \quad z = \pm h$$

The plate is simply supported along its edges $x = 0, a$. the condition of siding contact is given at the edges $y = 0, b$. The normal component of the induction of the electric field vanishes on the lateral surfaces of the plate.:

$$M_1 = 0, \quad W = 0, \quad \varphi = 0 \qquad \text{for} \quad x = 0, a$$

$$N_2 = 0, \quad \frac{\partial W}{\partial y} = 0, \quad D_y = 0 \qquad \text{for} \quad y = 0, b \tag{2.15}$$

We seek the solutions of the system (2.3) in the form

$$W = W_{mn}\sin\lambda_m x \cos\mu_n y$$

$$\varphi_c = \varphi_{mn}\sin\lambda_m x \cos\mu_n y$$

Then the boundary conditions are identically satisfied and the expressions (2.13) are obtained for the maximal deflexion and the maximal value of the electrostatic potential. These values coincide with those obtained in the case of the simply supported by its lateral surfaces plate.

However, in the considered case the maximal deflection and the maximal value of the electrostatic potential are reached in different from the previous case, points.

Expanding the function $\tanh(\chi Mh)$ into series and limiting ourselves to the first three terms of the expansion, we obtain the following expression for the maximal deflexion:

$$W_0 = \frac{\dfrac{2A}{3}\left[-1 + \dfrac{2}{5}(\chi Mh)^2\right]\varphi_0}{\dfrac{DB}{\varepsilon_1 h^3} + \dfrac{2A^2}{3\varepsilon_1}\left[1 - \dfrac{2}{5}(\chi Mh)^2\right]} \tag{2.16}$$

Thus, the problems of bending and generalised plane stress state are decoupled under the considered electric boundary conditions. They yield independent systems with respect to w, φ_c and u, v, φ_a respectively. Moreover, in the case when the symmetric part of the electrostatic potential is given on the facial surfaces of the plate, the displacements u and v of the middle plane do not depend on piezoeffect. The expressions have the same form as the ones, obtained for transversely-isotropic body. In the absence of the mechanical loading the plate is bent under the influence of the electric field.

In the bending problem of the plate the values for the maximal deflexion and the maximal value of the electrostatic potential are obtained to coincide for two cases of mechanical boundary conditions:1) in the case of a simply supported plate, 2) in the case of the plate, which is simply supported by its two opposite lateral surfaces and the sliding contact is prescribed on the other two. However, these values are reached in different points for different boundary conditions.

3. References:

1. Mktrchyan L. R. (1997) On separation of transverse and planar vibrations of piezoelectric plates, *Proceedings of NAS of RA Mechanika*. **50**, 3-4,(in Russian).

2. Belubekian M.V., Mkrtchyan L.R. (1996)Vibrations of transversely polarised thin piezoelectric plates, *Proceed of the Int. Conf. on "Smart Structures and Materials-95"*. San Diego, California,. Editor. A. Peter Jardine. SPIE, **2441**, 233-242.

3. Grinchenko V.T., Karlash V.M., Meleshko V.V., Ulitko A.F. (1976) Investigation of planar vibrations of rectangular piezoceramic plates, *Prilkladnaya Mechanika*, **Vol.13, No5**, 71-79. (in Russian)

4. Parton V.Z. Kudriavtsev B.A. (1988) *Electomagnetoelasticity of piezoelectric and electro-conductive plates*, Nauka, Moscow, 422p (in Russian).

QUASICRYSTALLINE MATERIALS. STRUCTURE AND MECHANICAL PROPERTIES

Yu.V. MILMAN, D.V. LOTSKO, A.M.BILOUS
I.M.Frantsevych Institute for Problems of Material Science of the
National Academy of Sciences of Ukraine
3 Krzhizhanovsky Str, o3142 Kyiv, Ukraine
S.M. DUB
Institute for Superhard Materials of the NAS of Ukraine
2 Avtozavodska Str, 04074 Kuiv, Ukraine

A brief description of the structure of quasicrystalline materials, their behavior under mechanical load and the mechanism of their high-temperature plastic deformation on the base of literary data are given. Original data about the investigation of the deformation of AlCuFe quasicrystal by a complex of micro- and nanoindentation techniques in a wide temperature interval are presented.

1. Introduction

1.1. STRUCTURE OF QUASICRYSTALS

Quasicrystals are solids with perfect long-range order, but with no translational periodicity in atomic arrangement which is characterized by 5, 8, 10, and 12-fold rotation symmetry forbidden in crystalline materials [1]. Many scientists consider the discovery of quasicrystals together with amorphous metal glasses to be the highest achievement of material science in twentieth century.

Quasicrystalline phases are found in more than 100 systems, many of them aluminum based. Two types of quasicrystals are indentified: quasiperiodic in two dimensions (polygonal) with one periodic direction of 8, 10 or 12-fold symmetry; and quasiperiodic in three dimensions (icosahedral) with no periodic direction and 12×5-fold axes. Among the latter, AlPdMn and AlCuFe are the most investigated because they are stable phases and can be obtained in a rather perfect form, AlPdMn in the form of single crystals [2,3]. The formation of quasicrystalline phases obeys to Hume-Rothery rule of electron concentrations, in particular, in AlCuFe(Cr) system, these phases exist near to a line at constant electron per atom ratio $e/a = 1.86$ [4].

The structure of icosahedral quasicrystals is built of multiple shell clusters [5]. For Al-based systems, there are two types of icosahedral clusters: Frank-Kasper type (e.g. AlLiCu) and Mackay type (AlPdMn, AlCuFe etc.). Mackay clusters in AlPdMn consist of 51 atoms [3], and in AlCuFe - of 33 atoms [6]. These are based

289

M.-I. Baraton and I. Uvarova (eds.),
Functional Gradient Materials and Surface Layers Prepared by Fine Particles Technology, 289–296.
© 2001 *Kluwer Academic Publishers. Printed in the Netherlands.*

on elementary icosahedrons consisting of 12 Al atoms with covalent bonding [5]. Mackay type clusters of about 0.9 nm in diameter are found to be the main obstacle in plastic deformation and cleavage of AlPdMn quasicrystals [3,7], so in this sense quasicrystals can be regarded as nanostructural materials.

The crystal lattice of quasicrystals is interpreted as a cubic one (FCC or BCC) in 6-dimensional hyperspace [3]. Distortions in this lattice, including Burgers vectors of dislocations, have two components: a phonon one which is a projection to the real physical 3-dimensional space; and a phason one which is a projection to the complementary, or perpendicular, space.

1.2. MECHANICAL PROPERTIES OF QUASICRYSTALS

Unusual atomic arrangement are responsible for complex physical and mechanical properties of quasicrystals. They are characterized by high hardness (to 8-10 GPa), high Young modulus (to 200 GPa), high wear and corrosion resistance, low thermal conductivity etc. For practical use, quasicrystals are of a great interest in two ways: as thermal barrier and low friction coatings; and as hardening precipitates of nanoscale size in aluminum and steel alloys stable to high temperatures [8].

As to mechanical properties, quasicrystals reveal no macroplasticity to the temperature $T_{db} \approx 0.8-0.85 T_m$ [9] (600°C for AlCuFe [10] and 680°C for AlPdMn [3]). At temperatures higher than T_{db}, quasicrystals can deform to rather high strains without cracks (to 130% in compression test of AlCuFe [10]), and this deformation is accompanied by a feature never observed in crystalline and amorphous materials, namely the prominent deformation softening. An exception is the deformation of strongly deformed metals at high temperatures, when recrystallization and recovery processes take place during deformation.

The mechanism of plastic deformation at $T>T_{db}$ for AlPdMn was established by experiments in situ in a transmission electron microscope with single crystals [3,7]. The deformation was shown to occur by dislocation generation, gliding and climb which were also observed in polycrystals [11]. More than 90% of dislocations were found to have Burgers vectors along 2-fold quasilattice directions and to move on 5-fold planes. Strain softening was connected with the observed increase of phason components of dislocation Burgers vectors which was explained by energetically favourable dislocation reactions [3,7]. Moving dislocations shall circumvent or cut the Mackay clusters which are regions of higher mechanical strength. Phason components of Burgers vectors cause structural and chemical disorder which has no analogs in crystalline solids. This leads to the destruction of clusters and to weakening of obstacles to dislocation motion. High-temperature deformation by dislocation mechanism was observed also in AlCuFe [12], in this case it was concentrated in deformation bands exhibiting a high dislocation density and networks of small-angle boundaries.

At temperatures $T<T_{db}$, quasicrystals reveal no macroplasticity, and while deformation their behavior is similar to ceramic materials [2]. Thus, fracture toughness in quasicrystals $K_{1C} = 0.5-1.5$ MPa·m$^{1/2}$ which is close to the K_{1C} value in silicon. In investigations by indentation [9,13-16], scratching [17] and friction [18] quasicrystals at room temperature revealed a prominent microplasticity: hardness prints were clear, without any sign of destruction under the indenter; and

plastic swellings and extrusions in edges of scratching and friction tracks were observed. Dubois with co-authors [18,19] had made a supposition that microplasticity is caused by a phase transition stimulated by high pressure under the indenter. The possibility of such transitions was demonstrated in experiments for anisotropic compression of AlPdMn under a pressure higher than 20 GPa [20]. But, direct evidences of a phase transformation with indentation were absent.

2. Experiment

A complex technique for studying mechanical properties of materials by local loading including measuring hardness by Vickers indenter in the temperature interval from -196 to 950°C (the load P=234N), the load dependence of hardness, hardness by penetration depths with the estimation of Young modulus, nanoindentation technique, and a new technique of the construction of the deformation curve by using a set of trihedral indenters with various angles at the tip were developed and reported [21-23]. The last method is a unique means to obtain a deformation curve for materials that are brittle in the usual testing methods.

Experiments in nanoindentation were carried out in Nano Indenter II (Nano Instrument Inc., Oak Ridge, USA) using a Berkovich indenter. The displacement was measured with the accuracy of 0.04 nm, and the load - with the accuracy of 75 μN. Registrations were made 3 times a second.

AlCuFe quasicrystals obtained by compacting air-atomized powder under a quasihydrostatic pressure of 5 GPa at 700°C were used for the experiments. Structure characterization of specimens was done using an X-ray diffractometer with CuK$_\alpha$ radiation. Specimen density was determined by hydrostatic weighing.

3. Results and discussion

3.1. ROOM-TEMPERATURE PROPERTIES

As reported [24], after atomization, the powder consisted of two phases - quasicrystalline AlCuFe and crystalline β-phase of FeAl type. To obtain a single-phase quasicrystal, it is enough to make an annealing at 700°C 2 h. The annealing was carried out before (specimen 1) and after pressing (specimen 2). For specimen 1, X-ray investigation revealed a significant line broadening, and the hardness was much lower (see Table). Thus, high temperature deformation defects induced by pressure have facilitated the deformation at room temperature under the indenter. Lower hardness in specimen 1 is not caused by larger porosity because its density is even a little higher than of specimen 2.

TABLE. Properties of AlCuFe quasicrystal at room temperature

Specimen No.	Hardness HV, GPa	Young modulus E, GPa	Density ρ, g/cm³	X-ray line half-width b, deg
1	4.37		4.709	0.247
2	7.53	113	4.693	0.075

3.2. TEMPERATURE DEPENDENCE OF HARDNESS

In the temperature dependence of hardness of Al-Cu-Fe, the investigation revealed a low-temperature athermal part to about $0.55T_m$ (Fig.1). In silicon, as was shown [25,26], it is undoubtedly connected with the phase transition under the indenter to a plastic metallic phase, and the level of hardness corresponds to the pressure of the phase transition. In silicon indentations, plastic swellings were also observed and extrusions [27] similar to ones reported by us [14-16]. An alternative explanation of the athermal part at low temperatures in HV(T) is due to the intensive development of brittle fracture under the indenter [22,23], but in our case, prints are fairly smooth without cracks. The formation of radial and lateral cracks around the indentation does not lead to the appearance of the athermal part in the HV(T) curve [22].

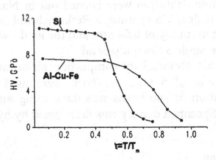

Figure 1. Temperature dependence of hardness for AlCuFe quasicrystal and silicon

It should be noted that the characteristic deformation temperature (knee temperature) in a quasicrystal ($\sim 0.9T_m$) is much higher than in covalent silicon crystal ($0.8T_m$) that depicts a very high value of activation energy of dislocation motion in quasicrystals [2, 28].

3.3. STRESS-STRAIN CURVES FOR QUASICRYSTALLINE AlCuFe

In Fig.2, stress-strain curves are presented in coordinates σ-ε where σ is the flow stress estimated from the measured Meier hardness by Tanaka's formula [29], and ε is the total deformation under the indenter that is determined by the angle at the tip [21,22]:

$$\frac{HM}{\sigma} = \frac{2}{3}\left[1 + \frac{3}{2}f + \ln\left(\frac{E\sqrt{\pi}\,ctg\gamma_1}{12\sigma(1-v)}\right)\right].$$

Here HM is Meier hardness, γ_l is the angle of indenter grinding, $\nu=0.28$ is the Poisson coefficient, $f=1$ for metals, and $f\approx1/3$ for ceramics. For calculations $f=1$ was taken.

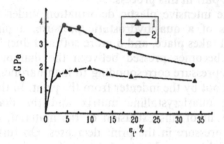

Figure 2. Stress-strain curves constructed by measuring Meier hardness by a set of indenters with various angles at the tip for specimens 1 and 2

In curves for both specimens, there is seen a prominent strain softening, especially strong in specimen 2. It leads to a supposition that dislocation motion which leads to the destruction of clusters may play a part in the deformation under the indenter, and in specimen 1 which contains destroyed clusters, this deformation is much easier.

3.4. NANOINDENTATION OF AlCuFe QUASICRYSTAL

A characteristic feature of nanoindentation curves is the appearance of steps at a certain value of the load, which is increased with the growth of the loading rate (Fig.3).

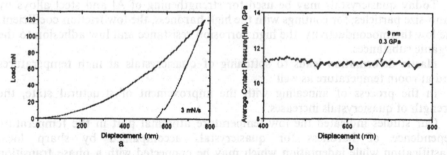

Figure 3. Depth-penetration curves for nanoindentation of AlCuFe specimen 2, loading rate 3 mN/s: a - in coordinates load-displacement; b - in coordinates HM - displacement

Evidently, Wolf and Paufler [30] did not observe these steps because they used loads up to 0.5 mN, and in our experiments with the loading rate as low as 1 mN/s, steps appeared only at a load of 30 mN. The step-like process of quasicrystal nanoindentation is expressed more clearly after the reconstruction of loading curve with the determination of the average contact pressure, i.e. Meier hardness [26] (Fig.3 b). It is seen that after rather gradual growth of the load, a sudden drop takes place that evidently corresponds to the formation of a small portion of more plastic phase that facilitates the deformation. Maybe, a local destruction of Mackay clusters occurs in a thin layer under the indenter, and dislocations may take part in this process.

As a result of the intensive plastic deformation under the indenter (5-8 % [19,20]) in conditions of a quasihydrostatic pressure, a phase transition from quasicrystal to crystal takes place, and a more soft and ductile crystalline phase is formed. This phase becomes pressed between the diamond indenter and the quasicrystal. Under a pressure corresponding to high hardness of the quasicrystal, this phase is squeezed out by the indenter from the print. In this process, the forces of friction with the quasicrystalline matrix and the diamond indenter are surmounted. As a result of this extrusion of the material, the indenter falls by several nm, and the pressure in the print decreases. On further increase of the load, new portions of the more ductile phase are formed, and, when its thickness will be sufficient for overcoming friction forces, the extrusion of the material from under indenter occurs again. With the further increase of the load on the indenter, the pressure in the print grows again, till a new portion of the quasicrystal in the print sufficient for overcoming friction forces will transform in a crystalline state.

Steps of similar kind were observed while nanoindentation of germanium [31], in which a low-temperature phase transition under the indenter was also found [25]. But, if in semiconductors the phase transition from semiconductor to metal is completely controlled by the value of pressure [25] in quasicrystals, it is observed in the case of a combination of high pressure and some critical plastic deformation.

4. Conclusions

Quasicrystals are a new perspective class of materials with unusual properties, which can be used for creating nanocomposite materials and coatings.

Today quasicrystals may be used for strengthening of Al and steel alloys by nano-size particles, for coatings with the high hardness, the low friction coefficient, the low thermoconductivity, the high corrosion resistance and low adhesion to the organic substances.

Plastic deformation leads to softening of quasicrystals at high temperatures and at room temperature as well.

In the process of annealing with the improvement of structural state, the strength of quasicrystals increases.

Our studies indicated the low temperature athermal part in the temperature dependence of hardness for quasicrystals accompanied by sharp local plastification while indentation which may be connected with a phase transition under the indenter.

For the first time, we are reporters of stress-strain curves for quasicrystals at room temperature.

5. Acknowledgements

The authors are grateful to Dr. I.I. Timofeeva and Dr. A.I. Bykov for the help in producing quasicrystalline compacts and to Dr. V.L. Rupchev for help in producing Al-Cu-Fe powders.

6. References

1. Dubois, J.M. (1998) *Introduction to Quasicrystals*, Springer Verlag, Berlin, 392 p..
2. Takeuchi, S. (1994) Physical properties of quasicrystals - an experimental review, *Mater. Sci. Forum* **150&151**, 35-52.
3. Feuerbacher, M., Metzmacher, C., Wollgarten, M., Urban, K., Baufeld, B., Bartsch, M., and Messerschmidt, U. (1997) The plasticity of icosahedral quasicrystals, *Mater. Sci. and Eng.* **A233**, 103-110.
4. Dong, C., Perrot, A., Dubois, J.-M., and Belin, E. (1994) Hume-Rothery phases with constant e/a value and their related electronic properties in Al-Cu-Fe-(Cr) quasicrystalline systems, *Mater. Sci. Forum* **150&151**, 403-416.
5. Kimura, K., Matsuda, H., Tamura, R., Fujimori, M., Schmechel, R., and Werheit, H. (1995) Interrelation between icosahedral aluminum-based quasicrystal and boron-rich solids, in Ch. Janot and R. Mosseri (eds), *Proc. of the 5th Internat Conf. on Quasicrystals, Avignon, 22-26 May 1995*, World Scientific, Singapore, p.p. 730-738.
6. Katz, A., and Gratias, D., Chemical order and local configurations in AlCuFe-type icosahedral phase, *ibid.*,p.p. 164-167.
7. Urban, K., Ebert, Ph, Feuerbacher, M., Franz, V., Wollgarten, M., Bartsch, M., Baufeld, B., and Messerschmidt, U. (1997) Mechanical properties of quasicrystals, in S.Takeuchi and T.Fujiwara (eds), *Proc. of the 6th Internat. Conf. on Quasicrystals, Tokyo, Japan, 26-30 May 1997*, World Scientific, Singapore, p.p. 493-500.
8. *Quasicrystals* (1999), J.-M. Dubois, P.A. Thiel, A.-P. Tsai, and K. Urban (eds), *Proc. of 1998 MRS Fall Meeting, Boston, MA,*, 553 p.
9. Wolf, B., and Paufler, P. (1999) Mechanical properties of quasicrystals investigated by indentation and scanning probe microscopes, *Surface and Interface Analysis* **27**, 592-599.
10. Bresson, L., and Gratias, D. (1993) Plastic deformation in AlCuFe icoahedral phase, *J. Non-Cryst. Solids* **153&154**, 468-472.
11. Shield, J.E., Kramer, M.J., and McCallum, R.W. (1994) Plastic deformation in icosahedral AlPdMn alloys, *J. Mater. Research* **9**, 343-347.
12. Köster, U., Ma, X.L., Greiser, J., and Liebertz, H. (1997) Plastic deformation of AlCuFe quasicrystals, in Ch. Janot and R. Mosseri (eds), *Proc. of the 6th Internat. Conf. on Quasicrystals, Tokyo, Japan, 26-30 May 1997*, World Scientific, Singapore, p.p. 505-508.
13. Köster, U., Liu, W., Liebertz, H., and Michel, M. (1993) Mechanical properties of quasicrystalline and crystalline phases in Al-Cu-Fe alloys (1993) *J. Non-Cryst. Solids* **153&154**, 446-452.
14. Trefilov, V.I., Milman, Yu.V., Lotsko, D.V., Belous, A.N., Chugunova, S.I., Timofeeva, I.I., and Bykov, A.I. (2000) Study of mechanical properties of quasicrystalline Al-Cu-Fe phase by indentation tecnique, *Reports of Russian Academy of Sciences*, in press.
15. Belous, A.N., Milman, Yu.V., Lotsko, D.V., Chugunova, S.I., Ivashchenko, R.K., Bykov A.I., Timofeeva, I.I., and Rupchev V.L. (1998) Investigation by indentation technique of mechanical properties of a quasicrystalline compact of Al-Cu-Fe system manufactured by hot pressing, in S.A. Firstov (ed), *Electron Microscopy and Strength of Materials*, IPMS of NASU, Kyiv, 9, 189-212.
16. Milman, Yu.V., Lotsko, D.V., Belous, A.N. (1999) Mechanical properties of quasicrystalline materials, in V.I. Betekhtin, S.P. Beliayev, Yu.M. Dal, Z.P. Kamentseva, A.I. Slutsker (eds.) *Deformation and Fracture Mechanisms of Advanced Materials, Proc. of the XXXV Seminar "Actual Problems of Strength", 15-18 September 1999, Pskov*, p.p. 463-470.
17. Hornbogen, E.. and Shandl, M. (1992) Probing mechanical properties of quasicrystalline aluminum alloys, *Zs. Metallkunde* **83**, 128-131.
18. Von Stebut, J., Strobel, C., and Dubois, J.M. (1997) Friction response and brittleness of polycrystalline AlCuFe icosahedral quasicrystals, in Ch. Janot and R. Mosseri (eds), *Proc. of the 5th Internat Conf. on Quasicrystals, Avignon, 22-26 May 1995*, World Scientific, p.p. 704-713.

296

19. Kang, S.S., and Dubois, J.M. (1992) Compression testing of quasicrystalline materials, *Phil. Mag A* **66**, 151-163.
20. Yan, Y., Baluc, N., Peyronneau, J., and Kleman, M. (1995) Pressure-induced transformations in icosahedral AlPdMn, in Ch. Janot and R. Mosseri (eds), *Proc. of the 5th Internat Conf. on Quasicrystals, Avignon, 22-26 May 1995*, World Scientific, Singapore, p.p. 668-671.
21. Milman, Yu.V., Galanov, B.A., and Chugunova, S.I. (1993) Plasticity characteristic obtained through hardness measurement, Overview No. 107, *Acta metall. mater.* **41**, 2523-2532.
22. Milman, Yu.V. (1999) New methods of micromechanical testing of materials by local loading with a rigid indenter, in I.K.Pokhodnia (ed.) *Advanced Materials Science: 21ˢᵗ Century*, Cambridge International Science Publishing, Cambridge, p.p. 638-659.
23. Milman, Yu.V., Chugunova, S.I., Goncharova I.V., Chudoba, T., Lojkowski, W., and Gooch, W. (1999) Temperature dependence of hardness in silicon - carbide ceramics with different porosity, *Refractory Metals & Hard Materials* **17**, 361-368.
24. Sordelet, D.J., Besser, M.F., and Anderson, I.E. (1996) Particle size effects on chemistry and structure of Al-Cu-Fe quasicrystalline coatings, *J. Thermal Spray Technology* **5(2)**, 161-174.
25. Gridneva, I.V., Milman, Yu.V., and Trefilov, V.I. (1972) Phase transition in diamond-structure crystals during hardness measurements, *Phys. stat. sol. (a)* **14**, 177-182.
26. Novikov, N.V., Dub, S.N., Milman, Yu.V., Gridneva, I.V., and Chugunova, S.I. (1996) Application of nanoindentation method to study a semiconductor-metal phase transformation in silicon, *J. of Superhard Materials* **18**, 32-40.
27. Suzuki, T. and Ohmura, T. (1996) Ultra-microindentation of silicon at elevated temperatures, *Phil. Mag. A* **74**, 1073-1084.
28. Trefilov, V.I., Milman, Yu.V., and Gridneva, I.V. (1984) Characteristic temperature of deformation in crystalline materials, *Crystal Res. Technol.* **19**, 413-421.
29. Tanaka, V. (1987) Elastic/plastic indentation hardness and indentation fracture toughness: the inclusion core model, *J. of Mater. Science* **22**, 1501.
30. Wolf, B. and Paufler, P. (1999) Mechanical Properties of quasicrystals studied by scanning probe mocroscopy, *European Microscopy and Analysis*, **62**, 21-23.
31. Pharr, G.M.,Oliver, W.C., Cook, R.F., Kirchner, P.D., Kroll, M.C., Dinger, T.R., and Clarke, D.R. (1992) Electrical resistance of metallic contacts on silicon and germanium during indentation, *J. Mater. Res.* **7**, 961-972.

MD SIMULATION OF THE ION-STIMULATED PROCESSES IN SI SURFACE LAYERS

A.E. KIV, T.I. MAXIMOVA, V.N. SOLOVIEV
Department of Materials Engineering,
Ben-Gurion University of the Negev, Beer-Sheva, 74895, Israel

1. Introduction

The progress in understanding of Si (001) surface structure has been well described in a series of review papers and conference proceedings [1,2] This surface has two dangling bonds per surface atom which move towards each other in pairs leading to (2x1) unit cell formation. Higher-order periodicities have been observed as well as disordered atomic configurations. Ab initio theoretical calculations of relaxation processes in semiconductor surface layers are limited by the great complexity of the phenomena involved, and thus forcing the use of simulation tools.

The Molecular Dynamics (MD) method gives a useful insight into the problem. Many results (for example [3]) show that the MD method can provide an important guide for the exploring models for Si surface, and what's more the simulation scheme may be optimized by simultaneous electronic calculations [4].

A widespread technological process in microelectronics is radiation treatment of Si surface. This process is used for surface cleaning and ion-assisted dry etching, but it has the inherent drawback that the particle – surface interaction produces a distorted and disordered surface layers. It is of great interest to study destruction and relaxation processes induced by low-energy ion bombardment of Si surface, in particular by self-ion implantation [5]. According to the ion energy dose, dose rate, and the temperature conditions Si surface can be damaged, amorphized, recrystallized [6], etc.

The problem is to clarify the conditions of the ion irradiation and to determine parameters of ion beams which stimulate relaxation processes leading to the best surface characteristics.

In this paper we described calculations of relaxation processes of Si surface layers at elevated temperatures and caused by ion beam bombardment. New details in microstructure of relaxed Si surface layers are obtained. The energy dependencies of ion-stimulated atomic processes show that the most expressed effect of improvement of Si surface layers takes place in vicinity of the energy threshold for elastic atom displacement in Si lattice.

M.-I. Baraton and I. Uvarova (eds.),
Functional Gradient Materials and Surface Layers Prepared by Fine Particles Technology, 297–303.
© 2001 *Kluwer Academic Publishers. Printed in the Netherlands.*

2. Model and Method

The starting configuration was taken as a parallelepiped containing 864 atoms: 12 layers with 72 atoms in each one. Periodic boundary conditions were used in two dimensions.

At first all atoms were in normal lattice positions. MD method was applied in its standard form [7] i.e. the equations of motion were solved by using the central difference scheme. The time-step was 10^{-14} s.

Simulations were performed with Stillinger-Weber(SW) [8] potential. Many other potentials for Si exist in the literature. But experience with SW potential indicates that it is a reasonable presentation of Si for the study of ion beam processing, accurately describing many properties of small Si clusters, bulk and surfaces.

The scenario of pulse ion irradiation was as follows. After equilibration of the system, one atom is given the velocity corresponding to the chosen energy and beam angle of incidence that needs to be simulated. Then we were waiting when the system reaches the equilibrium with the surrounding thermal bath. In the case of continuous ion irradiation the next ion pulse was done immediately after the previous one.

We have chosen new routes to investigate relaxation processes in Si surface layers. In this work we have used the new MD approach for investigations of full Si surface relaxation with taking into account the possibility of re-building and re-hybridization of chemical bonds as it was done in [9]. This approach allowed us to get new structural peculiarities of the relaxed Si surface layers.

3. Results

3.1. QUASI-DISORDERED PHASE

The results of computer simulation of Si surface relaxation processes have shown that a quasi-disordered phase (QDP) arises as a result of free Si surface relaxation. Only four near-surface layers form the QDP. The model has emphasized that in particular the microstructural constituents in relaxed Si surface layers are nodes with one or more dangling bonds. One can see that the Radial Distribution Functions and Angle Distribution Functions (the angles between chemical bonds are implied) are similar to those for a-Si (Fig.1). As a result of relaxation non-hexagonal polygons and dangling bonds were discovered in Si near-surface layers. The fifth layer does not differ practically from more deep layers.

Each of four layers which forms the QDF has his specific structural characteristics. The third layer has less structural and electronic defects in comparison with other layers of QDP. This layer almost has not dangling bonds. But at the same time the third layer is the most distorted one. Atoms of the third layer are characterized by the largest displacements upwards relatively to the normal position of the corresponding crystallographic plane. Analysis of microstructure of near-surface layers of Si indicates that the third layer plays a special role in stabilization of QDP. This layer is a transitional one between the crystal volume and the surface layers.

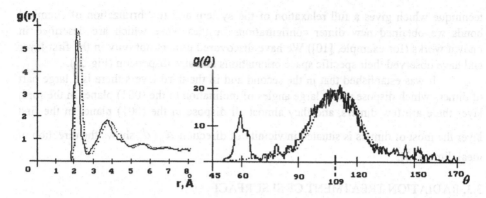

Figure 1. The Atomic Radial Distribution Function (left) and the Angle Distribution Function (right) for the relaxed (001) Si surface layers (unbroken curve). The doted lines correspond to a-Si.

3.2. STRUCTURE OF DIMERS IN RELAXED SI (001) SURFACE

Because of the structure of relaxed Si surface layers contains dangling bonds, the conditions arise for formation of dimers not only in the first layer. In near surface layers new space configurations of dimers were obtained.

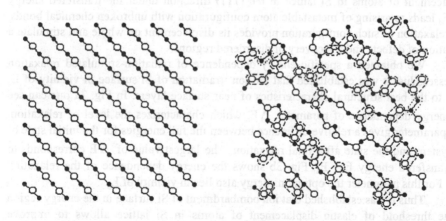

Figure 2. The dimer structure in the relaxed (001) Si surface layers
(dimers in the first layer on the right). On the left is a nonrelaxed surface.

As a rule dimers are investigated by moving only atoms of the first layer by equal and opposite distances along the given direction. By using the simulation technique which gives a full relaxation of the system and rehybridization of chemical bonds we obtained new dimer configurations besides those which are described in known works (for example, [10]).We have discovered dimers not only in the first layer and have observed their specific space orientations and size dispersion (Fig. 2).

It was established that in the second and in the third layers there is a large part of dimers which dispose under large angles of inclination to the (001) plane. In the forth layer there are few dimers, and they almost all dispose in the (001) plane. In the first layer the most of dimers is situated in vicinity of direction $\vec{d}_0$ ($\vec{d}_0$ shows the direction of ideal dimers [3].

3.3. RADIATION TREATMENT OF SI SURFACE

We investigated radiation effects in Si surface layers caused by low-energy ion bombardment. The results were obtained for the transfered energies near the threshold E_d for atom displacements in Si lattice under the elastic collisions [11]. The following energies of bombarding particles were chosen: 10, 20, 30, 40 and 50eV. The irradiation flux was modeled so that one pulse corresponded to $2 \cdot 10^{12}$ particles/sm$^2 \cdot$s.

For monoenergetic monoisotopic ion bombardment the induced structural changes depend on the bombardment angle. Therefore we have compared the results for different energies of ion beams at the same bombardment angles, ion doses and dose rates.

There are structural processes in this energy region of E_d which are known as "grasshopper effect" in diamond-like lattices [12]. It was shown in [12] that the displacement of atoms in Si lattice in the (111) direction under the transfered energy $E \cong E_d$ leads to arising of metastable atom configuration with unbroken chemical bonds. The relaxation of such configuration provides its displacement on whole and stimulate a migration of defects and a recovery of disordered regions.

We obtained a specific energy dependence of radiation-stimulated relaxation processes. Just it was established that the ion irradiation of Si surface in vicinity of E_d leads to the best structural characteristics of near surface layers. In Fig. 3a one can see the energy dependence of parameter ΔE, which characterizes the level of relaxation. This parameter gives a relative difference between the full energies for the initial state of the system and the state after final relaxation. The largest value of ΔE corresponds to the transfered energy $E \sim E_d$. Fig. 3b shows the energy dependence of the relaxation time. For this parameter the optimum energy also lies in vicinity of E_d.

Thus it was established that ion bombardment of Si surface in the energy region of the threshold of elastic displacement of atoms in Si lattice allows to improve structural characteristics of surface layers and decrease the relaxation time. The Fig.4a,b illustrates the influence of ion-beam treatment of Si surface on the each of near-surface layers separately and the results concerning the pulse and the continuous irradiation.

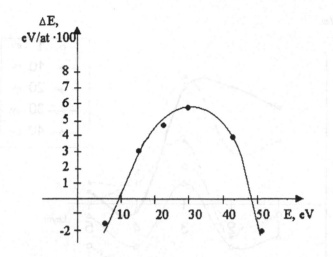

Figure 3a. Energy dependence of the level of relaxation for the pulse ion treatment of Si surface layers.

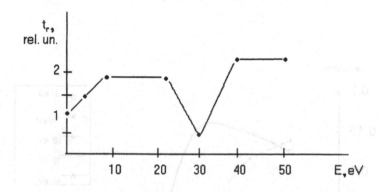

Figure 3b. Energy dependence of the relaxation time for the pulse ion of Si surface layers.

4. Conclusion

Molecular Dynamics technique gives a useful insight into the problem of radiation-stimulated semiconductor surface relaxation.

The high-temperature relaxation of the free Si surface leads to formation of quasi-disordered phase (QDP). Re-building and re-hybridization of chemical bonds may occur in the first four layers.

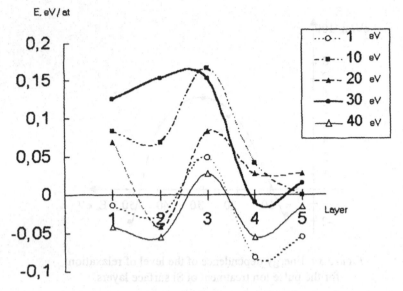

Figure 4a. Effect of radiation-stimulated restoration
of different Si surface layers.

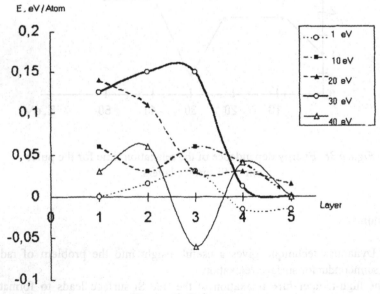

Figure 4b. Pulse treatment

The QDP of Si surface is characterized by non-usual polygons and dangling bonds. Radial distribution function and angle distribution function are similar to those characteristics of a-Si.

The bombardment of Si surface by low energy particles leads to radiation-induced relaxation of surface layers.

Energy dependencies of radiation-induced relaxation have shown that the radiation stabilization of Si surface layers takes place in the near-threshold energy range $E < 30eV$.

The simulation results open the important ways to influence on the atomic configurations of each layer separately by using low energy ion beams in radiation treatment of Si surface

References

1. Srivastava, G.P. (1997) Microstructure of the silicon surface layers, *Rep. Prog. Phys.* **60**, 561-613.
2. Bechstedt, F., Enderline, R. (1988) *Semiconductor Surfaces and Interfaces (Their Atomic and Electronic Structures*), Akademie-Verlag, Berlin.
3. Abraham, F.F., Batra, I.P. (1985) Molecular dynamics simulation of semiconductor surface layers, *Surf. Sci.* **163**, L752-L758.
4. Khakimov, Z.M., (1994) Ab initio calculation of electronic structure of a-Si, *Computer Mater. Sci.* **3**, 94-108.
5. Zhang, P.X., Mitchell, P.X., Tong, B.Y. et al. (1994) Ion-beam treatment of Si surfaces, *Phys. Rev.* **B50**, 17080-17084.
6. Huang, L.J., Lau, W.M., Tang, H.T. et al. (1994) Diffusion processes in ion-bombarded Si surfaces, *Phys. Rev.* **B50**, 18453-18468.
7. Allen, M.P., Tildesley, D.J. (1987) *Computer simulations in liquids*, Oxford University Press, Oxford.
8. Stillinger, F.H., Weber, T.A. (1984) New interatomic potential for silicon, *Phys. Rev.* **B31**, 5262-5267.
9. Jacobs, P.W., Kiv, A.E., Balabay, R.M. et al. (1998) Computer Modelling of borbarded Si layers at high ion-doses, *Computer Modelling & New Technologies.* **3**, 15 – 21.
10. J. Dabrovski, J., Scheffler, M. (1992) Dimer structure in silicon surface, *Appl. Surf. Sci.***56-58**, 15–22.
11. Vavilov, V.S., Kiv,A.E., Niyazova, O.R. (1981) *Mechanisms of formation and migration of defects in semiconductors*, Nauka, Moskow.
12. Kiv, A.E., Soloviev, V.N. (1979) Grass-hopper effect in diamond like lattices, *Phys. Stat. Sol. (b)* **94**, K91-K95

PHASE RELATIONS AND NANOCRYSTALLINE ALLOYS IN THE TERNARY SYSTEMS ZrO$_2$ (HfO$_2$)-Y$_2$O$_3$-La$_2$O$_3$ (Eu$_2$O$_3$, Er$_2$O$_3$)

E.R. ANDRIEVSKAYA
Frantsevich Institute of Materials Science NAS of Ukraine, Kiev.
3, Krzhizhanivs'ky St., 03142 Kiev, Ukraine

1. Introduction

The overall purpose of the present paper is the overview of phase equilibria in the binary and ternary systems based on ZrO$_2$, HfO$_2$ and oxides of the III B subgroup (Y$_2$O$_3$, Ln$_2$O$_3$) and properties of the phases potentially useful for nanostructured materials, in particular for nanograined ceramics. Double doped zirconia or doped pyrochlore intermediate, the objects of ternary phase diagrams, which in turn were not studied well. Our efforts for last decade to fill this space of knowledge resulted in several ternary diagrams developed.

The present work is a quint-essence of the pioneer studies of phase relations in the ternary systems HfO$_2$-Y$_2$O$_3$-Ln$_2$O$_3$ and ZrO$_2$-Y$_2$O$_3$-Ln$_2$O$_3$ (Ln=La,Eu,Er) in the wide range of temperature and concentrations using XRD, thermal analysis in air (including solar furnace up to 3000 °C) [1], DTA in He at temperatures up to 2500 °C [2], petrography and electron microscopy. The comparison of the phase diagrams based on zirconia and hafnia permits elucidation the difference in phase diagrams of the systems with crystallographic analogs with different ionic radii.

Basic interest to this research originates from the diversity of polymorphic modifications inherent to the mentioned oxides, intermediate phases and solid solutions stable or metastable, as well as from the effect of ionic radii of lanthanides on phase stability and boundaries of phase fields. Polymorphous transformations and mutual solubility of oxides are different in nanograined ceramics compared to coarsegrained one.

From the general view point, the nanostructured materials based on zirconia, hafnia, rare-earth oxides, yttria and some intermediate phases might be obtained through the following technical approaches:

- solidification of eutectic or peritectic alloys on quenching of melts;
- compacting of nanosized powders by solid state sintering or under external pressure;
- precipitation of the secondary phases (ordering or martensite transformation for instance) in solid solutions; deposition of thin films.

As for current and potential applications of ceramics based on the systems with different lanthanides ZrO$_2$(HfO$_2$)-Y$_2$O$_3$-Ln$_2$O$_3$ we have to mention the following: IC circuit substrates, solid electrolytes in fuel cells, oxygen sensors, gas generators, membrane reactors, chemical sensors, catalysts and catalyst supports, desiccants, automotive heat exchangers, thermal barrier coatings, diesel port liners, artificial valves for hearts etc.

M.-I. Baraton and I. Uvarova (eds.),
Functional Gradient Materials and Surface Layers Prepared by Fine Particles Technology, 305–312.
© 2001 *Kluwer Academic Publishers. Printed in the Netherlands.*

2. Quenched alloys

Rapid quenched and amorphous materials based on zirconia doped by rare-earth oxides are interested for many useful applications such as chemically stable refractory and toughened ceramics. Quenched eutectic alloys are attractive because of unusual and anisotropic behavior and ultrafine grain structure. Such behavior is prospective for layered fianites demonstrating unique optical properties.

Ternary eutectic alloys might consist of ordered and disordered phases, which are easily transformed into the stable state with demanded structure. Zirconia stabilized by elder lanthanide oxides forms lamellar structure consisting of parallel fibers as a result of solidification from melts. Grain size in such structures depends on cooling rate, chemical composition and ratio of ionic radii of zirconium and lanthanide. At the same time, Tiller and Hunt had proposed the relation $\lambda^2 R$=const, where λ is interlamella spacing and R is the crystallization rate [3]. From this relationship the cooling rate can be estimated to get nanograined solidified microstructure. Indeed, if direct liquid quenching is not quite rapid enough to give a fully amorphous product for the given composition, a nanocrystalline product may be obtained. Such data might be useful for thermal barrier materials, reinforced structural ceramics and applications of laser cladding process. In addition to solidification rate one must know phase relations at supersolidus and subsolidus temperatures. Information on liquidus and solidus of the ternary systems considered here is useful in preparation of nanosized powders or ceramic thin films by laser ablation, evaporation under electron beam or CVD [4].

Liquidus and solidus surfaces are combined together giving the scheme of alloys crystallization, which is helpful for selection of prospective alloys of definite phase composition and definite supersolidus temperature difference. Projection of the liquidus surface for the HfO_2-Y_2O_3-La_2O_3 system are presented in (Fig. 1).

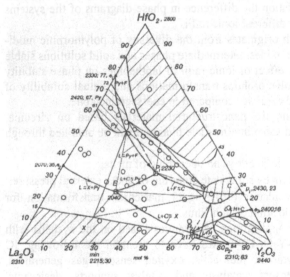

×5000

Figure 2. Microstructure of solidified: pyrochlore solid solution 58 mol % HfO_2-10 mol % Y_2O_3-32 mol % La_2O_3

Figure 1. Crystallization paths of the alloys in the system: HfO_2-Y_2O_3-La_2O_3;

o — experimental points

The liquidus surface formed by five fields of primary crystallization of the most high-melting phases H, C, F, X and Py. There are three non-variant equilibria with liquid phase: in the system ZrO_2-Y_2O_3-La_2O_3 [5] all of them are peritectics but two peritectics and one eutectic are revealed in the system HfO_2-Y_2O_3-La_2O_3 [6]. When doped by yttria, the pyrochlore phase originally congruently melted compound becomes incongruently melted and therefore, the sections $La_2Hf_2O_7$-Y_2O_3 and $La_2Zr_2O_7$-Y_2O_3 can not be considered as quasi-binary. Fluorite solid solutions form the largest fields on the both liquidus and solidus surfaces. The uniform grain structure of primaryly solidified pyrochlore phase $La_2Hf_2O_7$ of the composition $50HfO_2$-$10Y_2O_3$ (mol% here and bellow) is shown in Fig. 2.

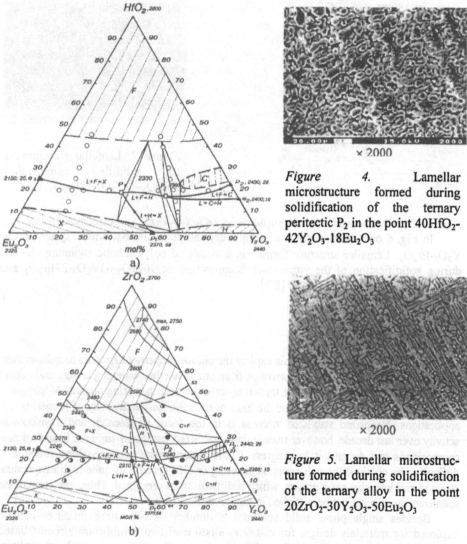

Figure 4. Lamellar microstructure formed during solidification of the ternary peritectic P_2 in the point $40HfO_2$-$42Y_2O_3$-$18Eu_2O_3$

Figure 5. Lamellar microstructure formed during solidification of the ternary alloy in the point $20ZrO_2$-$30Y_2O_3$-$50Eu_2O_3$

Figure 3. Crystallization paths of the alloys in the systems: HfO_2-Y_2O_3-Eu_2O_3 (a); ZrO_2-Y_2O_3-Eu_2O_3 (b); o — single phase, o — two phases, • — three phases.

308

Liquidus and solidus of the systems with europia (Figs. 3 a,b) are simplier compared to ones with lanthana: two non-variant points are on the liquidus and two tie-lines corresponding them are on the solidus [7]. Formation of lamellar structures was observed in the alloys, which are located along the monovariant line e_1p_2 or belongs to the tie-line triangle.

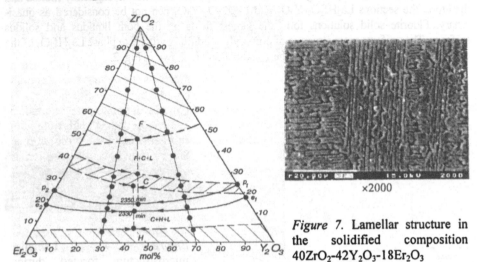

×2000

Figure 7. Lamellar structure in the solidified composition $40ZrO_2$-$42Y_2O_3$-$18Er_2O_3$

Figure 6. Crystallization paths of the alloys in the system ZrO_2-Y_2O_3-Er_2O_3; • — experimental points

In Fig. 6 one can find paths of alloy's crystallization in the ternary system ZrO_2-Y_2O_3-Er_2O_3. Lamellar structure formed is a result of polysynthetic twinning (C→H) during solidification of the yttria-reach composition of $10ZrO_2$-$63Y_2O_3$-$27Er_2O_3$ and $40ZrO_2$-$42Y_2O_3$-$18Er_2O_3$ (Fig. 7) [8,9].

3. Sintered ceramics

Ceramics based on stabilized zirconia explore the unique property of zirconia to exist in three polymorphic forms and their transformation from one to another with temperature and dopant concentration. Rare-earth oxides are known to serve the best stabilizing dopants of tetragonal and cubic zirconia, which in turn, are the basis of such impressive diversity of properties and applications. Nanosized stabilized zirconia is in the focus of research and development activity over last decade, however mainly yttria-stabilized tetragonal zirconia. Search of new compositions and dopants is in progress. Cubic zirconia with nanocrystalline structure is difficult to be obtained through common sintering or HIP because of quite high temperature required and intensive grain growth, which is difficult to be prevented. Thus, rare-earth oxide stabilized zirconia has not been studied in nanocrystalline state.

Besides single phase solid solutions a number of two-phase mixtures can be explored for materials design, for instance, mixture of two equilibrium zirconia-based solutions of cubic and tetragonal modifications as well as tetragonal with monoclinic. Our recent data on martensitic transformations in YSZ and rare-earth oxide stabilized zirconia demonstrated different behavior in stable and metastable ceramics [10]. Conditions of phase metastability are defined by the pyrochlore-type phase's ($Ln_2Zr_2O_7$) ordering rate, which depends on lanthanides ion radius. The smaller is the ion radius the slower is the ordering rate. This rule is fair for the lanthanides of ceria subgroup, but for yttria subgroup the pyrochlore compounds are unknown.

It has been established that 3 mol% of Y_2O_3 is enough to stabilize the tetragonal zirconia, transformation of which to monoclinic phase proceeds at 560 °C on cooling. The percentage of the monoclinic phase usually depends on grain size because in nano-grained state one can observe overstabilization effect due to both doping ions and forces of grain boundary tension. Thus, in nanograined zirconia this content of Y_2O_3 can be reduced to 1.5% compared with standard 3% in order to obtain the optimally transformation-toughened ceramics with highest fracture toughness 16 MPa·m$^{1/2}$.

Undoubtedly the complex doping by rare-earth oxides is of great interest as soon as one enable to control flexibly the effective ionic radius of dopant and finally the microstructure and properties of nanoceramics. Phase equilibria in the zirconia-reach region of the phase diagrams with La_2O_3, Nd_2O_3, Sm_2O_3, Eu_2O_3, as well as ternary systems $HfO_2(ZrO_2)$-Y_2O_3-Ln_2O_3 at low temperatures in the range 1250-1600 °C were studied. Several diagrams are presented in Figs. 8 a-d.

Equilibrium temperature of eutectoid transformation increases with ionic radius of dopant and this fact is confirmed by temperature of martensitic transformation, however the metastable transformations proceed with wider loop of temperature hysteresis as a rule, that is similar to behavior in nanostructured yttria-doped zirconia ceramics.

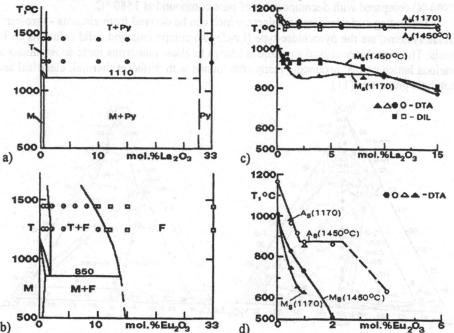

Figure 8. Fragments of the phase diagrams ZrO_2-Ln_2O_3 and corresponding parameters of martensitic transformation in equilibrium and non-equilibrium compositions [10].

310

These figures illustrate the change of martensitic transformation conditions with lanthanide's ionic radius in both equilibrium and non-equilibrium compositions, where the non-equilibrium one corresponds to the conditions of lower sintering temperature (1170 °C) and ultrafine-grained state of the ceramics. As soon as the strengthened and toughened zirconia ceramics commonly works at elevated or high temperatures, it becomes especially important to know how stable is the nanograined ceramics and what is the equilibrium composition of such ceramics.

Several isothermal sections of equilibrium phase diagrams are represented. The series of isothermal sections demonstrates stability of phases and solubility change with temperature. Attractive properties of fluorite, perovskite and pyrochlore solid solutions require careful study of the third component's solubility. Homogeneity field of the yttria doped pyrochlore phase, for instance, becomes wider compared to pure compound.

On the isothermal section of the system HfO_2-Y_2O_3-La_2O_3 at 1600 °C (Fig. 9a) we found solid solutions based on C-Y_2O_3, A- and B-La_2O_3, M- and F-HfO_2 as well as compounds $La_2Hf_2O_7$ and $LaYO_3$. Phase relations are determined by high thermodynamic stability of lanthanum hafnate ~100 kJ/mol. Phase equilibria at 1250 °C are similar to those at 1600 °C, though the homogeneity fields become narrower (Fig. 9b). Isothermal sections in the system ZrO_2-Y_2O_3-La_2O_3 have similar structure.

From the consideration of isothermal and polythermal sections the homogeneity fields of solid solutions and two-phase fields become wider with temperature rise while three-phase fields become narrower. Zirconia or hafnia dopants improve thermal stability of the perovskite $LaYO_3$ in ternary compositions and this phase was revealed at 1900 °C compared with decomposition of pure compound at 1580 °C.

Interesting and prospective objects, which can be derived from zirconia - rare-earth oxide systems, are the pyrochlore-type ($Ln_2Zr_2O_7$) compounds and solid solutions on its basis. The pyrochlore crystal structure is known to allow numerous ionic substitutions at various lattice points, producing many compounds with different thermal, electrical and catalytic properties [11].

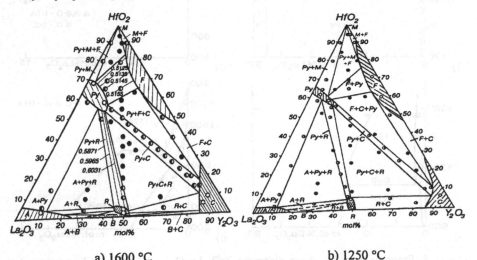

a) 1600 °C   b) 1250 °C

Figure 9. Isothermal sections in the system HfO_2-Y_2O_3-La_2O_3 at 1250 and 1600 °C.

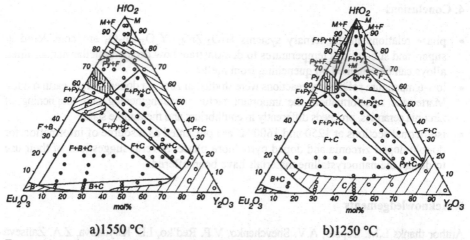

a)1550 °C   b)1250 °C

Figure 10. Isothermal sections in the system HfO₂-Y₂O₃-Eu₂O₃ at 1550 and 1250 °C

Pyrochlores are specified by extremely low thermal and ionic conductivity and low diffusion mobility, that is useful for graded thermal-barrier materials and nanostructured carrier of catalyst with thermally and chemically stable porous structure. Despite of these opportunities, a few works had been carried out to study nanograined doped rare-earth zirconates and hafnates.

Before characterization of the systems with europia, the variable oxidation degree of Eu ion (+2 and +3) in reducing and oxidizing media respectively should be taken into consideration. Isothermal sections in the $HfO_2(ZrO_2)$-Y_2O_3-Eu_2O_3 systems at 1550 °C and 1250 °C (Figs. 10a,b) are characterized by solid solutions based on M-and F-HfO₂, T-ZrO₂, C-Y₂O₃, B-Eu₂O₃, as well as pyrochlore $Eu_2Hf_2O_7(Eu_2Zr_2O_7)$. All the homogeneity fields are widen compared with those in lanthana-based systems due to smaller ionic radius of rare-earth metal. The most interesting detail of these sections is a solubility gap of the fluorite solid solutions due to high stability of europium zirconate. Solubility of dopants (yttria+europia) in tetragonal zirconia becomes much larger compared to one in boundary binary systems. This fact opens a possibility to control easier the volume of transformed monoclinic phase in toughened ceramics.

Finally, the phase diagrams with erbia (Fig. 11) are characterized by wide homogeneity fields of the solid solutions based on F-HfO₂ and C-Y₂O₃ and narrow field of solid solution based on M-HfO₂. Ion radii of yttrium and erbium are very close and, therefore they are easily substitute each other in solid solutions of wide extension. Using this property we can change the composition of the same phase in wide range and obtain new ceramics.

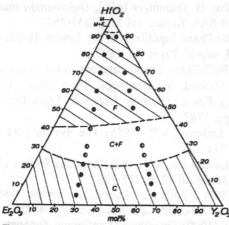

Figure 11. Isothermal section in the system HfO₂-Y₂O₃-Er₂O₃ at 1600 °C.

312

4. Conclusions

- phase relations in the ternary systems $HfO_2(ZrO_2)-Y_2O_3-Ln_2O_3$ are considered at super- and sub-solidus temperatures to demonstrate how do the lamellar nanograined alloys can be obtained on quenching from melts;
- low-temperature solid phase reactions were studied in zirconia doped by rare-earth oxides. Martensitic transformation, the important factor of strengthening and toughening of zirconia ceramics, proceeds differently in equilibrium and metastable states;
- isothermal sections at 1250 and 1600 °C are presented as a source of information for double-doped zirconia and doped pyrochlore phase. Several suggestions of their use to design the nanocrystalline materials have been offered.

5. Acknowledgements

Author thanks L.M. Lopato, A.V. Shevchenko, V.P. Red'ko, I.E. Kiryakova, Z.A. Zaitseva for their helpful advices and assistance in experimental study of presented phase diagrams.

6. References

1. Shevchenko, A.V., Tkachenko, V.D., Lopato, L.M., et al. (1986) Methods of identification of temperatures of phase transformations under solar heating, *Powder Met. and Metal Ceram.* **1**, 91-94.
2. Shevchenko, A.V., Lopato, L.M., Kuschevskiy, A.E. (1972) Investigation of systems of high-refractory oxides, *Powder Met. and Metal Ceram.* **1**, 88-92.
3. Yoshimura, M., Yashima, M., Noma, T., and Somiya, S. (1990) Formation of Diffusionlessly Transformed Tetragonal Phases by Rapid Quenching of Melts in ZrO_2 -$RO_{1.5}$ Systems (R=rare earths), *J. Mater. Sci.* **25**, 2011-2016.
4. Brock J.R. (1998) Nanoparticles Synthesis: A Key Process in the Future of Nanotechnology, in G.-M. Chow and N.I. Noskova (eds.), *Nanostructured Materials. Science and Technology*, Kluwer Academic Publishers, Dordecht, pp. 1-14.
5. Andrievskaya, E.R., Lopato, L.M., Duran, P (1999), The System ZrO_2-Y_2O_3-La_2O_3 and Field of its Applications, in L. Parilak, H. Danninger (eds.), *Deformation and Fracture in Structural PM Materials*, IMR SAS, Kosice, vol.2, pp. 251-257.
6. Andrievskaya, E.R., Lopato, L.M. (2000) Phase Equilibria in the System Hafnia-Yttria-Lanthana, *J. Amer.Ceram. Soc.*, **83**, in print Paper No. 189226.
7. Andrievskaya, E.R. and Lopato, L.M. (1997) Phase Transformations in the Ternary Systems $HfO_2(ZrO_2)-Y_2O_3-Eu_2O_3$ in P.Abelard, M. Boussuge, Th. Chartier, G. Fantozzi, G. Lozes, and A. Rousset (eds.), *Key Engineering Materials*, Trans Tech. Publications, Switzerland, vol. 132-136, pp. 1782-1785.
8. Andrievskaya, E.R., Lopato, L.M., and Smirnov, V.P. (1996) The System HfO_2-Y_2O_3-Er_2O_3, *J. Amer. Ceram. Soc.* **79**, 3, 714-720.
9. Andrievskaya, E.R., Lopato, L.M., and Shevchenko, A.V.(1996) Liquidus Surface in the System ZrO_2-Y_2O_3-Er_2O_3, *Inorganic Materials* **32**, 6, 721-726.
10. Andrievskaya, E.R. and Lopato, L.M., (1995) Influence of Composition on the T-M Transformation in the Systems ZrO_2-Ln_2O_3 (Ln=La, Nd, Sm, Eu), *J. Mater. Sci.* **30**, 2591-2596.
11. Nair, J., Nair, P., Doesburg, G.B.M. et al. (1999) Sintering of Lanthanum Zirconate, *J.Amer. Ceram. Soc.* **82**, 8, 2066-2072.

Author Index

314

Subject Index

316